D0025392

THE EYE OF HEAVEN
Ptolemy, Copernicus, Kepler

Masters of Modern Physics

Advisory Board

Dale Corson, Cornell University
Samuel Devons, Columbia University
Sidney Drell, Stanford Linear Accelerator Center
Herman Feshbach, Massachusetts Institute of Technology
Marvin Goldberger, Institute for Advanced Study, Princeton
Wolfgang Panofsky, Stanford Linear Accelerator Center
William Press, Harvard University

Series Editor

Robert N. Ubell

Published Volumes

The Road from Los Alamos by Hans A. Bethe
The Charm of Physics by Sheldon L. Glashow
Citizen Scientist by Frank von Hippel
Visit to a Small Universe by Virginia Trimble
Nuclear Reactions: Science and Trans-Science
 by Alvin M. Weinberg
In the Shadow of the Bomb: Physics and Arms Control
 by Sidney D. Drell
The Eye of Heaven: Ptolemy, Copernicus, Kepler
 by Owen Gingerich

(THE EYE OF HEAVEN)
Ptolemy, Copernicus, Kepler

OWEN GINGERICH

AIP

The American Institute of Physics

Copyright and permissions notices for use of previously published material are
provided in the Acknowledgments section in the back of this volume.

American Institute of Physics
335 East 45th Street
New York, NY 10017-3483

Library of Congress Cataloging-in-Publication Data

Gingerich, Owen.
    The Eye of Heaven: Ptolemy, Copernicus, Kepler / Owen Gingerich.
      p. cm. -- (Masters of modern physics)
    Includes index.
    ISBN 0-88318-863-5
    1. Astronomy--History. 2. Ptolemy, 2nd cent. 3. Copernicus,
Nicolaus, 1473–1543. 4. Kepler, Johannes, 1571–1630. I. Title.
II. Series.
QB15.G563 1992   91-26227
520'.9--dc20   CIP

This book is volume seven of the Masters of Modern Physics series.

Contents

Preface   vii

INTRODUCTION

1. Ptolemy, Copernicus, and Kepler   3

PTOLEMY AND THE GEOCENTRIC UNIVERSE

2. Was Ptolemy a Fraud?   55

3. Ptolemy Revisited   74

4. Zoomorphic Astrolabes: Arabic Star Names Enter Europe   81

5. The 'Abd al-A' imma Astrolabe Forgeries   102
(with D. King and G. Saliba)

6. Alfonso X as a Patron of Astronomy   115

7. The 1582 "Theorica Orbium" of Hieronymus Vulparius   129

8. The Search for a Plenum Universe   136

COPERNICUS AND THE HELIOCENTRIC UNIVERSE

9. *The Astronomy and Cosmology of Copernicus* 161

10. *Did Copernicus Owe a Debt to Aristarchus?* 185

11. *"Crisis" versus Aesthetic in the Copernican Revolution* 193

12. *Early Copernican Ephemerides* 205

13. *Erasmus Reinhold*
    and the Dissemination of the Copernican Theory 221

14. De revolutionibus:
    An Example of Renaissance Scientific Printing 252

15. *The Censorship of Copernicus's* De revolutionibus 269

16. *Heliocentrism as Model and as Reality* 286

KEPLER AND THE NEW ASTRONOMY

17. *Johannes Kepler and the New Astronomy* 305

18. *Kepler as a Copernican* 323

19. *Kepler's Place in Astronomy* 331

20. *The Origins of Kepler's Third Law* 348

21. *The Computer versus Kepler* 357

22. *The Computer versus Kepler Revisited* 367

23. *The Mercury Theory from Antiquity to Kepler* 379

24. *Kepler, Galilei, and the Harmony of the World* 388

25. *Circumventing Newton* 407

Epilogue 419

Acknowledgments 425

Index 431

Preface

The twenty-five papers in this collection were originally published throughout a quarter of a century, beginning in 1964 with "The Computer versus Kepler," which marked my entry into history of astronomy as a professional activity. The essays come from a wide variety of publications, not all readily available, ranging from the *Annali* of the History of Science Institute in Florence and an exhibition catalog of the Landesmuseum in Linz to the *Quarterly Journal of the Royal Astronomical Society* and the *Journal for the History of Astronomy*. I am grateful to all the original publications for permission to reprint the articles.

Here they have been arranged in a chronological sequence, not according to the date of publication, but approximately by historical period. After an overview that gives the subtitle to this volume come essays on Ptolemy and on instruments, next a group on Copernicus and early Copernicans, and then several papers on Kepler. The final paper considers the question of how the development of celestial physics might have differed had Newton never lived.

As for the title of this collection, it seemed appropriate to select a phrase from Kepler's English contemporary, William Shakespeare, whose Sonnet XVIII includes the lines

Sometime too hot the eye of heaven shines,
And often is his gold complexion dimm'd
And every fair from fair sometimes declines,
By chance or nature's changing course untrimm'd.

For purposes of this anthology, the papers have been lightly edited. The first article, for example, originally appeared in *Great Ideas Today*

1983 and contained numerous cross-references to the *Great Books of the Western World,* which have here for the most part been omitted. Elsewhere, obvious typographical errors have been corrected, as have occasional misstatements of fact. Some redundancy of material, especially illustrations, has been avoided. The footnotes have in general been cast into a uniform style, and the orthography has been brought to uniform American standards.

Certain major blunders still stand, however, and the reader is warned especially about the conclusion of the first article in the middle section, "The Astronomy and Cosmology of Copernicus," wherein I announce the discovery of a major series of annotations in a copy of Copernicus's *De revolutionibus* then believed to be in the hand of Tycho Brahe. It turns out that these important and revealing manuscript pages were written by one Paul Wittich, whose adventures have now been detailed in a monograph entitled *The Wittich Connection: Priority and Conflict in Late Sixteenth-Century Cosmology.* Published in the *Transactions of the American Philosophical Society* (volume 78, number 7, 1988), that book, by me and Robert S. Westman, must be considered a companion to this anthology.

For readers unfamiliar with sexagesimal notation, I hasten to point out that 45;43,15 equals 45°43′15″. Also, two works that figure prominently in the ensuing pages, Ptolemy's *Almagest* and Copernicus's *De revolutionibus,* are divided into books (XIII and VI, respectively) and these in turn are divided into chapters. *Almagest* XII,1 means *Almagest,* Book 12, Chapter 1, and a similar convention applies to Copernicus's *magnum opus.*

I have taken the opportunity afforded by this collection to place at the end an epilogue to reflect on the unity and diversity of the researches described in these articles.

INTRODUCTION

Ptolemy, Copernicus, and Kepler

O n the title page of the first edition of Copernicus's *On the Revolutions* appears this text, the equivalent of the modern publisher's dust jacket blurb:

"You have in this newly created and published book, diligent reader, the motions of the fixed stars and planets restored from both the old and new observations, and furthermore, furnished with new and wonderful hypotheses. You also have the most convenient tables from which you can with the greatest ease calculate their positions for any time. Therefore buy, read, and profit!"

Fourteen centuries earlier, Claudius Ptolemy's publisher (if indeed there had been "publishers" in the ancient world) could have made an even stronger claim: With his tables you could *for the first time* calculate the positions of the planets for "any time."

To understand the relative contributions of Ptolemy, Copernicus, and Kepler, we must appreciate the fact that each was tackling a fairly esoteric problem—how to find the positions of the heavenly bodies in the sky—and that each had a cosmological framework in which the problem was solved. I say a "fairly esoteric problem" because most people have rather little use for such specific information as to where

Selection 1 reprinted from *The Great Ideas Today 1983*, ed. by M. J. Adler and J. Van Doren (Chicago, 1983), pp. 137–80.

Mars is to be found at a particular moment, and even less use for a position accurate within two minutes of arc, which was part of Kepler's great achievement.

Nevertheless there are subtle reasons why this quest has been important to mankind. The motions of the planets have a demonstrable, predictable regularity which raised the hope that other aspects of our world might also have underlying regularities, if only the secrets of nature could be teased from the seemingly capricious patterns of weather, catastrophe, or even human personality. In Ptolemy's day (the second century A.D.) there was the lure that the regularity of the stars was reflected in mundane events. Tycho Brahe, the renowned Danish observer and Kepler's sometime mentor, expressed it in the motto *Suspiciendo despicio*—"By looking upwards, I see below." It was a false scent, to find in the clockwork of the stars the key to personal affairs, yet the knowledge of the planetary rhythms did unlock Newtonian physics and with it the fundamental understanding of the physical universe. The value of accurate planetary positions has come full circle: in their precision came an appreciation of elliptical orbits, then gravitation and planetary perturbations, and now the trajectories of spacecraft exploring the distant new worlds of the solar system—and requiring accuracies undreamed of by Ptolemy or Copernicus or even Kepler.

In the ancient world it was surely the astrological motivation that provided much of the impetus for Ptolemy's treatise and perhaps most of his financial support, just as it supported Kepler in the seventeenth century. But beyond that was the sheer intellectual curiosity that drove all of these men. In some of the manuscripts Ptolemy's treatise begins with a memorable epigram:

> I know that I am mortal by nature, and ephemeral; but when I trace at my pleasure the windings to and fro of the heavenly bodies I no longer touch earth with my feet: I stand in the presence of Zeus himself and take my fill of ambrosia, food of the gods.

Ptolemy's epoch-making book is generally called by its Arabic name, the *Almagest*, meaning literally "The Greatest." Written in Alexandria around A.D. 150, it is indeed the greatest surviving astronomical work from antiquity. What Ptolemy has done is to show for the first time in history (as far as we know) how to convert specific observational data into the numerical parameters for his planetary models, and with the models to construct some ingenious tables from which the solar, lunar, and planetary positions and eclipses can be calculated for any given time. Altogether it is a remarkable achievement, combining in a brilliant synthesis a treatise on theoretical astronomy with a practical handbook for the computation of ephemerides.

Ptolemy accomplished his task within the cosmological framework almost universally accepted in his day, the Earth-centered or geocentric world view. As a consequence of the revolutionary work of Copernicus and Kepler, of Galileo and Newton, the geocentric Ptolemaic system has by now been tossed into the trashcan of discarded theories. Euclid's *Elements*, with its logically ordered series of geometrical proofs, is still honored as a timeless classic, but why should we take time to consider the *Almagest*? Because, more than any other book, it demonstrated that natural phenomena, complex in their appearance, could be described by relatively simple underlying regularities in a mathematical fashion that allowed for specific quantitative predictions.

Ptolemy admits at the outset that it might perhaps be simpler, from a strictly celestial viewpoint, to have the Earth spinning daily on its axis rather than the entire heavens rotating about the Earth. But this, he says, fails to take into account the terrestrial physics: if the Earth moved, "animals and other weights would be left hanging in the air, and the Earth would very quickly fall out of the heavens. Merely to conceive such things makes them appear ridiculous." And so, after only a few pages devoted to his cosmological assumptions, Ptolemy marches on to the "practical" matter of a mathematical description of the heavens. Ptolemy's model is Euclid, and even the Greek title of his volume, *Syntaxis*, or "Mathematical Treatise," reflects his desire to give a tidy geometrical and numerical account of his subject. As such, it must be considered more as a masterful pedagogical textbook than as an account of discovery.

Ptolemy's attempt to write the astronomical equivalent of Euclid's geometry was all too successful: nobody bothered to copy any competing treatises that might have been available, leaving us with the *Almagest* and rather little else for reconstructing how Ptolemy might have arrived at the main structures of his system. And in retrospect it has become increasingly obvious that, in writing his textbook, he has imposed a logical order on the material that is markedly different from the one he actually used to arrive at his conclusions. This in turn has today embroiled Ptolemy in such a heated controversy that one critic has recently called him "the greatest fraud in the history of science."

Despite objections that have been raised against Ptolemy in the past few centuries, it is clear that his work did, in fact, provide the fundamental text for well over a millennium, and it served as the model for Copernicus's *On the Revolutions* in 1543. For better or for worse, Copernicus followed Ptolemy's pattern so well that historians of science learn rather little from Copernicus's book as to why he arrived at the principal features of *his* system—that is, the Sun-centered or heliocentric cosmology. It is not true, as many secondary accounts would have us believe, that the Ptolemaic system was by then falling apart from complex additions that it had accrued, or that it was hopelessly failing to predict the celestial phenomena. By Kepler's standards the tables of both the *Almagest* and *On the Revolutions* were embarrassingly deficient, but the fact of the matter is that in the sixteenth century there was astonishingly little information for choosing between them, as far as accuracy was concerned. Copernicus had tried to effect a cosmological revolution without really raising the standards of prediction; his was a vision of the mind's eye, not the revolt of an observer with his quadrant or armillary sphere.

In contrast to Ptolemy or Copernicus, Kepler gives us an abundance of autobiographical detail. Thereby we can reconstruct his progress in the reform of astronomy, a prodigious feat in which he raised the accuracy of prediction (take Mars, for instance) from several degrees to within a few minutes of arc—that is to say, from ten times the diameter of the Moon to near the limit of naked-eye astronomy. This magisterial accomplishment was profusely described in his *Astronomia nova* of 1609. Seldom was a scientific treatise better named. The "New Astronomy" broke with two millennia of tradition (a) by introducing elliptical orbits to replace the time-honored combination of uniform circular motions, (b) by showing how to extract intricate details from vast quantities of conflicting observations, and (c) by arguing from physical principles in a way both novel and foreign to astronomical expectations. Kepler's greatest work, the *New Astronomy* is simultaneously a formidably mathematical and exasperatingly obscure book. It is easier to get the flavor of Kepler at work from his later *Harmonice mundi* (1619), the "Harmony of the World," in which he stumbled across his important "harmonic law" and his *Epitome of Copernican Astronomy* (1618–21), in which are given the final and far-better-digested results. Nevertheless, there are several quite accessible portions of the *New Astronomy* that show Kepler forging his new celestial physics.

In the ensuing sections I shall attempt to lead the reader through the technical positional astronomy as well as the cosmology of these three pioneering astronomers, and, for those who want to see how it is *really* done, I shall show how the tables in Ptolemy's *Almagest* can in fact be used to find a planetary position.

Ptolemy's Invention

Ptolemy's goal is nothing less than the calculation of planetary positions at any time—past, present, or future. This differed in a major way from the received Babylonian astronomy, which attempted to find the time and place only of specific phenomena, such as the disappearance of a planet in the Sun's rays, its reappearance, or its opposition to the Sun in the sky. To be precise, Ptolemy wanted to calculate the planetary positions with respect to the ecliptic—that is, the Sun's great circle route through the zodiac. (The ecliptic is the line along which lunar eclipses can take place, whence its name.) He wanted to specify the longitude measured eastward along the ecliptic, as well as the latitude, measured in degrees north or south of the ecliptic.

The *Almagest* describes Ptolemy's complete program. On the one hand, it uses geometrical mechanisms, based in spirit on Euclid (and Apollonius), and on the other, specific numerical information, coming largely from the Babylonians (and in part transmitted via Hipparchus in the second century B.C.). For the latter, however, Ptolemy had no earlier example to govern his presentation. The origins of his geometrical models are as mysterious and perplexing as the origins of his numerical data. Nevertheless, we can attempt to reconstruct this unwritten background to his treatise.

Ptolemy's *Almagest* is divided into 13 books, and he does not get to the critical planetary longitude tables until Book XI, nor the planetary latitude tables until the very end, so it is obvious that a considerable foundation must be laid into place before he reaches his goal. Yet it is unlikely that he started as systematically in his researches as his great textbook lays them out.

The problem at hand, already posed by Plato to his students, was finding some explanation for the windings to and fro of the planets. We can take as a prototype the behavior of Mars. This ruddy and not very bright planet moves eastward throughout the zodiacal signs for months on end, but then it slows, comes to a stop against the background stars, brightens, and, now quite conspicuous, moves westward for several weeks before stopping, fading, and finally resuming its direct motion. How could this retrograde motion be accounted for, particularly within the constraint of unending and repetitive circular motions?

In Plato's day his contemporary, Eudoxus, proposed a series of cleverly nested spheres for turning the planets. It was an admirable scheme; in fact, it was far more to be admired than used. Not only did it fail to reproduce the motions except in the most general way, but (because all the spheres were completely concentric to the Earth) there was no way for the planets to approach or recede, and hence no way to account for the changes in brightness.

Five centuries later, by the time of Ptolemy, the works of Euclid and Apollonius were at hand, and also a vast array of specific astronomical data from the Babylonians, who had been making crude but systematic observations since before 700 B.C. For example, the Babylonians knew that in 79 years Mars made almost exactly 42 complete revolutions through the zodiac, and that it moved 40 percent faster when it was in Capricorn than when it was opposite in the sky in Cancer. Furthermore, the astronomer Hipparchus had made some progress in representing the motion of the Sun and Moon by circles not centered exactly on the Earth. His tables sufficed to get the dates of eclipses. Presumably, all these materials were readily available to Ptolemy in the great library at Alexandria.

The challenge facing Ptolemy was to find some geometric model that could reproduce not only the retrograde motion of Mars but also its faster motion in the half of the sky toward Capricorn. The model needed to take into account the observation that the retrograde loops were about half as long when they were in Capricorn as when they were in Cancer. The task of building a geometric model to predict the detailed positions of Mars had defeated Hipparchus as well as Eudoxus before him, and by A.D. 135 could be considered the outstanding unsolved problem of astronomy.

In seeking a solution to the problem of planetary motion, Ptolemy had available a now-lost work of Apollonius of Perga (ca. 200 B.C.). Apollonius had considered the problem of a secondary circle moving on a larger one—an epicycle on a deferent, as it came to be called in the Ptolemaic scheme—and this provided a suggestive clue. A point on the rim of the epicycle traced out a looped path swinging near and far from the center of the deferent, and moving in reverse (as seen from the center) each time it approached the center (Figure 1). The problem was, could the model be made to represent the apparently irreg-

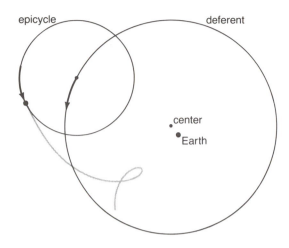

FIGURE 1. *Retrograde motion generated by an epicycle.*

ular behavior of Mars while retaining uniform motions for both the deferent and epicycle?

Now the period of Mars to get around the entire sky was known pretty well from the Babylonians, as was the period between retrogressions. Hence Ptolemy could calculate the period of the uniform motion in longitude as well as the uniform motion of what is called the anomaly, in this case referring to the anomalous retrograde motions, to be represented in the model by the epicycle. In fact, Ptolemy tabulates both of these in terms of the motion per unit length of time in the *Almagest*, IX,4. Armed with a theorem from Apollonius, he could decide how large the epicycle had to be with respect to the deferent to reproduce the average retrograde loop.

But the observations showed two troublesome details. As mentioned above, Mars appeared to move (after averaging out the effect of the retrogression) 40 percent faster on one side of the orbit compared with the other. Second, the retrograde loops themselves varied in size from one retrogression to another. How could Ptolemy produce the necessary variations in the loops as well as the variations in the speed of the center of the epicycle?

Here the previous work of Hipparchus provided the way. To get a nonuniform motion of the Sun, Hipparchus had placed the solar orbit eccentrically with respect to the Earth. The Sun, in moving along its deferent at uniform speed, actually appeared from the Earth to spend more time in the summer quadrant than in the winter one. The actual motion remained uniform, but the apparent motion matched the observed view from the Earth.

In order to account for the 40 percent faster speed of Mars on one extreme of the orbit compared to the other, the deferent circle could simply be placed eccentric by 20 percent, so that the planet would be 20 percent closer (and faster) than the average in one direction and 20 percent farther (and slower) than the average in the other direction. Alas! While representing the change of speed quite well enough, the computed size of the epicycle came out different at the two extremes; or alternatively, with the epicycle held at a constant size, the sizes of the retrograde loops were absurdly out of line with those observed.

How many solutions Ptolemy may have tried to overcome this dilemma we shall never know. Perhaps the epicycle could be made to expand and contract—not a very nice solution, but perhaps possible. Or another smaller epicycle might ride upon the first. We know, from the work of the Islamic astronomers of the fourteenth century that such a procedure is feasible.

Ptolemy found that he could solve the problem of the size of the retrograde loops simply by making the eccentricity 10 percent instead of 20 percent. This permitted a constant size to the epicycle and correctly varied the loops provided he retained their angular speeds. However, the previous solution to the nonuniform motion of the epicycle center was now spoiled, for the variation in speed about the mean was now ±10 percent, for a total of 20 percent, half the requisite amount, and not the amount he had assumed in his calculation. If only some way could be found to vary the actual speed of the epicycle center along the deferent without disturbing the distances, which were yielding the correct sizes! Here is where Ptolemy's solution turned out to be both elegant and unexpectedly accurate.

Ptolemy proposed to insert a seat of uniform angular motion for the epicycle's center at a point equal and opposite from the Earth along the line through the deferent center (Figure 2). The arrangement, called an equant, splendidly solved this combination of problems, though it did violate the principle of uniform circular motion, because the motion of the epicycle center along the deferent was now smoothly *nonuniform*. As seen from the equant point, however, there was equal angular motion in equal times; that is, uniformity was preserved from this viewing point although not along the deferent itself.

The foregoing account makes Ptolemy's invention seem all too obvious. Yet we must remember that the observations at his disposal were incomplete and rough. Only for Mars are the effects large enough to show readily. As Kepler was to remark in a similar situation many centuries later (in his *New Astronomy*, Chapter 7), "for us to arrive at the secret knowledge of astronomy, it is absolutely necessary to use the motion of Mars; otherwise it would remain eternally hidden."

But the *Almagest* is not an account of discovery; rather, it is a textbook, and as such it carefully hides the process of finding the eccentric-epicycle-equant model. In fact, it takes the planets in order, beginning with the awkward, overly complicated mechanism for Mercury, next Venus, whose eccentricity is too small to allow this effect ever to be discovered from the sort of observations Ptolemy had, and finally to the critical case of Mars.

Given the discovery of this model for Mars, Ptolemy must have immediately tried next to work it out in full detail. However, the motion of Mars's epicycle is closely connected with the motion of the Sun. Rather than backtrack to the solar theory, Ptolemy may have simply been content to use the Hipparchan model, which, after all, predicted the eclipses pretty well. To get the precise period of Mars required some knowledge of the solar theory and a long temporal baseline. In the *Almagest*, Ptolemy uses a Babylonian observation from 272 B.C. together with those from his own day, around A.D. 135. This is more easily said than done, because the dates for the observations are on disparate calendars. The first is recorded in a Dionysian year with curious zodiacal months, and the others in Egyptian months in years taken with respect to the Emperors Hadrian and Antoninus. Thus, any deep analysis of Ptolemy requires a certain familiarity with chronology, a point that would later cost Copernicus much time before he could safely connect his own age with Ptolemy's.

It was one thing for Ptolemy to establish a successful theory of Mars and another to cast it in a form so that calculations could be made in a routine fashion. He proceeded along the following lines. First, consider the planet moving uniformly around its deferent. One can easily make a table of uniform motions, such as those already cited. What is needed next is to correct these mean motions on account of the eccentric placement of the deferent; this could be simultaneously combined with the effect of the equant. For Mars, these combined corrections can reach a maximum of nearly $11\frac{1}{2}°$. Second, there must be a correction for the effect of the epicycle, the amount of which depends on the degree of rotation of the planet within the epicycle. For Mars, the epicycle can advance or retard the mean motion by something over $41°$.

It is here that the real complications begin. If the epicycle is at its farthest point from the Earth, clearly its correction is going to be smaller than if the epicycle were closer. If we wish to tabulate these corrections for every degree, we would need 180 positions for the deferent (symmetry will duplicate the entries for 181° to 360°), and for each of them values for the 180 positions within the epicycle, or a grand double-entry table with 32,400 values! (Actually, Ptolemy did not tabulate as closely as every degree.) Ptolemy very cleverly gets around such a double-entry table monster by a multiplicative combination of two single-entry tables, thereby using only 360 values instead of 32,400. His is one of the neatest tricks of practical mathematics from all of antiquity.

Those who do not wish to see in detail how the tables of the *Almagest* work should simply skip past the next section, in which I will show how the Ptolemaic theory can be used to find the longitude of a planet, specifically Mars. The date will be chosen in Ptolemy's system of Egyptian years.

The Longitude of Mars—An Example

As an example, we shall calculate the longitude of Mars used by Ptolemy in the *Almagest*, Book X, Chapter 8. [The best translation, in any language, is Gerald J. Toomer's *Ptolemy's Almagest*, 1984; an earlier and more widely available edition is in Vol. 16 of *Great Books of the Western World* (*GBWW*).] The longitude is measured eastward throughout the zodiac, beginning at the northbound intersection of the ecliptic and the equator. Ptolemy breaks the ecliptic into twelve 30° segments, each named for a sign of the zodiac. Thus, when he specifies an observation of Mars at "$1\frac{3}{5}°$ within the Archer," we note that the Archer (Sagittarius) is the ninth zodiacal sign, making the longitude 241°36'. Ptolemy specifies that this observation was made in the second year of Antonine, in the Egyptian month Epiphi, in the night of 15–16; with the tables on p. 467 of *GBWW*, we can readily establish that Epiphi is the eleventh Egyptian month (so that 300 days have already elapsed in the year).

The base date or epoch for Ptolemy's tables is midday of the first day of the first month (Thoth 1) of the first year of the reign of King Nabonassar of Babylon. Hence the first task is to discover how much time has elapsed between the epoch and the date of the observation.

This is facilitated by a King List provided elsewhere by Ptolemy, and conveniently appended in *GBWW*, p. 466. There it may be seen that

From Nabonassar to the death of Alexander	424 Egyptian years
From the death of Alexander to Antonine 1	460 Egyptian years
From Antonine 1 to Antonine 2	1 Egyptian year
	885 Egyptian years

We now wish to calculate the mean motion for Mars and its epicycle (the "anomaly") over this elapsed time, starting with the positions at the epoch. This is conveniently done by combining the numbers for the various intervals tabulated by Ptolemy in IX,4. Mars moves about half a degree per day, or 191°16′54″ per year. In 810 years (the largest interval Ptolemy tabulates), Mars circumnavigates the sky 430 times with 138° left over, and of course it is only the remainder that is of interest in locating a later position. Note that because Ptolemy begins with Thoth 1 (rather than Thoth 0), it is necessary to take one day less than the day of the month to get the elapsed days; note also that, since he specifies the observation as three hours before midnight, nine hours have elapsed since midday. The epoch positions are tabulated at the head of the tables.

	Longitude		*Anomaly*
At epoch	3° 32′		327° 13′
810 years	138 15		24 49
72 years	92 17		250 12
3 years	213 51		145 26
300 days	157 13		138 28
14 days	7 20		6 28
9 hours	0 12		0 10
Mean longitude	252° 40′	Mean Anomaly	172° 46′

These results are obtained after taking out entire cycles of 360°. It is next necessary to establish where the mean longitude has carried the epicycle center with respect to the apogee (see Figure 2) of the planet. The apogee is, according to Ptolemy, fixed with the stars and therefore precessing with them at the rate of 1° per century. For the 885 years since the epoch, the precession amounts to 8°51′, a number that must be added to the specified apogee at epoch:

Apogee at epoch = Crab 16°40′ =	106° 40′
Precession	8° 51′
Apogee at date of observation	115° 31′

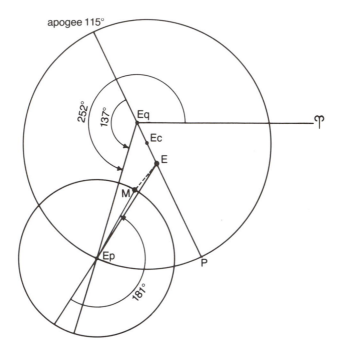

FIGURE 2. *Longitude of Mars according to Ptolemy.*

We next wish to compute the eccentric anomaly or the "longitude from the eccentric's apogee," as Ptolemy calls it; this is found by differencing the mean longitude and the apogee:

Mean longitude	252° 40′
Apogee	115° 31′
Eccentric anomaly	137° 09′

This value agrees well with the 137°11′ specified in X,8 and will be used in the correction tables in the next step.

    For the remaining steps it is worthwhile to consult Figure 2 in order to understand why the corrections are added or subtracted. The angles are exaggerated on the diagram for clarity. The Earth is at *E*, the center of the eccentric deferent at *Ec*, the equant at *Eq*, the perigee at *P*, the epicycle center at *Ep*, and the position of Mars at *M*.

The corrections or "prosthaphaereses," which may be translated as "addition-subtractions" or "equations," are tabulated in XI,11. The first correction, for the equant and eccentric, is found in columns 3 and 4 combined. We enter with 137°09' in column 1 and interpolate, noting that column 4 is in the negative range and needs to be subtracted from column 3. On the diagram, this correction is angle $EqEpE$, and its effect is to *decrease* the longitude, but to *increase* the angle in the anomaly, which needs to be measured from the line of sight to the Earth E, rather than the equant Eq. Hence:

Mean longitude	252° 40'	Anomaly	172° 46'
$c_3 + c_4 =$	− 8° 21'		+ 8° 21'
Corrected longitude	244° 19'	Corrected anomaly	181° 07'

We now enter with the corrected anomaly to find from column 6 the correction for the movement within the epicycle, the angle $EpEM$. Because this observation was deliberately chosen by Ptolemy to be a few days after an opposition and within the retrogression, the value is near the bottom of the table where linear interpolation is not especially accurate but will suffice. Note from the diagram that this correction must decrease the angle and thus be subtracted.

$$
\begin{array}{rl}
 & 244° \; 19' \\
c_6 = & - 2° \; 10' \\
\hline
 & 242° \; 09'
\end{array}
$$

The final correction adjusts for the fact that the epicycle is closer than its average position, and therefore an additional amount must be subtracted. This depends both on how much closer the epicycle is and on the position in the epicycle. The former part is given as a proportion with respect to 60 in column 8, whereas the latter part is given in column 5 if the epicycle is farther than the average position, or in column 7 if the epicycle is closer than the average, as in this case. We get the product of these corrections as follows:

$$
\begin{array}{rl}
c_8 \; (\text{entering with } 137° \, 09') & 37/60 \\
c_7 \; (\text{entering with } 181° \, 07') & \times 53' \\
\hline
 & 33'
\end{array}
$$

so that we can make the final correction, obtaining:

$$
\begin{array}{r}
242° \ 09' \\
c_8 \times c_7 = \quad - \quad 33' \\
\hline
\text{True longitude} \quad 241° \ 36'
\end{array}
$$

This agrees exactly with the Ptolemy's value in the *Almagest* X,8.

Did Ptolemy Cheat?

Disappointingly for the analyst of Ptolemy's achievement, the epicyclic model for the planets appears full grown in the *Almagest*. There are no hints as to what motivated its invention. Ptolemy simply gets on with the business of finding the required numerical parameters. They are

- the ratio of the epicycle to the deferent,
- the eccentricity of the orbit and equant,
- the direction in which the equant-eccentric line lies,
- the period of revolution in the deferent,
- the period of revolution in the epicycle,
- the position in the deferent for a given starting date,
- and the position in the epicycle for a given starting date.

Thus Ptolemy requires altogether seven numbers to specify the model for longitude. Two of these parameters, the period of revolution and position in the epicycle, are connected to the solar theory, so that actually only five numbers are needed independently. In order to establish five parameters, it is necessary to have five observations, and for Mars these are given as follows:

Dionysius 13, Aigon 25	[272 B.C. Jan 18]	Sco 2° $\frac{1}{4}$
Hadrian 15, Tybi 26–27	[A.D. 130 Dec 15]	Gem 21°
Hadrian 19, Pharmouthi 6–7	[A.D. 135 Feb 21]	Leo 28° $\frac{5}{6}$
Antonine 2, Epiphi 12–13	[A.D. 139 May 29]	Sgr 2° 34'
Antonine 2, Epiphi 15–16	[A.D. 139 May 30]	Sgr 1° $\frac{3}{5}$

It is instructive to recompute each of these from Ptolemy's tables; we have done it for one of them in the preceding section, and in carrying out the same procedure for the other four, we obtain an almost perfect match with the stated observations. However, when we use a modern computer to calculate where the planet really was on each of these occasions, we find the following discrepancies:

	Ptolemy	Modern	Difference
272 B.C. Jan 18	212° 15'	212° 31'	0° 16'
A.D. 130 Dec 15	81°	81° 25'	25'
A.D. 135 Feb 21	148° 50'	150° 15'	1° 25'
A.D. 139 May 29	242° 34'	242° 49'	15'
A.D. 139 May 30	241° 36'	241° 55'	19'

In other words, although Ptolemy's stated observations are not particularly good and in one case err by more than a degree, they do match his tables in an uncanny way. In fact, this situation prevails throughout the entire *Almagest*, which leads to the suspicion that something is going on that does not meet the eye, something craftily concealed in the writing of the *Almagest*. Here we must remember, above all, that the *Almagest* is not like a modern research paper. It is a well-honed textbook, modeled as much as applicable on Euclid's *Elements*. It was no doubt Ptolemy's intention to present the procedures with sample data so that future astronomers could see how it was done and could introduce their own observations over a longer temporal baseline in order to derive even better parameters. For such a purpose he surely wanted to exhibit mutually consistent observations, the best he could find.

There are several ways that Ptolemy could have obtained such agreeable data. He could have taken five well-chosen observations and fit his parameters to them. Then necessarily the tables would yield back precisely these same five data points. I say "well-chosen" because the final parameters for his models turn out to be remarkably accurate, especially for Mars, and this would hardly be the case if he had picked his five observations more or less at random. Perhaps he used multiple observations and found the average of what seemed to be the most frequent solutions. (The mind boggles to think how many graduate students he might have had sweating over the calculations!) Given a fairly stable solution based on a series of observations, he could then use his tables to choose which were the best observations. However, I am inclined to think that, if he got this far, he may no longer have felt it necessary to find perfect observations, for he could simply have used his theory to judge how much observational error an observation contained, and he could then correct it accordingly. By modern standards, to do this without mentioning it would be considered reprehensible, but in the second century A.D. there were no models and no agreed-upon method for handling redundant data. Remember that any extra data beyond the minimum required were very likely to be self-contradictory on account of the inevitable errors of observation.

For at least a couple of centuries, astronomers have been to some degree aware of this problem with Ptolemy's data, but they have generally swept it under the carpet. In recent years it has been rediscovered by the geophysicist R. R. Newton, who has reacted with such alarm that in his *The Crime of Claudius Ptolemy* he has written that it would have been better for astronomy if the *Almagest* had never been written. This judgment is so extreme as to be silly, but nevertheless it has reminded astronomers that the data in the *Almagest* cannot necessarily be taken at face value.

It is generally not, in fact, the task of historians of science to cast moral judgments or to brand a pioneer of the past as a criminal. Rather, it is to study the evolution of ideas, to try to reconstruct *how* and *why* particular ideas were presented in a particular way by a particular person at a particular time. Ptolemy, like astronomers of today, undoubtedly built his edifice on a great array of traditional materials, rejecting, adjusting, or incorporating them as he saw fit, and molding them into a new theoretical framework of geometric planetary models. To examine how these elements can blend together, we now return to the presentation of the *Almagest* itself, to see the difficulties that some of Ptolemy's borrowings have caused him and his successors.

The Arrangement of the Almagest

The philosophical or cosmological assumptions are tersely stated at the outset of Book I, whereupon Ptolemy turns immediately to the mathematical preliminaries. For the Alexandrian astronomer, this means not only presenting a trigonometrical table but explaining in full detail how to compute it. Ptolemy does not use our familiar sines and cosines, but the chord function, which is directly related to them. Following this, he must show how to solve for angles on the celestial sphere, no mean trick considering that spherical trigonometry as we know it had yet to be invented! Although Ptolemy uses the most advanced mathematics of the time, discovered in the preceding generation by Menelaus, the resulting procedures are certainly clumsy. The Menelaus theorem connects six quantities, and, to solve for one unknown, the problem must be set up so that the other five are known. Nevertheless, it has been shown that all problems of spherical trigonometry can be solved by this technique.

Given the Menelaus theorem, what problem might require a solution? Ptolemy first picks a simple one: the relation between the equator, which slices the celestial sphere into two polar hemispheres, and the ecliptic, which is the great circle annual path of the Sun, tilted with respect to the equator by an amount called the obliquity of the ecliptic. Ptolemy wishes to relate the angular distance along the ecliptic to the distance of the ecliptic north or south of the equator. He sets up the problem by using the obliquity, the angular distance along the equator as well as its complement, and two other segments known to be right angles; with these five given, the unknown sixth can be calculated and tabulated.

The geometry is impeccable, but the wedding of abstract geometry with particular numerical values immediately poses a conundrum for connoisseurs and critics alike. How can the obliquity be determined? Presumably in a straightforward manner, by measuring the altitude of the Sun at noon when it is highest in the sky in summer, the Sun "standing" (= *sol stice*) at the top of the ecliptic, and then by repeating the measurement at the lowest point, the winter solstice. Ptolemy obligingly gives some details of setting up a calibrated brass circle, or an inscribed stone or wood block whose flat surface is oriented precisely in the north–south–zenith plane. The total angle (twice the obliquity) always lies between 47°40' and 47°45', which, he reports, agrees nearly with Eratosthenes (ca. 200 B.C.) and Hipparchus (ca. 135 B.C.); half of this gives the adopted obliquity of 23°51'20". The problem is that, to the best of our modern calculations, the obliquity should have been 23°40'50" in Ptolemy's time.

Apparently Ptolemy has settled on a traditional value, although conceivably he really did try to confirm the obliquity and something subtle went wrong, fooling him into agreement with his predecessors (whose value was just as wrong for them as for Ptolemy). Where could such a value for the obliquity come from? Ptolemy remarks that the double obliquity, 47°42'40", is almost exactly $\frac{11}{83}$ of the circumference of a circle, and since we know that the Greeks with their elementary mathematics particularly liked ratios of integers, we can guess that in some circuitous way the traditional obliquity stems from that ratio.

In the third book of the *Almagest*, Ptolemy takes on the problem of the length of the year and the seasons. This seemingly simple problem has vexed astronomers of every age, including our own. Roughly speaking, the year is the time required for the Sun to make its circuit through the sky (or the Copernican converse, the time required for the

Earth to make its circuit around the Sun). But how do we know when the circuit has been completed? Is it when the Sun has come back to the same position against the starry background? This is not so easy to measure, because of the Sun's brilliance, which blots out the stars in the daytime. However, during a lunar eclipse we know that the Moon is precisely 180° away from the Sun, so this geometry provides a clever method to pin down the Sun's position among the stars. A year defined this way, with respect to the stars, is called the sidereal year.

Another way to measure a year is to record the Sun's seasonal movement, to document the time that passes between the moment when the Sun crosses the celestial equator northward at the vernal (spring) equinox, and when the phenomenon repeats again the following year. This is called the tropical year, which is about 20 minutes shorter than the sidereal year.

Ptolemy chooses to use the tropical year as his fundamental base, and so he requires observations of the Sun at the equinoxes. Again, Ptolemy describes his observational procedures in some detail; again he claims his results confirm Hipparchus, and again there are problems. Ptolemy reports, for example, that the Sun could be observed as its shadow changed sides on an equatorial ring on 22 March 140 (of course, he specified this in the current Egyptian calendar, which was Pachon [Pactom] 7 of the year Antonine 3). Now this happens to be exactly the date that would have been predicted by Hipparchus on the basis of the equinox he had observed and with a length of the tropical year of $365\frac{1}{4}$ days minus $\frac{1}{300}$ of a day. However, the actual date of the equinox in A.D. 140 was March 21, not March 22.

Probably Ptolemy had some trouble making this observation. He actually complains that the old bronze rings in the gymnasium square had settled and sometimes gave spurious results. Furthermore, the phenomenon of atmospheric refraction can also make the observation ambiguous. So far, however, no one has been able to give a completely convincing explanation of how Ptolemy could have misread his instruments to come up with figures that agree so well with the calculated numbers and disagree a day from the actual phenomena.

There is a further hint in the *Almagest* III, however, that his numbers are traditionally based on, and ultimately rooted in, Babylonian parameters. In addition to finding the length of the year Ptolemy seeks the length of the seasons, because these reveal the nonuniform motion of the Sun and tell him the eccentricity of the Sun's orbit. It is comparatively easy to get the times of the equinoxes because in March the Sun is moving northward quite perceptibly each day, and in September the Sun is going southward quite perceptibly. But, when the Sun rounds the northernmost portion of its path along the ecliptic in June, a week or two can pass with barely any change in the noonday height of the Sun; the same is the case in December when the Sun rounds the southernmost portion of its path. Yet Ptolemy comes up with a pretty good date for the solstice (a day wrong but agreeing with the error in the equinoxes), and hence with a reasonable value for the eccentricity of the Sun's deferent. Interestingly enough, however, the length of summer in the *Almagest* exactly matches the motion of the Sun on the slow arc in the Babylonian solar theory. It is possible that Ptolemy, in borrowing Hipparchus's solar theory, was completely unaware of this. In any event, it seems likely that the rather good value Ptolemy got for the solar eccentricity via Hipparchus stemmed at least indirectly from the Babylonians.

But how could the Babylonians find the length of the seasons so well, since it would have been no easier for them than for Ptolemy to find the time of the solstice by direct observation? The answer seems to lie in the idea that the Babylonian astronomy was thoroughly dependent upon lunar observations, and particularly on a long series of lunar eclipses. Over the past century the astronomical cuneiform tablets have gradually been deciphered, and one of the most surprising things that has emerged is the relatively high accuracy with which parameters can be extracted from very approximate observations. Provided there are enough records over a considerable period of time, even crude measurements furnish quite reliable figures for planetary periods and for their nonuniform motion along the ecliptic.

In particular, the Babylonians discovered that the lunar eclipses repeated in certain patterns, and that the possible eclipse positions were more crowded together in the direction of Sagittarius than in Gemini. This meant that the Sun was moving more slowly when it was in Gemini, and more rapidly in Sagittarius. From this observation it was possible to work backward and establish when the seasons began without actually making daytime measurements of the solstices.

There is, indeed, a third kind of year in Babylonian astronomy that is related to the Moon. The Babylonians had set the length of the average lunar synodic month, that is, the time between successive full moons, as

$$29^d 31'50''8'''20''''$$

or precisely the number later adopted by Ptolemy. The Babylonians (and the Greeks as well) knew that there were 235 lunations (full moons) in almost precisely 19 years, something now called the Metonic cycle. If one assumes this relation is exact and multiplies $\frac{235}{19}$ times the lunar synodic month above, the length of what I shall call the Metonic year is

$$365^{d}14'48''$$

which just happens to be Hipparchus's length of the year, the one quoted approvingly by Ptolemy and adopted by him. Hipparchus had found that this year differs from the sidereal year by about fifteen minutes (of time), leading to a very accurate value of the difference between the two systems of 1° per century. Unfortunately, the year of $365^{d}14'48''$ is about $5\frac{1}{2}$ minutes (of time) longer than the correct tropical year, a difference apparently too small to be noticed by Hipparchus but building up to a full day in the interval between Hipparchus and Ptolemy.

Thus it appears that Ptolemy, by believing that Hipparchus's year was essentially a correct tropical year, and by not observing times of the equinoxes carefully enough, spoiled the numerical foundations of his theory. As we shall see presently, instead of the 1° per century precession given by Ptolemy (which is accurate for the difference between the sidereal and Metonic years but not between the sidereal and tropical years) he should have had 1° per 70 years, an error that was to cause astronomers of the Middle Ages and Renaissance almost endless grief.

After considering the solar motion, Ptolemy turns to the Moon. In Book IV of the *Almagest* he presents the lunar theory received from Hipparchus, which, while quite satisfactory at times of new and full moons (when eclipses can take place), was pretty wretched at the quarters. Ptolemy ameliorates this deficiency in Book V, after detecting a new feature in the lunar motion that has come to be known as *evection*. This phenomenon, a periodic change in the *range* of the speeds of the Moon, can be interpreted as a regular variation in the eccentricity itself. Ptolemy coped with evection by placing the center of the lunar deferent on its own circle, and as it cranked the deferent nearer and farther from the Earth, the effective eccentricity varied. I suspect that Ptolemy worked unusually hard untangling the lunar motions, for the parameters are astonishingly accurate, better, for example, than those for Jupiter or Saturn.

The task of sorting out the Moon's motion was rendered especially difficult because the Moon, unlike the Sun or other planets, is close enough to the Earth to have a sensible parallax—that is, a change in its position depending on the location of the observer. Hence it is necessary to consider the lunar motion *as seen from the center of the Earth*.

When the Moon is directly overhead, so that the observer is on the line between centers of the Earth and Moon, there is no parallactic effect. On the other hand, when the Moon is on the horizon, the effect is the greatest and the apparent position of the Moon is raised by a degree— that is, twice its apparent diameter. Precisely how large this effect is depends not only on the Moon's altitude above the horizon but on its varying distance from the Earth. In trying to sort out the Moon's distance, Ptolemy had a real chicken-and-egg problem: he needed the observations to establish the model, but he needed the model to correct the observations for the distance.

Given the model of the Sun, established in Book III, and of the Moon, completed in Book V, Ptolemy is ready to take on eclipses, which follow in Book VI. Ptolemy's eclipse theory was a pioneering accomplishment, for this was the first time in history that solar eclipses could be calculated for a particular geographical location.

Only after the relatively complicated eclipse procedures are in hand does the Alexandrian astronomer move on to the much simpler topic of stars (Books VII and VIII), and here it becomes quickly apparent why this order is crucial to his presentation. Ptolemy has made the tropical year the essential basis for his astronomy, and from this follows the definition of his coordinate system. Celestial longitude is measured eastward along the ecliptic, so that the Sun and Moon always move to progressively higher numerical values. But where does the numbering begin? By definition, from the intersection of the ecliptic with the equator. Needless to say, this intersection is invisible in the sky and can only be deduced by the motion of the Sun. To transfer the invisible coordinates traced out by the Sun onto the nighttime sky, Ptolemy used lunar positions. Today we could use accurate clocks in conjunction with precise transit instruments to connect the coordinate system from the Sun with the nighttime sky, but for Ptolemy, who had only the most primitive clocks, it was essential to have the lunar theory available before discussing the positions of the stars.

Within Ptolemy's textbook everything is, for better or worse, tightly knit; in this case for worse, because his error of a day in the date of the equinox put the Sun behind by about a degree (since the Sun moves 360° in 365 days, and therefore approximately a degree a day). Hence the entire coordinate frame was zeroed about one degree beyond the actual intersection point of the ecliptic and equator. This meant that all the stars listed by Ptolemy had systematically erroneous longitudes, too low by about 1°12'.

Now the errors begin to compound. O unlucky Ptolemy! As I have said, the length of the sidereal year, dependent on the stars, differs from the length of the tropical year, dependent on the time of the equinoxes, by an amount called the precession of the equinoxes. Because Ptolemy's star positions are wrong, when he calculates their differences with respect to the positions in Hipparchus's day ($2\frac{2}{3}$ centuries earlier), the result is too small, only 2°40′ instead of 3°52′, and Ptolemy ends up with a precession of 1° per century instead of the correct 1° per 70 years. It is no wonder that by the time of Copernicus there were some obvious problems with astronomy, although this was surely not one that cried for a heliocentric solution!

The source of Ptolemy's star catalog has long been a matter of contention. Ptolemy describes his observing procedures, but, had an existing catalog been conveniently available, it is likely that he would have made good use of it, just as any subsequent astronomer would have used Ptolemy's as a starting point for his own work. Where Ptolemy has leaned heavily on earlier astronomers such as Hipparchus (in the solar and lunar theories) or Apollonius (in the theorem for retrogressions), he gives them credit, and more generously than most ancient writers who reprocessed material from their predecessors. He credits Hipparchus and Timocharis with certain specific observations, but not with respect to the catalog itself, where he says, "We have not used for each of the stars altogether the same formulations as our predecessors, just as they did not use the same as their predecessors." I suppose Ptolemy does not wish to attribute his specific ordering of the stars to them, but he may well have taken many of the positions from a catalog made for the epoch of Hipparchus and simply updated it by adding 2°40′ to each position.

In the final five books of the *Almagest*, Ptolemy addresses the central problem of his work, the modeling of the planets. We have already examined in detail how Ptolemy may have gone about this and how the tables work for the longitude of Mars. It remains to comment on the peculiarly complicated model for the planet Mercury and to mention briefly his latitude theory.

Because Mercury stays comparatively close to the Sun in the sky, it is difficult to get the proper observations to establish its motions. In particular, Ptolemy failed to get observations of the morning elongations from the Sun in the spring and the evening elongations in the fall, something he mentions explicitly. Had Ptolemy stuck with the model he had already achieved for Mars, his results with Mercury would have been ultimately more satisfactory, but this might have forced him to approximate some of his observations. Instead, he took some comparatively poor observations much too seriously and was therefore obliged to invent a second, inner wheel (similar to the scheme for the Moon) in order to reproduce these data. Since Mercury was the second fastest moving object after the Moon, it was placed immediately beyond the Moon's orbit and inside the model for Venus; perhaps Ptolemy thought it might be appropriate for such a transition planet to have a more complicated mechanism something like the Moon's. Like Ptolemy, Copernicus found it impossible to get those key observations of Mercury, and so he simply transferred Ptolemy's model to the heliocentric arrangement. Not until Kepler was the Mercury mechanism made similar to the other planets, resulting in a highly successful prediction of the first observed transit of Mercury across the face of the Sun.

Ptolemy's *Almagest* concludes in Book XII with the theory of planetary stationary points and retrogressions, using (as mentioned earlier) the theorem of Apollonius, and in Book XIII with the theory of latitudes. It is fascinating to note that he uses somewhat different modeling for the latitudes, which reinforces the view that Ptolemy was more interested in the predictive results than in the physical reality of his mechanisms.

Subsequently Ptolemy wrote a more cosmologically oriented work, the *Planetary Hypotheses*, in which he described how the individual mechanisms for the various planets could be assembled into a whole, truly a "Ptolemaic system"; the numbers therein suggest that he was still carrying out improvements to his theory. In the ensuing years, the criticisms of the *Almagest* centered primarily on the philosophical aspects, such as whether Ptolemy correctly preserved the principle of uniform circular motion, and very little on his specific choice of parameters. It is some of these philosophical critiques that we next examine.

Prelude to Copernicus

In the centuries that followed Ptolemy's *Almagest*, astronomy fell into a serious decline. Among the Islamic astronomers were a number who commented intelligently on the *Almagest* and who even criticized it, but no one systematically reworked all the parameters or amended the models to make them more accurate. Part of the reason was a great dearth of suitable observations. From the death of Ptolemy until the birth of Copernicus in 1473 there exist in the West not more than a dozen records with accurate planetary positions.

Nevertheless, it gradually became apparent that Ptolemy's value of the precession was untenable or at least incomplete. Observations revealed that the stars were drifting through the coordinate system more rapidly than 1° per century as specified by Ptolemy. Yet no one had the audacity to challenge the observations recorded by the Alexandrian astronomer, for that would have undermined the entire foundation of astronomy. Instead, an ingenious system of variable precession was introduced in just such a way as to preserve the values between the times of Hipparchus and Ptolemy, but to give a faster rate in the centuries that followed. Called *trepidation*, the new feature was invented by Islamic astronomers and incorporated into astronomical procedures prior to A.D. 1000.

By the thirteenth century, when the astronomers of King Alfonso X of Spain prepared new and handier tables for finding planetary positions, trepidation had been codified into a standard method. A famous but somewhat dubious anecdote survives from that time: Alfonso is reputed to have told his astronomers, after looking over their work, that if he had been around at creation he could have given the Good Lord some hints. From this has grown a legend that the Alfonsine tables were encumbered with epicycles on epicycles, and that as a consequence the entire Ptolemaic system was crying for reform in order to reduce its incomprehensible complexity. Nothing could be farther from the truth. In the absence of systematic observations during the Middle Ages, there was no basis for making the planetary models more complicated, except for adding the table of trepidation. Furthermore, the clever method of the Ptolemaic tables in coping with the corrections by single-entry tables would have been vitiated by the addition of any independent epicycles.

In the eastern part of the Islamic world, however, astronomers did begin to experiment with new planetary models involving one or more additional epicycles. But, contrary to what we might imagine from a modern viewpoint, such models were not invented to introduce more accuracy into the procedures. Their entire purpose was to satisfy certain philosophical requirements. These little epicycles, or epicyclets, were designed to replace the equant, partly in order to preserve the ancient requirement of *uniform* circular motion, and partly to set up a strictly mechanical model of the planetary system.

In the world view of Aristotle, each planet occupied its own concentric zone of aethereal material, and these spheres were tightly nested together. Ptolemy attempted to preserve certain features of this scheme, and in his *Planetary Hypotheses* he calculated the distances of the planets on the assumption that the epicyclic mechanisms of the successive planets were stacked together as compactly as possible, yet spaced just far enough apart so they would not collide.

By the Middle Ages the Aristotelian aether was envisioned as a hard crystalline substance, smooth and transparent, out of which the planetary spheres were constructed. It was easy to conceive of the epicycle as made of hard crystal, and gliding without friction in a rigid eccentric circular sleeve surrounded by more hard crystal, but it was impossible to construct a mechanical linkage to the equant point without conflicting with other crystalline material. For example, if all the mechanisms were nested one inside another, then the equant point of the large outer assembly for Saturn would fall within the crystal spheres for Mars. Clearly if the Ptolemaic system was to be reconciled with an Aristotelian picture of physical reality, some modification was needed. It was the failure of Ptolemy's model to provide a philosophically acceptable picture that brought his astronomy under increasing criticism throughout the Islamic period, and it was for this reason that astronomers of the thirteenth-century Maragha school (in present-day Iran) explored the epicyclic alternatives to the Ptolemaic equant. (The Maragha astronomers likewise worried about the other two cases in which Ptolemy had used an interior mechanism: for Mercury and for the Moon.) Incidentally, the additional epicyclets adopted by these astronomers did not introduce a further independent motion because they were constrained to duplicate the effects of the equant; hence the appearance of the resulting tables was the same as before.

Precisely how much this philosophical attack on Ptolemy filtered into the Latin West is still difficult to assess, but undoubtedly it helped shape the climate of opinion at the end of the fifteenth century when Copernicus was a student. In addition, the temporary fix provided by the trepidation was coming undone; the trepidation had been set up to provide a variable rate of precession, and by 1500 its effect was diminishing, contrary to the witness of the stars. Some kind of reform was desirable on both accounts, yet neither the criticism of the equant nor the failure of precession theory called for anything as radical as the introduction of an entirely new cosmology. In order to come to grips with that larger issue, we now turn to Copernicus himself.

Copernicus and the Heliocentric Cosmology

Nicolaus Copernicus was born in Toruń, Poland, on February 19, 1473. After his father died, his maternal uncle provided for his education and later arranged for him to become a canon and lawyer at the Cathedral of Frombork (Frauenburg), the northernmost Catholic diocese in Poland. At the historic and flourishing university in Cracow, Copernicus presumably studied the standard medieval curriculum, which included the *Sphere* of Sacrobosco. This beginning textbook gave a very elementary account of spherical astronomy. Probably Copernicus also encountered the second-level text, Peurbach's *Theoricae novae planetarum*, the "New Theories of the Planets," which was a relatively new textbook but not a new theory. While in Cracow, Copernicus took up astronomy with sufficient enthusiasm to buy his own printed copy of the *Alfonsine Tables*, and his interest continued when he went to Italy for graduate work, even though his official studies were canon (church) law and medicine.

How early Copernicus developed his radically new heliocentric cosmology we do not know for certain. He might have got the first inkling already during his undergraduate days in Cracow, or perhaps in Renaissance Italy, or possibly not until he had returned to northern Poland to work as private secretary and physician to his uncle, who had become bishop. The first indirect hint of his novel ideas comes down to us from 1514 in the inventory of the books of a Cracow scholar, whose library contained a manuscript by "someone who held that the Sun stands still while the Earth moves." Entitled the *Commentariolus* or "Brief Treatise," it represents Copernicus's earliest known foray into astronomy. Closely modeled on Peurbach's *Theoricae*, it criticizes the standard geocentric astronomy and outlines a very different heliocentric arrangement. Without discussing specific observations or presenting tables of planetary motions, the *Commentariolus* includes enough technical details to show key differences from the final version of the Copernican system as given at the end of his life in his major treatise, the *De revolutionibus orbium coelestium*, "On the Revolutions of the Heavenly Spheres."

Copernicus opens the *Commentariolus* with a specific complaint against the equant, a theory that appeared "neither sufficiently perfect nor pleasing to the mind." He then lists seven assumptions, the two most important being

> (3) All the spheres surround the Sun as though it were in
> the middle of all of them, and therefore the Sun is near the
> middle of the universe,

and

(7) What appears as the direct and retrograde movements arises not from the planets themselves, but from the
Earth. This motion alone therefore suffices to explain many
apparent irregularities in the heavens.

These premises form both the basis of, and one of the chief arguments
for, the new cosmology; they are important for understanding Copernicus's defense, but they tell little about the genesis of the idea in his
mind. Displeasure with the equant may have turned him toward an
examination of the accepted planetary modeling; similarly, the problems with the Ptolemaic coordinate system (that is, precession and
trepidation) may also have played a role, since immediately after a
short discussion of the solar theory Copernicus turns to a section
entitled "That uniform motion should be referred not to the equinoxes
but to the fixed stars." Yet neither problem has as its solution a heliocentric cosmology as such.

In the case of Ptolemy, we have only the surviving works themselves
as evidence for his thought processes. For Copernicus we do have a
few background materials: the much worked-over manuscript of his
De revolutions survives in Cracow, plus some working notes in books
now preserved in Sweden where they were taken during the Thirty
Years' War. One of the manuscript pages gives some highly suggestive
clues as to the route traveled by Copernicus in his examination of the
existing planetary models. The argument is subtle, but a modern figure
helps us understand the path Copernicus may have taken (Fig. 3).

In the figure the solid lines represent the Ptolemaic model for Mars.
The diagonal arrow represents the direction from the Earth to Mars,
which is determined by the motions of the two circles. Of course the
primary goal of a planetary model is to give such a direction (i.e., the
geocentric longitude), and the procedure can rest on purely fictitious
intermediate mechanisms. It is instructive to consider an alternative
construction, represented by the shaded lines in the diagram; it can be
quickly verified that this variant arrangement will continue to give
identical answers for the direction to Mars.

The basic difference in the two schemes is a philosophical one: if the
smaller circle is placed in the center, with the larger one riding on it,
then the larger circle will cut through the smaller in a way that makes
it difficult to think of the circles as rigid crystalline spheres. This
interpenetration of the two spheres is a mighty obstacle to anyone
taking the idea of solid spheres seriously. We know that Copernicus
considered this arrangement, but we suspect that he considered the
interpenetration of the spheres such a drawback that he quickly dis-

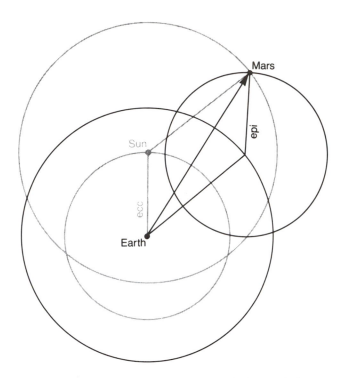

FIGURE 3. *The Ptolemaic system (solid lines) and the geo-heliocentric system with its interpenetration of circles for Mars (shaded lines).*

missed it as a viable possibility, even though he does not explicitly say so. In the generation after Copernicus, Tycho Brahe made the arrangement the basis of a new cosmology, and he is quite explicit in saying that for a long time he was hung up on precisely this point about the larger circle cutting through the smaller.

The alternative arrangement (the one shown in the shaded lines) has one feature of great interest: when the large and small circles are interchanged, then the direction from the Earth to the center of the larger circle (labeled "ecc") always turns out to be the direction to the Sun! And since the Ptolemaic system gives only relative distances, it is possible to scale the Sun's orbit so that the Sun occupies this point. Now, instead of having two separate mechanisms, one for Mars and another for the Sun, both have been combined into a single system. This is an extremely powerful insight, the crucial beginning of the entire Copernican system. Surely such a detail had been noticed by Ptolemy, and if by some incredible stroke he overlooked it, then it must have been well known by the Middle Ages. Indeed, the *Alfonsine Tables* take implicit advantage of this possibility in their very organization. But to accept this combination as a physically real picture of the universe meant giving up the whole substance that supported and turned the heavens, the crystalline spheres. This was a formidable leap, one not easily taken.

If Mars and the Sun can be combined, is it possible to combine the Sun with other planets? And if so, would it be possible to increase or diminish the sizes of the circles so as to fit, say, Mars, Jupiter, and the Sun all together? The answer is yes, but it was philosophically very troubling for astronomers accustomed to considering each planet separately.

At this crucial juncture we pause, recognizing that this step must have been psychologically exceedingly difficult, or else it would have been explored much earlier by other astronomers. And we pause to look at the evidence suggesting that Copernicus got to this point and hesitated, appalled at the idea of destroying the crystalline structure of the universe.

The two alternative arrangements in the diagram carry a differing description of the circles. The small solid circle is of course the *epicycle* of the Ptolemaic model. But when it is moved to the interior position, it now delineates the movable *eccentric* center of the larger circle. And remarkably, on a single page in one of Copernicus's books we find a series of numbers labeled "ecc" agreeing exactly with the numbers that Ptolemy sets for the epicycle sizes. There follows a second table, in which each planetary mechanism is scaled so that the Earth-Sun distance (the "ecc") is the same, and all can be superimposed.

At once a common measuring stick appears: each large planetary circle is now scaled and fixed *with respect to the Earth-Sun distance.* "We find that the world has a wonderful commensurability," declares Copernicus in one of his most poetic passages, referring to this Earth-Sun standard, the "common measure" of the universe.

But nowhere does Copernicus mention this Earth-centered arrangement in which the Sun, with its entourage of planets, marches around the Earth in annual progression. It remained for Tycho Brahe, the Danish observer and cosmographer, to propose it as a serious alternative in 1588. Why does not Copernicus settle for such a scheme, which preserves not only the ancient physics but a literal reading of the Holy Scriptures as well? For Tycho to defend this geo-heliocentric arrange-

ment, he had to take the iconoclastic stand that the crystalline substance of the heavens was a mere fiction invented by the ancients. For
Copernicus, who followed his fellow Renaissance humanists in looking
back to a golden antique age when knowledge was still pure and un-
forgotten, such a rejection of ancient wisdom was perhaps too much to
consider. And without the spheres, what could keep the planets in
their fixed paths? Possibly Copernicus was also an idealist who saw in
his mind's eye something still more compelling, something so remarkable and beautiful that he was willing to accept it despite the resulting
conflict with traditional physics.

What Copernicus noticed was that the amalgamated arrangement
placed Mercury, the swiftest planet, in the smallest orbit; Saturn, the
slowest, fell much the farthest from the Sun, and the others were
placed in a harmonious progression in between, as shown below:

Planet	Relative distance	Sidereal period in days
Mercury	0.387	88
Venus	0.723	225
Earth	1.000	365
Mars	1.52	687
Jupiter	5.20	4333
Saturn	9.54	10759

Did it not make sense to assign to the Earth rather than to the Sun the
365-day period, which fit so nicely between the 225 days for Venus and
the 687 days for Mars? Besides, to place the Earth in motion around
the Sun, rather than vice versa, immediately cleared up the problem of
the interpenetration of the spheres!

Copernicus never breathes a hint of that intermediate stepping stone
in the path from the geocentric to the heliocentric system. Rather, he
proclaims the obvious uniqueness of the Sun: "Who would place this
lamp. . .in another or better place than this, from which it can illumi-
nate everything at the same time?" He sings of the harmonious arrangement of larger orbs for slower planets: "There is a sure bond of
harmony for the movement and size of the orbital spheres that can be
found in no other way." And he proceeds to explain how the different
sizes of the retrograde loops for the various planets now have a natural
explanation. Copernicus's disciple, Georg Joachim Rheticus, waxes
rhapsodic about this point: "For all these phenomena appear to be
linked most nobly together, as by a golden chain; and each of the
planets, by its position and order and every irregularity of motion,
bears witness that the Earth moves."

It is perhaps not surprising that Kepler, who comes after Tycho, argues more forcefully against the geo-heliocentric arrangement. Tycho, in defending this intermediate arrangement that he had adopted, had remarked, "Copernicus nowhere offends the principles of mathematics, but he throws the Earth, a lazy, sluggish body unfit for motion, into a motion faster than the aetherial torches (the stars)." In the introduction to his *New Astronomy*, Kepler argues from a newly emerging physics:

> Furthermore, it is very likely that the source of the Earth's motion is at the same place as that of the other five planets, namely, in the Sun. For this reason it is therefore very probable that the Earth moves, since there is an apparent, plausible cause for its motions. . . . But let us consider the bodies of both the Sun and the Earth: which would be the better source of motion for the other? Does the Sun, which moves the other planets, move the Earth, or does the Earth move the Sun, itself the mover of the other planets and so many times larger than the Earth? Let us not be forced to concede that the Sun is moved by the Earth, which is absurd. We must concede immobility to the Sun and motion to the Earth.

Regardless of the path to the new cosmology or the arguments raised in its defense, Copernicus had taken a bold and provocative step, one that required both imagination and courage. He also had the intellectual perception to know that its serious consideration would require more than a simple statement of the new blueprint for the planetary system. Unlike others who may have entertained that same speculation, Copernicus went on to place his ideas into a complete context—that is, a book from which one could actually calculate planetary positions from the heliocentric arrangement.

Copernicus's *On the Revolutions of the Heavenly Spheres*

In the *Commentariolus* Copernicus remarked that "I have decided here, for the sake of brevity, to omit the mathematical demonstrations, which are planned for a larger book." There is no way to know precisely what Copernicus then had in mind, but by the time he had acquired his copy of the *Almagest*, first printed in 1515, he must have realized that for his new theory to rival Ptolemy's, it would have to be set forth in a treatise of comparable structure and magnitude. He labored over his *magnum opus* for almost 30 years, and when he was nearly 70 he was finally persuaded to let it go off to a printer in Germany, even though he was still in the process of making its numbers more self-consistent.

It is possible that Copernicus originally intended to call his larger work by the full title *De revolutionibus orbium coelestium*, "On the Revolutions of the Heavenly Spheres," although there is good evidence that he later changed his mind and preferred to omit *orbium coelestium*. The "heavenly spheres" may have played a key role at first when he opted for a heliocentric system (rather than the geo-heliocentric version Tycho was later to adopt) in order to avoid the interpenetration of the spheres of Mars and the Sun. But evidently his interest in the spheres faded, for his *On the Revolutions* contains neither a description of the crystal spheres nor an indictment of the equant (although he worked systematically to eliminate it by means of a small epicyclet included for each planet).

Now, in defending his new cosmology in his *On the Revolutions*, Copernicus returns to the power of unity, the idea that everything is linked together by the common measure of the Earth-Sun standard:

> Moreover, they [the ancients] have not been able to discover or to infer the chief point of all, i.e., the form of the world and the certain commensurability of its parts. But they are in exactly the same fix as someone taking from different places hands, feet, head, and the other limbs—shaped very beautifully but not with reference to one body and without correspondence to one another—so that such parts made up a monster rather than a man.

One of the essential differences between *On the Revolutions* and the *Almagest* is the vigorous defense of its cosmology, something that Ptolemy could rather easily dismiss because his readers would have automatically assumed a geocentric framework. Copernicus's arguments rise to a brilliant and glorious polemic in favor of the Sun-centered system by Chapter 10, where, in what is easily the finest passage of the entire book, he proclaims, "In the center of all rests the Sun. . . as if on a kingly throne, governing the family of stars that wheel around." Nevertheless, this opening cosmological section constitutes a mere 4 percent of the treatise, extending through only eleven chapters of the first of its six books. The remaining 96 percent is a technical treatise as formidable as the *Almagest* and in many places parallel to it.

Like the *Almagest*, *On the Revolutions* includes near the beginning a section on mathematics, complete with a trigonometrical table; but Copernicus can rely on more modern and much handier spherical trigonometry, and instead of chords (despite the table headings) he presents the sine function. In Book II, however, a fundamental distinction occurs: because for Copernicus the "fixed stars" really are fixed and are providing the basic reference frame, he at once provides a star catalog. Unlike Ptolemy, Copernicus makes no claims of having observed the catalog himself; in fact, he states right out that he has for the most part borrowed it from Ptolemy.

Only after discussing the stars does Copernicus turn (in Book III) to the complex threefold motion of the Earth. The Earth not only spins on its axis and revolves about the Sun, its axis slowly changes direction, producing the precession-trepidation phenomenon. Copernicus combines the revolution and the precession somewhat differently than we would today, but the results are the same apart from the fact that we no longer accept an oscillatory trepidation superimposed on the steady precessional motion. More interesting is Copernicus's almost mystical linking of the changes in the eccentricity of the Earth's circular orbit with the direction of that displacement, with the change in obliquity of the ecliptic, and with the period of trepidation. The actual numbers were not well known and were affected as well by the problems of Ptolemy's reference framework. The fact that Copernicus could link all these phenomena together by the single remarkable number of 1717 years is an astonishing display of his intuitive belief in the unity of nature. Such details impressed all the astronomers of the century, whether or not they accepted the heliocentric hypothesis. There was a touch of relativity here: what worked for the Earth's orbit would by reflex also work for the Sun's orbit about a fixed Earth.

In Book IV Copernicus takes up the orbit of the Moon, and here he brings against Ptolemy a perceptive criticism that had already been noticed in the Islamic period. In the *Almagest*'s model, the Moon was sometimes as far as 64 Earth radii and sometimes as close as 34. This should have caused the diameter of the Moon to appear half as large in diameter when it was far compared to when it was close, a variation never observed. "Experience and sense-perception teach us that the parallaxes [i.e., distances] of the Moon are not consonant with those which the ratio of the circles promises," remarks Copernicus in Book IV,3. Although his own model in no substantial way improves the longitude or latitude predictions of Ptolemy, it does greatly ameliorate the problem of the apparent size. In the Copernican model, the Moon varies in size by about ± 10 percent, approximately twice the actual change, but a vast improvement over the range in Ptolemy's lunar mechanism.

In both the lunar theory and in the planetary longitudes that follow in Book V, Copernicus takes considerable pains to eliminate the detested Ptolemaic equant. The equivalence between the equant and Copernicus's epicyclet is diagramed in Fig. 5 of selection 9, p. 175 in this volume. If the distance between the Earth and equant in the Ptolemaic arrangement is twice the eccentricity e, then in the equivalent Copernican layout the center of the orbit is placed at $\frac{3}{2}e$ from the Sun and the radius of the epicyclet is $\frac{1}{2}e$. The motion in the epicyclet is uniform, so that a regular trapezoid is generated as shown, with the dashed line always behaving the same way as would the line from the equant. The effective orbit, shown by the dashed curve, is not exactly a circle, but slightly fattened.

In discussing the planetary longitudes in Book V, Copernicus relies on the observations of the *Almagest*, checked by four modern ones of his own. Thus Copernicus, like Ptolemy, gives only the minimum number of observations for finding the parameters. His own observations are not particularly good; in fact, they can be quite dreadful, such as one of Mars that is $2°$ off the mark! For Mercury, which hovers quite close to the Sun, Copernicus borrowed observations from his near contemporary Bernard Walther. Since these observations were later independently published, we can see that Copernicus gently adjusted them to get them to fit the model.

The advantage of a longer temporal baseline enabled Copernicus to reset the hands of his clockwork, as it were, without changing the mechanism itself. That is, his heliocentric framework with circular orbits and epicyclets simply mapped onto Ptolemy's geocentric framework with its use of the equant, and these changes made essentially no difference with respect to the technical machinery. Indeed, it requires only minor adjustments to use the tables of *On the Revolutions* to predict planetary positions compared to the procedure outlined earlier for the *Almagest*.

Because there were few accurate observations before the time of Tycho Brahe, contemporary astronomers could not, in fact, tell if Copernicus's tables were giving better results than Ptolemy's. Despite this, most of the sixteenth-century astronomers calculating almanacs did convert to the more modern tables. Today we can use electronic computers to calculate where the planets were in the 1500s, and we find that although Copernicus actually improved most positions, the improvement was not enough to get excited about, especially considering how long it had been since the *Almagest* was written. As for the planetary latitudes, discussed in Book VI, Copernicus's work was a disaster. He neither took advantage of the heliocentric geometry to straighten out the *Almagest*'s theory of latitudes, nor did he reobserve the lines of nodes that determine when a planet is north or south of the ecliptic. To be sure, Copernicus here inherited the worst part of the *Almagest*, which was without specific observations, but it was nonetheless an opportunity lost. It would remain for Kepler to use the latitudes in an integrated way with the theory of longitudes, thereby producing a powerful lever for advancing planetary theory.

The Reception of Copernican Theory

Victorian rationalists, who created such concepts as "martyrs of science" to describe the effects of the religious opposition eventually encountered by the ideas of Copernicus and Galileo, assumed that Copernicus hesitated to publish his book because of an expected adverse reaction from the Catholic Church. But there were at least two other reasons, probably much more likely, that prevented him from trying to disseminate his ideas. On the one hand, he was far from the major international centers of printing that could profitably handle a book as large and technical as *De revolutionibus*. On the other, his manuscript was still full of numerical inconsistencies, and he knew very well that he had not taken complete advantage of the opportunities that the heliocentric viewpoint offered. For example, the theory of Mercury retained curious motions with the period of a year because not all the geocentric elements had been purged from its construction.

Furthermore, Copernicus was far from academic centers, thereby lacking the stimulation of technically trained colleagues with whom he could discuss his work. Then, in 1539, the situation changed: a young professor from the Lutheran University of Wittenberg arrived to learn about the new cosmology. Georg Joachim Rheticus became Copernicus's first and only disciple, and it is owing to his enthusiasm that Copernicus allowed a copy of the manuscript to be taken to Germany for printing. In 1542 Rheticus took the copy to Nuremberg, where the typesetting began. Presently Rheticus received a tempting offer to become professor at Leipzig, so he left the work in the hands of the printer Petreius and his clergyman-proofreader, Andrew Osiander.

The last pages to be printed in the spring of 1543 were the first leaves of the book, and here Osiander conspired with the printer to add an anonymous introduction to safeguard its contents. "These hypotheses need not be true nor even probable," he wrote, "as long as they provide a calculus that fits the observations. . . . Let no one expect truth from astronomy, lest he leave a bigger fool than when he entered." When Copernicus, then on his deathbed, saw these pages, he was much agitated. But whether he was well enough to appreciate their contents is not at all clear, and perhaps he was only excited to realize that his life work was at last on the verge of publication.

In any event, this anonymous introduction certainly spared the book the theological criticism it might otherwise have attracted. Throughout the Lutheran university system the book was systematically studied by the advanced students; in the elementary courses Copernicus was held in high regard, but students were carefully shielded from his heliocentric doctrine. The book had been dedicated to Pope Paul III, but because it was published within the Lutheran sphere, relatively few copies of the first edition found their way to England or to Italy and the other Catholic countries. Probably fewer than 500 copies were originally printed, and after the Nuremberg edition went out of print, a second edition was published in 1566 in Basel. This edition was well distributed both in England and the Catholic countries, so by the beginning of the seventeenth century Copernicus's ideas were readily available, even though the heliocentric cosmology was rarely taught openly and then almost always as a hypothesis, just as the anonymous introduction declared.

The outstanding astronomer in the interval between Copernicus and Kepler was Tycho Brahe, the eminent Danish observatory builder. Like the other leading astronomers of the day, he was attentive to Copernicus's innovations, and he found the interlinking of the planets and the Sun particularly appealing. But he had considerable difficulty in accepting it as a physically real system because it contradicted the accepted Aristotelian physics and apparently also the Bible. Unwilling and unable to forge a new physics, he chose instead to modify the cosmology to the geo-heliocentric form. Of course, this shattered the crystalline spheres, a radical step he eventually espoused. That left the heavens with no obvious means of generating their motions, something previously supposed to have been supplied by God at the outside edge of the nested spheres and transmitted inward through the crystal machinery. It is at this point in the story that the young German astronomer Johannes Kepler arrived on the scene, eager to restore some kind of celestial physics.

Kepler

Johannes Kepler was conceived on the 16th of May, 1571, at 4:37 A.M. and born on the 27th of December at 2:30 P.M. Such minutely kept dates remind us that Kepler lived in an age when "astronomer" still meant "astrologer" and when the word *scientist* had not yet been invented. Like many of the world's greatest scientists, including Ptolemy and Copernicus, Kepler had a profound feeling for the harmony of the heavens; although he rejected most of the traditional details of astrology, he believed in a powerful accord between the cosmos and the individual.

There was little in Kepler's youth to indicate that he would become one of the foremost astronomers of all time. A weak and sickly child, but intelligent, he easily won a scholarship to the nearby Tübingen University so that he could study to become a Lutheran clergyman. In recommending him for a scholarship renewal, the University Senate noted that Kepler had "such a superior and magnificent mind that something special may be expected of him."

Yet Kepler himself wrote in his *New Astronomy*, Chapter 7, that, although he had done well enough in the prescribed mathematical studies, nothing indicated to him a special talent for astronomy. Hence he was surprised and distressed when, midway through his third and last year as a theology student at Tübingen, he was summoned to Graz, far away in southern Austria, to become an astronomy teacher and provincial mathematician. It was there in one of his astronomy lectures that he hit upon what he believed to be the secret key to the construction of the universe.

This key hung upon a crucial thread: at Tübingen, Kepler had become a Copernican. The astronomy teacher at the University, Michael Maestlin, was remarkably knowledgeable about Copernicus's *On the Revolutions*. In his lectures, Maestlin explained how the new Copernican system accounted for the retrogradations in a most natural way, and how the planets were laid out in an elegant harmonic fashion both with respect to their spacing from the Sun and to their periods.

It was undoubtedly the beautiful harmonic regularities so "pleasing to the mind" that appealed strongly to Kepler's sense of the aesthetic and induced him to become such an enthusiastic Copernican. To Kepler the theologian, such regularities revealed the glory of God. When he finally hit upon that secret key to the Universe, he attributed it to Divine Providence. "I believe this," he wrote, "because I have constantly prayed to God that I might succeed if what Copernicus had said was true." Later, in writing to his teacher Maestlin, he said, "For a long time I wanted to become a theologian; for a long time I was restless. Now, however, behold how through my effort God is being celebrated in astronomy."

Because of his preoccupation with the Copernican system, Kepler had begun to ask himself three unusual questions: Why are the planets spaced this way? Why do they move with these regularities? Why are there just six planets? All these questions are very Copernican, the last one particularly so because a traditional geocentrist would have counted both the Sun and the Moon, but not the Earth, thereby listing seven planets.

In illustrating to his class how the great conjunctions of Jupiter and Saturn fall sequentially along the ecliptic, Kepler drew a series of quasi-triangles whose lines began to form an inner circle half as large as the outer ecliptic circle. The proportion between the circles struck Kepler's eye as almost identical with the proportions between the orbits of Saturn and Jupiter. Immediately he began a search for a similar geometrical relation to account for the spacing of Mars and the other planets, but his quest was in vain.

"And then it struck me," he wrote. "Why have plane figures among three-dimensional orbits? Behold, reader, the invention and the whole substance of this little book!" He knew that there were five regular polyhedra, that is, solid figures each with faces all of the same kind of regular polygon. By inscribing and circumscribing these five figures between the six planetary spheres (all nested in the proper order), he found that the positions of the spheres closely approximated the spacings of the planets. Since there are five and only five of these regular or Platonic polyhedra, Kepler thought that he had explained the reason why there were precisely six planets in the solar system.

Kepler published this scheme in 1596 in his *Mysterium cosmographicum*, the "Cosmographic Secret." It was the first new and enthusiastic Copernican treatise in more than 50 years, since *De revolutionibus* itself. Without a Sun-centered universe, the entire rationale of the book would have collapsed.

The young astronomer realized, however, that Copernicus had made the Sun immobile without actually using it as his central reference point; rather, he had used the center of the Earth's orbit. Although the Sun was nearby, it played no physical role. But, Kepler argued, the Sun's centrality is essential to any celestial physics, and the Sun itself must supply the driving force to keep the planets in motion. Not only did he propose this critically significant physical idea, he attempted to describe mathematically how the Sun's driving force diminished with distance. Again, his result was only approximate, but at least the important physical-mathematical step had been taken. This idea, which was to be much further developed in his *New Astronomy*, establishes Kepler as the first scientist to demand physical explanations for celestial phenomena. Although the physical explanation of the *Mysterium cosmographicum* was erroneous, never in history has a book so wrong been so seminal in directing the future course of science.

Kepler sent a copy of his remarkable treatise to the most famous astronomer of the day, Tycho Brahe. Unknown to Kepler, the renowned Danish astronomer was in the process of leaving his homeland. Tycho had boasted that his magnificent Uraniborg Observatory had cost the king of Denmark more than a ton of gold. Now, however, fearing the loss of royal support at home, Tycho had decided to join the court of Rudolf II in Prague.

Kepler describes the sequence of events in his greatest book, the *Astronomia nova* (or *New Astronomy*). Tycho had been impressed by the *Mysterium cosmographicum*, though he was unwilling to accept all its strange arguments; then, Kepler writes,

> Tycho Brahe, himself an important part in my destiny, did not cease from then on to urge me to visit him. But since the distance of the two places would have deterred me, I ascribe it to Divine Providence that he came to Bohemia, where I arrived just before the beginning of the year 1600, with the hope of obtaining the correct eccentricities of the planetary orbits. . . . Now at that time Longomontanus had taken up the theory of Mars, which was placed in his hands so that he might study the Martian opposition with the Sun in 9° of Leo [that is, Mars near perihelion]. Had he been occupied with another planet, I would have started with that one. That is why I consider it again an act of Divine Providence that I arrived at the time when he was studying Mars; because for us to arrive at the secret knowledge of astronomy, it is absolutely necessary to use the motion of Mars; otherwise it would remain eternally hidden.

Kepler's *Astronomia nova* was not to be published until nine years later, in 1609. His greatest work, it broke the two-millennium spell of perfect circles and uniform angular motion—it was truly the "New Astronomy." Never had there been a book like it. Both Ptolemy in the

Almagest and Copernicus in *De revolutionibus* had carefully dismantled the scaffolding by which they had erected their mathematical
models. Kepler's book is nearly an order of magnitude more complete
and complex than anything that had gone before, but he himself admits that he might have been too prolix.

Kepler had quickly perceived the quality of Brahe's treasure of
observations, but he had realized that Tycho lacked an architect for
the erection of a new astronomical structure. A devoted Copernican,
he nevertheless recognized certain shortcomings in the *heliostatic* system described in *De revolutionibus*, and he was determined to derive a
truly *heliocentric* system in which the Sun played a vital physical role
in keeping the planets in motion.

In the first three months in Prague, he established two fundamental
points: first, the orbital plane of Mars had to be referred to the Sun
itself, and not to the center of the Earth's orbit, as Copernicus had
assumed; and second, the traditional eccentric circle of the Earth-Sun
relation had to be modified to include an equant or its equivalent.
Although the other planetary mechanisms had traditionally employed
the equant (or the equivalent Copernican epicyclet), the Earth-Sun
system did not. Hence, it was of paramount importance to Kepler's
physics to prove that the Earth's motion resembled those of the other
planets, and this he accomplished by an ingenious triangulation from
the Earth's orbit to Mars.

In contrast to Copernicus, Kepler had no objections to using the
equant as a mathematical tool, but unlike all previous astronomers, he
decided to allow the equant to fall an arbitrary distance beyond the
center of the circular orbit. In this way by the spring of 1601 he had
been able, with the help of Tycho's accurate observations, to achieve a
far better solution for Mars's longitudes than any of his predecessors.
By a series of iterations he established a model accurate to 2′; "If you
are wearied by this tedious procedure," he implored the readers of the
New Astronomy, "take pity on me who carried out at least 70 trials."

From the predicted latitudes, however, Kepler realized that his
scheme gave erroneous distances; again, unlike previous astronomers,
who were satisfied with separate models for longitudes and latitudes,
Kepler sought a unified, physically acceptable description. To obtain
the correct distances his physical model demanded, he was obliged to
reposition his circular orbit with its center midway between the Sun
and the equant, but this move destroyed the excellent results he had
previously found for the longitudes. The errors now rose to 8′ in the
octants of the orbit, and in the *New Astronomy* Kepler goes on to say,

"Divine providence granted us such a diligent observer in Tycho Brahe that his observations convicted this Ptolemaic calculation of an error of 8′; it is only right that we should accept God's gift with a grateful mind, because these 8′ have led to a total reform of astronomy."

While Kepler's admiration for Tycho's achievements always remained high, the imperious and high-handed Dane exceedingly frustrated the young German, almost driving him to a nervous breakdown. How well a collaboration between the two very different personalities would have worked in the long run is now impossible to assess, but what actually happened after Kepler had worked with Tycho for only ten months was that his mentor unexpectedly took ill and died. Suddenly Kepler inherited both the use of Tycho's observations and his position as Imperial Mathematician (although at only a third of the salary!). Kepler took time out to complete Tycho's nearly finished book—the *Progymnasmata astronomiae instauratae*, "Exercises for the Reform of Astronomy"—and then he returned to his warfare on Mars.

Kepler now revived his earlier speculations on a planetary driving force, something like magnetism, emanating from the Sun. He envisioned a rotating Sun with rotating emanations that continuously pushed the planets in their orbits. His revised model, with the center midway between the Sun and equant, enabled him to formulate what we can call his distance law, that the velocity of a planet is inversely proportional to its distance from the Sun. Finding the angular motion from the distances immediately raised a computational problem of some proportion that could only be solved by tedious numerical summations. Here he had the fortunate inspiration to replace the sums of the lines between the Sun and the planet with the area swept out by the line between the Sun and the planet. This relation is formulated by Chap. 40 of his *New Astronomy*. Neither in the chapter itself nor anywhere else in the *New Astronomy* is this relation, the so-called law of areas, clearly stated. By 1621, however, Kepler finally understood its fundamental nature and clearly stated both the area law and a revised distance law.

At this point Kepler had an accurate but physically inadmissable scheme for calculating longitudes and an intuitively satisfactory physical principle (the distance law) that worked well for the Earth's orbit, but which left an unacceptable 8′ error when applied to Mars. In order to preserve simultaneously both his accurate longitude prediction and the properly centered circular orbit, Kepler next added a small epicycle to his circle. This was a time-honored procedure, but one that left him distressed by its absurdity. Just as sailors cannot know from the sea alone how much water they have traversed, he argued, so the Mind of the planet will have no control over its motion in an imaginary epicycle except by watching the apparent diameter of the Sun.

Kepler had difficulty in preserving the circular motion when he adopted an epicycle; it is therefore not surprising that he next turned to a closer examination of the shape of Mars's path. Having established the proper position of the Earth's orbit by triangulation of Mars, he was able to turn the procedure around and investigate a few points in the orbit of Mars itself. Although the method did not yield a quantitative result, it clearly showed that Mars's orbit was noncircular. Kepler recognized that observational errors prevented him from getting precise distances to the orbit. Because of this scatter, he had to use, as he picturesquely described it, a method of "votes and ballots."

Armed with these results, Kepler found in the epicycle a convenient means for generating a simple noncircular path. The resulting curve was similar to an ellipse, but was slightly egg-shaped with the fat end containing the Sun. In working with this ovoid curve, Kepler got himself into a very messy computational problem when he tried to apply his area rule to various segments. As an approximation he used an ellipse, but rather different from the one he was finally to adopt. Then, in the course of his calculations, he stumbled upon a pair of numbers that alerted him to the existence of another ellipse, a curve answering some of his requirements almost perfectly. "It was," he wrote, "as if I had awakened from a sleep."

Nevertheless, Kepler was also searching for a *physical* picture of planetary motion, something quasi-magnetic connected with the Sun that would explain not only the varying speed of Mars but also its varying distances. He fervently hoped that the oscillations of a hypothetical magnetic axis of Mars would satisfy his requirements. "I was almost driven to madness in considering and calculating this matter," he wrote in Chap. 58 of his *New Astronomy*. "I could not find why the planet. . .would rather go on an elliptical orbit as shown by the equations. O ridiculous me! As if the oscillation on the diameter could not be the way to the ellipse! So this notion brought me up short, that the ellipse exists because of the oscillation. With reasoning derived from physical principles agreeing with experience there is no figure left for the orbit except a perfect ellipse."

With justifiable pride he could call his book the *New Astronomy*; its subtitle emphasizes its repeated theme: "Based on Causes, or Celestial Physics, Brought out by a Commentary on the Motion of the Planet Mars." Although his magnetic forces have today fallen by the wayside, his requirement for a celestial physics based on causes has profoundly influenced contemporary science, which takes for granted that physical laws operate everywhere in the universe.

The work was completed by the end of 1605, but publication did not follow immediately, for Tycho's heirs demanded censorship rights over materials based on his observations, and they were displeased that Kepler had chosen a Copernican basis rather than Tycho's fixed-Earth, geo-heliocentric arrangement, which of course made little sense in the framework of Kepler's physical ideas. Eventually a compromise was reached, primarily with respect to the dedicatory materials at the beginning of the volume, and the book was at last printed in 1609. Kepler added to the work a long introduction defending his physical principles, and he described how the Copernican system could be reconciled with the Bible. This latter part he had already written in 1596 for inclusion in his *Mysterium cosmographicum*, but the Tübingen University Senate, which had been asked to referee his first publication, had objected to such theological material. Now independent, Kepler had no such restrictions; of all Kepler's writings, this section of the introduction was the most frequently reprinted during the seventeenth century, and was the only part to be translated into English.

It is difficult to gauge what impact Kepler's new astronomy would have had if he had stopped at this point. His cleansing and reformulation of the heliocentric system had been worked out in theory for only a single planet, Mars, and he had not provided any practical tables for calculating its motions. As Imperial Mathematician to Rudolf II, Kepler had been explicitly charged with the preparation of new planetary tables based on Tycho's observations, an arduous task that he still faced. "Don't sentence me completely to the treadmill of mathematical calculations," Kepler replied to one correspondent. "Leave me time for philosophical speculations, my sole delight."

The Harmony of the World

Soon after completing his *Mysterium cosmographicum*, Kepler had drafted an outline for a work on the harmony of the universe, but his plan had lain dormant while he grappled with the intricacies of Mars. Then, in the fall of 1616, after he had completed the first in a long series of ephemerides based on his work, he began to work intermittently on his *Harmonice mundi*, the *Harmony of the World*. A major work of 250 pages, it was finally completed in the spring of 1618.

Max Caspar, in his biography *Kepler*, gives an extended and perceptive summary of the *Harmony*, concluding:

> Certainly for Kepler this book was his mind's favorite child. Those were the thoughts to which he clung during the trials of his life and which brought light to the darkness that surrounded him.... With the accuracy of the researcher, who arranges and calculates observations, is united the power of shaping of an artist, who knows about the image, and the ardor of the seeker for God, who struggles with the angel. So his *Harmonice* appears as a great cosmic vision, woven out of science, poetry, philosophy, theology, mysticism....

Kepler developed his theory of harmony in four areas: geometry, music, astrology, and astronomy. It is the latter treatment, in Book V, which commands the primary attention today.

In the *Mysterium cosmographicum* the young Kepler had been satisfied with the rather approximate planetary spacings predicted by his nested polyhedrons and spheres; now, imbued with a new respect for data, he could no longer dismiss its 5 percent error. In the astronomical Book V of the *Harmony*, he came to grips with this central problem: By what secondary principle did God adjust the original archetypal model based on the regular solids? Indeed, Kepler now found a supposed harmonic reason not only for the detailed planetary distances but also for their orbital eccentricities. The ratios of the extremes of the velocities of the planets corresponded to the harmonies of the just intonation. Of course, one planet would not necessarily be at its perihelion when another was at aphelion. Hence, the silent harmonies did not sound simultaneously but only from time to time as the planets wheeled in their generally dissonant courses around the Sun. Swept on by the grandeur of his vision, Kepler exclaimed:

> It should no longer seem strange that man, the ape of his Creator, has finally discovered how to sing polyphonically, an art unknown to the ancients. With this symphony of voices man can play through the eternity of time in less than an hour and can taste in some small measure the delight of God the Supreme Artist by calling forth that very sweet pleasure of the music that imitates God.

In the course of this investigation, Kepler hit upon the relation now called his third or harmonic law: The ratio that exists between the periodic times of any two planets is precisely the ratio of the 3/2 power of the mean distances. (This is equivalent to saying that the square of the periodic time is proportional to the cube of the mean distance.) Neither here nor in the few later references to it does Kepler bother to show how accurate the relation really is. Using his own data, we can calculate the table he failed to exhibit:

	Period (years)	Mean distance	Period squared	Distance cubed
Mercury	0.242	0.388	0.0584	0.0580
Venus	0.616	0.724	0.3795	0.3795
Earth	1.000	1.000	1.000	1.000
Mars	1.881	1.524	3.540	3.538
Jupiter	11.86	5.200	140.61	140.73
Saturn	29.33	9.510	860.08	867.69

The harmonic law pleased him greatly for it neatly linked the planetary distances with their velocities or periods, thus fortifying the *a priori* premises of the *Mysterium* and the *Harmony*. So ecstatic was Kepler that he immediately added these rhapsodic lines to the introduction to Book V:

> Now, since the dawn eight months ago, since the broad daylight three months ago, and since a few days ago, when the full Sun illuminated my wonderful speculations, nothing holds me back. I yield freely to the sacred frenzy; I dare frankly to confess that I have stolen the golden vessels of the Egyptians to build a tabernacle for my God far from the bounds of Egypt. If you pardon me, I shall rejoice; if you reproach me, I shall endure. The die is cast, and I am writing the book—to be read either now or by posterity, it matters not. It can wait a century for a reader, as God himself has waited six thousand years for a witness.

Kepler's *Epitome of Copernican Astronomy*

At the same time that Kepler was preparing his planetary ephemerides and his *Harmony of the World,* he also embarked upon his longest and perhaps most influential book, an introductory textbook for Copernican astronomy in general and Keplerian astronomy in particular. Cast in the catechetical form of questions and answers typical of sixteenth-century textbooks, the *Epitome* treated all of heliocentric astronomy in a systematic way, including the three relations now called Kepler's laws. Its seven books were issued in installments; the first three appeared in 1617 and the final three in 1621.

Although Book IV, on theoretical astronomy, came last conceptually, it was published in sequence in 1620. Subtitled "Celestial Physics, that is, Every Size, Motion, and Proportion in the Heavens Explained by a Cause Either Natural or Archetypal," it is the most remarkable section of the *Epitome*. To a large extent it epitomizes both the *Harmony* and Kepler's new lunar theory, completed just before this part was sent to press.

Book IV opens with one of his favorite analogies, one that had already appeared in the *Mysterium cosmographicum* and which stresses the theological basis of his Copernicanism: the three regions of the universe were archetypal symbols of the Trinity—the center, a symbol of the Father; the outermost sphere, of the Son; and the intervening space, of the Holy Spirit. Immediately thereafter Kepler plunges into a consideration of final causes, seeking reasons for the apparent size of the Sun, the length of day, and the relative sizes and distances of the planets. From first principles he attempts to deduce the distance to the Sun by assuming that the Earth's volume is to the Sun's as the radius of the Earth is to its distance from the Sun. His result is 20 times greater than that assumed by Ptolemy and Copernicus, but he shows from a perceptive analysis of the observations that such a size is not excluded. Subsequently he argues that the sphere of fixed stars must be 2,000 times larger than the orbit of Saturn, thereby advocating a size of the universe vastly greater than that considered by the ancients or even by Copernicus.

Kepler's third or harmonic law, which he had discovered in 1619 and announced virtually without comment in the *Harmony*, receives an interesting and extensive treatment in the second part of Book IV. His explanation of the $P = r^{3/2}$ law (where P is the period of revolution of a planet and r is the mean distance from the Sun) is based on the relation

$$\text{Period} = \frac{\text{Path length} \times \text{Matter}}{\text{Magnetic strength} \times \text{Volume}}$$

Clearly, the longer the path, the longer the period; the greater the magnetic emanation reaching the planet from the Sun (which furnished the driving force), the shorter the period. The matter in the planet itself provides a resistance to continued motion: the more matter, the more inertia, and the more time required. Finally, with a larger volume of material, the magnetic emanation or "motor virtue" can be soaked up more readily and the period proportionately shortened. According to Kepler's distance rule, the density of each planet depends upon its distance from the Sun, a requirement quite appropriate to his ideas of harmony. To a limited extent he defends this arrangement from telescopic observations, but he generally falls back on vague archetypal principles.

In the third part of Book IV Kepler continues his discussion of the physical causes of planetary motions, and in particular the irregularities of speed and shape of orbit. "But the celestial movements are not the work of mind," he states, and by implication he accepts that their motions, either elliptical or circular, are compelled by material necessity after having been so arranged by the Creator. In his *New Astronomy*, Kepler had been much more equivocal on this point, and certainly in his *Mysterium cosmographicum* he had endorsed the idea of animate souls as moving intelligences for the planets. Now, in the *Epitome*, he seeks an explanatory foundation based strictly on the solar magnetic emanations and their interactions with the magnetic poles of the Moon and planets. In this framework he introduces the magnetic influence of the Sun, likening it to the laws of the lever and balance, and then he uses the oscillations or "librations" of the magnetic poles to explain physically the elliptical orbits.

With the way prepared by the discussion of the magnetic forces and the elliptical orbit, Kepler turns to the most complex case of all, the lunar theory. Although he understood the complications of lunar motion well enough to offer reasonable predictions for the Moon's positions, for the most part, his efforts yielded an ad hoc scheme that failed to provide any foundation for further advances.

Book V of the *Epitome* treats certain practical geometrical problems arising from elliptical orbits, and here Kepler considers the ellipse in considerably more detail. It is in this section that he correctly formulates his "distance law" in a form equivalent to the modern law of conservation of angular momentum, and here he also candidly apologizes for the fuzzy statement of the law of areas in his *New Astronomy*, saying, "I confess that the thing is given rather obscurely there, and most of the trouble comes from the fact that there the distances are not considered as triangles, but as numbers and lines." Also introduced in this book is what is now called Kepler's equation, used in the practical application of the law of areas to an elliptical orbit.

Kepler's *Epitome* can be considered the theoretical handbook to his *Rudolphine Tables*, finally published in 1627; this monumental work furnished working tables based on his reforms of the Copernican system. Kepler's planetary positions were generally about 30 times better than those of any of his predecessors, and in 1631 (the year following Kepler's death) they provided the grounds for a dramatically successful observation, for the first time, of a transit of the planet Mercury across the face of the Sun. In 1632 a further impetus to the Copernican system came through the brilliant polemic of Galileo's *Dialogue Concerning Two Great World Systems*.

Precisely how influential Kepler's *Epitome* was is difficult to assess. His reputation was considerably enhanced by the publication of his *Rudolphine Tables*, and the Copernican idea was rapidly gaining in acceptability. Thus, by 1635, there was sufficient demand for the *Epitome* to warrant reprinting it, and for many years it remained one of the few accessible sources for the details of the revised Copernican system.

With Ptolemy and Copernicus, Kepler shared a profound sense of order and harmony. In Kepler's mind this was linked with his theological appreciation of God the Creator. Nowhere is this more movingly expressed than in the prayer near the end of his *Harmony*:

> I give thanks to Thee, O Lord Creator. . . . Behold! now, I have completed the work of my profession, having employed as much power of mind as Thou didst give to me; to the men who are going to read those demonstrations I have made manifest the glory of Thy works, as much of its infinity as the narrows of my intellect could apprehend. . . . If I have been allured into rashness by the wonderful beauty of Thy works, or if I have loved my own glory among men, while I am advancing in the work destined for Thy glory, be gentle and merciful and pardon me; and finally deign graciously to effect that these demonstrations give way to Thy glory and the salvation of souls and nowhere be an obstacle to that.

Repeatedly, Kepler stated that geometry and quantity are coeternal with God and that mankind shares in them because man is created in the image of God. From these principles flowed his ideas on the cosmic link between man's soul and the geometrical configurations of the planets; they also motivated his indefatigable search for the mathematical harmonies of the universe.

Contrasting with Kepler's mathematical mysticism, yet growing out of it through the remarkable quality of his genius, was his insistence on physical causes. In Kepler's view, the physical universe was not only a world of discoverable mathematical harmony but also a world of phenomena explainable by mechanical principles. The result of his work was the mechanization and the cleansing of the Copernican system, setting it into motion like clockwork and sweeping away the vestiges of Ptolemaic astronomy.

It remained for Isaac Newton to banish the last traces of Aristotelian physics and to place the heliocentric system on a consistent physical foundation. Although Newton indirectly owed much to Kepler's insistence on physical causes, he rejected the type of physical arguments so dear to Kepler's mind, and with it, he tried to withhold credit for Kepler's achievement. As a result, Kepler is nowhere mentioned in Book I of Newton's *Principia*. Nevertheless, Newton's contemporaries felt differently, and thus his work was introduced to the Royal Society as "a mathematical demonstration of the Copernican hypothesis as proposed by Kepler." Perhaps the fairest evaluation of Kepler has come from Edmond Halley in his review of the *Principia*: Newton's first 11 propositions, he wrote, were "found to agree with the *Phenomena* of the Celestial Motions, as discovered by the great Sagacity and Diligence of Kepler."

PTOLEMY
AND THE
GEOCENTRIC
UNIVERSE

Was Ptolemy a Fraud?

The greatest surviving astronomical work from antiquity is the *Almagest* or *Syntaxis*, a mathematical treatise written in Alexandria around A.D. 145 by Claudius Ptolemy. In some of the manuscripts, the *Almagest* begins with the epigram, "I know that I am mortal by nature, and ephemeral; but when I trace at my pleasure the windings to and fro of the heavenly bodies I no longer touch earth with my feet: I stand in the presence of Zeus himself and take my fill of ambrosia, food of the gods."[1] The epigram seems to place Ptolemy within the long series of scientists who have tasted the intoxicating pleasure of a splendid theory.

It is difficult to convey the elegance of Ptolemy's achievement to anyone who has not examined its details. Basically, for the first time in history (so far as we know) an astronomer has shown how to convert specific numerical data into the parameters of planetary models, and from the models has constructed a homogeneous set of tables—tables that employ some admirably clever mathematical simplifications, and from which solar, lunar, and planetary positions and eclipses can be calculated as a function of any given time. Altogether it is a remarkable accomplishment, combining in a brilliant synthesis a treatise on theoretical astronomy with a practical handbook for the computation of ephemerides.

Possibly for pedagogical reasons Ptolemy strove for even greater completeness by including descriptions of observational techniques,

Selection 2 reprinted from *Quarterly Journal of the Royal Astronomical Society*, vol. 21 (1980), pp. 253–66.

but in retrospect he has managed only to cast doubt on the veracity of his text. For example, the *Almagest* opens with the theory of the Sun's motion. Ptolemy describes how the time of the equinoxes can be measured by watching the shadows on a bronze ring accurately aligned to the celestial equator, and he goes on to specify his own "very careful" observation of an equinox on a date corresponding to 26 September A.D. 139.[2] Modern calculations show that the equinox actually fell some 30 hours earlier. Now we can imagine that Ptolemy was an excellent theorist but a clumsy observer; nevertheless, our suspicions are aroused when we discover that Ptolemy's reported time agrees precisely with an extrapolation from an observation by Hipparchus 278 years earlier, an extrapolation depending on the slightly long length of the year rather arbitrarily arrived at by Hipparchus. Did Ptolemy fabricate his purported equinox observation? Perhaps.

This situation has been known for a long time, and over two centuries ago, in 1753, Tobias Mayer discussed it in a letter to Leonard Euler. What happened, according to Mayer, was that Ptolemy had to start somewhere, and he knew that the time-honored Hipparchan parameters for the eclipse theory worked very well. Thus, despite the ordered presentation of the *Almagest*, Ptolemy actually took the eclipse theory as his fundamental base and from there proceeded to the planetary theory, adopting the Hipparchan solar and lunar parameters in order to work out the planetary parameters. Mayer wrote, "It can be that Ptolemy perceived the error of his solar tables in his observations of the equinoxes, which are the very last of his extant observations; but, because he had already built his whole system upon it, he perhaps preferred to discard his observations rather than to start all over again. Since, however, no one could object, he pretended that the erroneous equinoxes of his tables were true and observed. There are more recent examples of astronomers, from too great a love for their constructions, falsifying observations (in Lansberg and Riccioli for sure). Ptolemy, who perhaps did not imagine that anyone would ever be able to detect this deception, could easily have fallen into this error."[3]

Over the course of time Ptolemy's apparent fabrication of his equinox observations have been repeatedly rediscovered, for example, by the astronomer-historian Delambre,[4] and about a decade ago these data were thoroughly analyzed by John Britton in a doctoral dissertation at Yale. Britton noticed that atmospheric refraction would seriously confuse the equatorial ring method of establishing the equinox, so that in about half the cases the illumination would change from the upper- to the under-side of the ring on the day following the autumnal equinox, and in one quarter of the cases at vernal equinox the illumination would change on two successive days even if the ring were perfectly aligned. Ptolemy seemed aware of the problem, although not of all its causes, when he wrote "And anyone can see an example of this in the bronze rings in the Palaestra [square for gymnastics], which are supposed to be in the plane of the equator. For in making observations we find such a distortion in their placement, especially the larger and older they are, that at times their concave surfaces twice suffer a shift in lighting at the same equinox."[5]

Ptolemy himself does not specify precisely how he made his equinox observations, although a bronze equatorial ring is implicated. It is quite likely that he checked the equinoxes only for plausible agreement with Hipparchus rather than "most accurately" as he states. Britton summarizes the affair by saying, "The conclusion that Ptolemy's equinox observations can have been scarcely no more than the results of computations is unsatisfying, but I can find no other explanation of the errors in his report of times and their agreement with Hipparchus's observations in length of the year. On the other hand, if Ptolemy set out to determine the times of the equinoxes using an equatorial ring, he could not have avoided encountering the difficulties and irregularities [of refraction]. Thus he might easily have concluded that he could make no secure improvement on Hipparchus's solar parameters."[6]

Concerning the single recent solstitial date given by Ptolemy, he made no claims for its accuracy except to say that he had computed it with care, nor did he specify a method of observations. While it is possible to deduce the time of the solstice fairly accurately by making altitude observations of the Sun some days before and after the actual solstice, when the Sun is still changing perceptibly in declination, it is difficult to establish the time by observations adjacent to the date itself. It has been known since the end of the last century that the summer solstice date given by Hipparchus derives from the traditional parameters of the Babylonian system A solar theory. The solar motion on its slow arc[7] is 28;7,30°/month or, converting with the modern value of 29^d5306 for the synodic lunar month, 0.95241°/d. Since the Sun began its slow arc around the beginning of March (to use the Julian calendar anachronistically) and finished in August, the interval from vernal equinox to summer solstice lay entirely within the slow arc. Ninety degrees at 0.95241°/d yields 94^d50, the precise interval adopted by both Hipparchus and Ptolemy.

Now an error of a day in establishing the time of the equinoxes results in an error of about a degree in the solar position, that is, in the deduced position of the invisible intersection of the ecliptic and equator. Ptolemy used positions of the Moon (at lunar eclipses and otherwise) to transfer his solar reference frame into the starry nighttime sky. Hence his erroneous value in the solar position propagated into his fundamental coordinate system, including the positions given for the planets and for the stars in the catalog, and in turn the faulty stellar positions led to a defectively small value of precession.[8]

Some years ago the geophysicist R. R. Newton has rediscovered these underlying difficulties in the *Almagest*, but he has gone much further than previous workers in analyzing Ptolemy's data and even his motivations. Thus he has been led to conclude his recent book, *The Crime of Claudius Ptolemy*, by saying, "The *Syntaxis* has done more damage to astronomy than any other work ever written, and astronomy would be better off if it had never existed. Thus Ptolemy is not the greatest astronomer of antiquity, but he is something still more unusual: he is the most successful fraud in the history of science."[9]

In coming to the defense of Ptolemy's scientific reputation, I shall concede at once that the *Almagest* poses some curious problems to the historians of science, and that Ptolemy's statements regarding observed quantities cannot always be taken at face value. R. R. Newton has systematically examined the observations reported in the *Almagest* and he finds (as I also do) quite appreciable errors compared to the actual positions of the celestial bodies at those times.[10] These errors can be as much as a degree, and sometimes more. He also finds (and I generally confirm) that the reported positions agree very closely with Ptolemy's theory—generally well within 10 minutes of arc. Somehow, according to Newton, the match with theory appears just too good. In his opinion this means that Ptolemy simply made up these observations.

Yet it has always seemed to me that this hypothesis runs into a serious difficulty. How can Ptolemy's parameters, which seem generally more accurate than his data base, be derived from observations that are simply fabricated?

We can consider four hypotheses:

1. Ptolemy borrowed good hypotheses from elsewhere (Hipparchus? Babylonians? A lost civilization?) and made up his theory to look as if he had done it all.

2. Ptolemy selected from a large data bank only those observations that fit the theory.

3. Ptolemy fitted his theory to a few preferred observations.

4. Ptolemy "corrected" his observations so as to agree with a theory established from numerous observations not mentioned in his work.

R. R. Newton has suggested that the mean motions, for example, were simply borrowed from the Babylonians.[11] In fact, O. Neugebauer has shown that it is Hipparchus who adopted the Babylonian values and that Ptolemy's mean motions represent the refinement possible with a longer temporal baseline.[12] Whence, according to hypothesis 1, the apsidal lines, eccentricities, epicycle sizes, and so on could be borrowed remains a mystery. Perhaps they were brought by ancient astronauts, but until further evidence is forthcoming, the first hypothesis is unacceptable within our present framework of historical understanding.

Hypothesis 2, that the excellent agreement between the stated observations and the Ptolemaic theory arises from deliberate observational selection, leaves open the source of the theoretical parameters. Hypothesis 3 leaves open the question of choosing "preferred" observations. The reason for supposing that there was a large data bank and/or preferred observations rests on the fact already mentioned that Ptolemy's parameters seem better than the recorded observations. Before analyzing the hypotheses further, we must examine this point.

Let us first consider Ptolemy's lunar theory. The *Almagest* approaches the lunar theory in three stages. The first, which Ptolemy attributes to Hipparchus, employs an epicycle on a concentric deferent. The epicycle generates the basic orbital eccentricity, and also the advance of perigee. Figure 1 shows the outer envelope of errors in longitude for the three stages, and is adapted from the work of Viggo Petersen.[13] The systematic displacement in longitude is one of the manifestations of Ptolemy's problems with the fundamental coordinate system. As shown in the upper part of the figure, the Hipparchan model is tolerably successful at new and full moons, but the errors are unacceptably large in the intermediate parts of the orbit. The discrepancies at first and last quarters arise from what is now called *evection*, an effect discovered by Ptolemy, who understood it as a change in the effective eccentricity of the orbit and who accounted for it by introducing a crank mechanism in the center of the orbit. The errors in this second stage are shown in gray in the lower part of the figure, and as may be seen, the fit is now tolerably successful at the quarters, although some problems remain in the octants. Observing the moon in

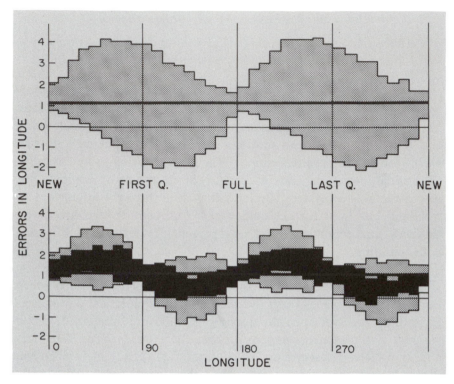

FIGURE 1. *The outer envelope of longitude errors in degrees for the three stages of Ptolemy's lunar theory, adapted from V. Petersen.*

the octants has apparently never been very popular with positional astronomers, and, for example, it was only very late in Tycho Brahe's observing career that he discovered the so-called "variation" with an amplitude of 40 minutes of arc.[14] In any event, Ptolemy, on the stated basis of only two observations, further modified his model to reduce the error envelope to the solid section in the lower part of the figure. The residual sinusoidal effect of Tycho's "variation" is still plainly evident.

Figure 2, from Apianus's *Astronomicum Caesareum* (Ingolstadt, 1540), shows the lunar mechanism. The direction to the Sun is toward the upper right. I have set the volvelles for the first of the two octant observations used by Ptolemy (from Hipparchus, 2 May 126 B.C.). In the second form of the model the motion in the epicycle is reckoned from the line drawn from the Earth through the center of the epicycle. Only a small change suffices to bring about the large improvement in the third form of the model, namely, measuring the motion in the

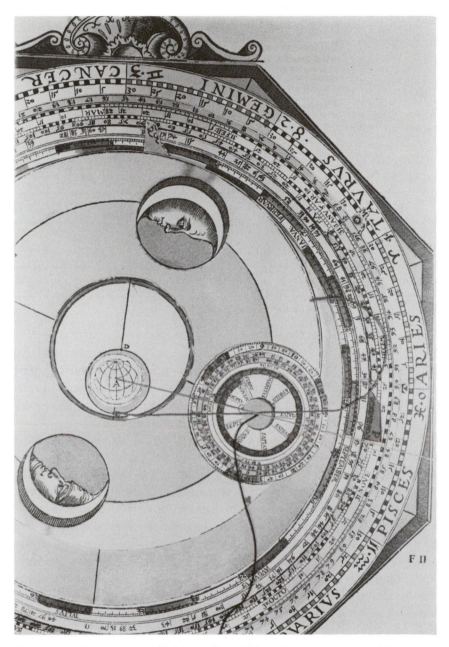

FIGURE 2. *The lunar mechanism from Apianus's* Astronomicum Caesareum *set for 2 May 126 B.C. In the second stage of Ptolemy's theory the position of the Moon in the epicycle is measured with respect to the line from the Earth's center,* A. *A considerable improvement is achieved in the third stage by measuring with respect to the line from* E.

TABLE 1. Lunar Parameters

	Modern		Ptolemy (Stage 3)	
	°	′	°	′
Equation of center	− 6	17.3 sin $\bar{\alpha}$	− 6	14
	+	12.8 sin $2\bar{\alpha}$	+	19
Evection	− 1	16.4 sin$(2\bar{\eta} - \bar{\alpha})$	− 1	16
Variation		39.5 sin $2\bar{\eta}$	—	

			(Stage 2)	
			°	′
Spurious term	—	sin $2\bar{\eta}$ cos $\bar{\alpha}$	− 1	16

($\bar{\alpha}$ = mean anomaly, $\bar{\eta}$ = mean elongation)

epicycle from the line that originates in a point equal and opposite from the equant. Note that the observation is chosen at a very special octant time, so that the line of apsides is perpendicular to the crank mechanism, and the difference in models is maximized.

The success of Ptolemy's final lunar model may be seen in Table 1 by comparing his derived parameters with the terms in the modern lunar theory.[15] Note the excellence of the fit for the equation of the center and the evection. Ptolemy's second theory differs from the third by the inclusion of a spurious term, sin $2\bar{\eta}$ cos $\bar{\alpha}$, which is eliminated by the small geometrical change in the models. I personally find it unbelievable that Ptolemy could have improved his theory so appreciably merely from these two octant observations. Ptolemy must have used many more observations, whereas those in the *Almagest* serve only to derive most directly the parameters after the inequality of motion had been extensively analyzed. Since he had no theory of errors, he could cope with the multiplicity of data by including in the *Almagest* only the minimum number of observations required to determine the parameters. It is possible that he picked these two as being representative of the average results from other observations. The fact that two somewhat erroneous observations fit the theory so well would then be explained by his use of carefully selected data according to

TABLE 2. Copernican Mars Positions

	Observed —Actual		Observed —Theory (Copernicus)	Observed —Theory (Reinhold)
	°	′	′	′
5 June 1512	—	20	− 9	− 2
12 December 1518	+ 1	17	− 9	− 3
22 February 1523	+ 2	16	− 2	− 5
1 January 1512	+	2	+ 4	+ 4

hypothesis 2, and not by fudging the observations. An alternative is that Ptolemy, who was unquestionably an able mathematician, skillfully fit his theory through the chosen observations, according to hypothesis 3.

Table 2 illustrates a clear case of fitting a theory precisely through very bad observations, done not by Ptolemy, but by Copernicus 1400 years later. Copernicus, like Ptolemy, gives only four Martian observations from his own century, and two of them are very bad indeed, far worse than Ptolemy's data. Nevertheless, Copernicus has succeeded in fitting his Mars model to this wretched data within 10 minutes of arc.[16] It is interesting to notice that his successor, Erasmus Reinhold, was able to perform the same geometric operation with slightly greater precision, but it apparently never occurred to Reinhold to question the underlying data itself.

Let us next look at Ptolemy's value for the perigee of Venus. Venus has a very low eccentricity, so it is not easy to find the orientation of its orbit. The model requires the vector sum of ae (semimajor axis times the orbital eccentricity) for the Earth and for Venus; because ae for the Earth's orbit is so much larger, it predominates, as may be seen in the small vector triangle inset into Figure 3. Ptolemy claims to find the perigee line from the two pairs of matched elongations. These are plotted on the figure (which is shown anachronistically in a heliocentric layout). Ptolemy's result of 235° (for ca. A.D. 130) is astonishingly good, with an error of only 3°.

The word "astonishingly" is used advisedly, for when we make a critical examination of these four observations, comparing Ptolemy's values with the actual positions as recomputed today,[17] we find noth-

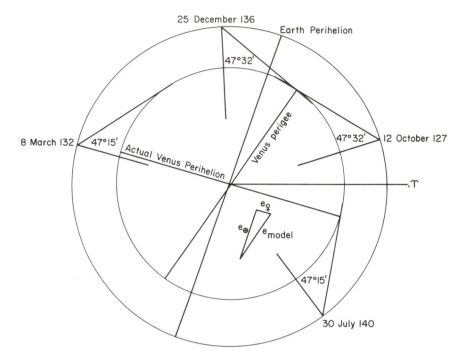

FIGURE 3. *Ptolemy's observations for the Venus perigee shown (anachronistically) on a heliocentric scheme. The inset shows an enlargement of the central vector triangle. The commensurability of the periods of Venus and the Earth prevented Ptolemy from picking arbitrary sighting points.*

ing like the aesthetic symmetry that marks the *Almagest*'s presentation. In fact, Ptolemy has sometimes reported his solar and Venusian observations rather roughly, often to an accuracy of only a quarter of a degree, and these do not even agree with the theoretical values derived from the tables of the *Almagest*. It stretches our confidence beyond its breaking point to suppose that Ptolemy established the very sensitive geometry and the perigee within 3° of the correct value by these two pairs of identical elongations. Either Ptolemy determined the perigee from other data and then by incredible luck was able to select from his data bank the pairs that just happened to match, or else the data have been "laundered." In other words, hypothesis 3 seems untenable (although I shall consider it further below), but hypotheses 2 and 4 are possible. However, if Ptolemy had adjusted his numbers for

TABLE 3. Positions of Venus

		Mean Sun		Venus		Elongation	
		°	′	°	′	°	′
	Ptolemy	344	15	31	30	47	15
132 March 8.75	*Almagest* theory	344	14	31	24	47	10
	Modern	345	18	32	9	46	51
	Ptolemy	125	45	78	30	47	15
140 July 30.25	*Almagest* theory	125	42	79	10	46	22
	Modern	126	48	80	40	46	08
	Ptolemy	197	52	150	20	47	32
127 October 12.25	*Almagest* theory	197	51	150	27	47	24
	Modern	198	54	151	50	47	04
	Ptolemy	272	4	319	36	47	32
136 December 25.75	*Almagest* theory	272	3	319	50	47	47
	Modern	273	8	320	2	46	54

the elegance of his presentation, according to hypothesis 4, then where was the basis for such a good determination of the perigee?

In my opinion the only reasonable explanation is that Ptolemy had a large number of elongation observations, none very precise, but sufficient to find the perigee line. Probably the pedagogic standards of his day (of which we know very little) dictated a simple presentation of his results, so he set forth a minimum number of symmetrically placed points. However, for this planet, Ptolemy was faced with a special problem in finding a symmetrical set of elongations, because there is a rhythmic interlock between the period of the Earth and Venus. As a consequence, Venus tends to repeat its pattern of greatest elongations with the same five terrestrial positions. In Ptolemy's day this pentagonal array was not symmetrical with the perigee line, and consequently he could not actually use greatest elongations. It was therefore not easy for him to obtain consistency, and for the dates chosen Venus was not quite at its greatest elongation. Hence, substantial discrepancies remain between the putative positions of Venus and those predicted by the tables in the *Almagest* itself (Table 3).

When we turn to Ptolemy's treatment of the superior planets, we find four observations per planet from his own lifetime. As illustrated in Figure 4, these data are reasonably accurate with respect to his

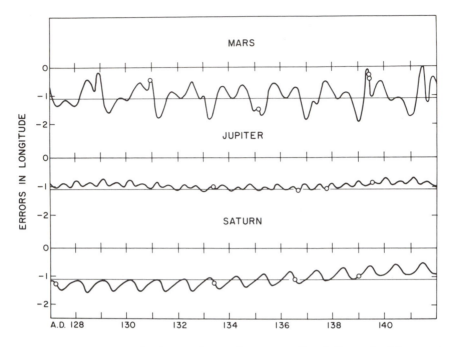

FIGURE 4. *Errors in the longitudes in degrees predicted from the* Almagest's *planetary theory, with the observations reported by Ptolemy shown as circles.*

coordinate system (that is, with the systematic precessional error of $1°.1$), but they agree even better with the predictions of Ptolemy's theory. The differences between the longitudes as computed from the tables in the *Almagest* and the modern calculations are shown by the continuous error curve on the diagram. The four observations suffice to define four orbital elements: the eccentricity, direction of perigee, time of perigee passage, and the size of the epicycle. There is no redundancy of data, and in principle Ptolemy, like Copernicus after him, could have derived the parameters for the orbits from these points, so that the error curve would necessarily pass through the data points, that is, according to hypothesis 3.[18]

For Venus (Figure 5) the situation is somewhat more complicated. Four observations (those diagrammed in Figure 3 and shown with solid circles in Figure 5) are used to demonstrate the direction of perigee, and five observations determine the three remaining orbital elements. As may be expected *if the observations are not fudged*, three of these points fall precisely on the error curve, and two do not.

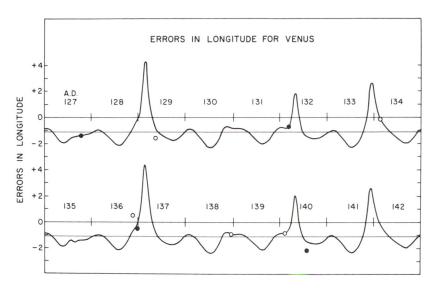

FIGURE 5. *Errors in the longitudes in degrees predicted from the* Almagest's *theory for Venus, with the observations reported by Ptolemy shown as circles. Solid circles represent the symmetrical observations used to find the perigee line. The large errors occur at inferior conjunction when Venus is lost in the Sun's glare.*

In the case of Mercury (Figure 6) six of Ptolemy's own nine observations fall within 6 minutes of arc of the error curve. Six observations are used to determine the apogee and (spurious) double perigee. Without dwelling on the details of this latter bizarre situation,[19] I can remark that Ptolemy must surely have put credence in some specific observations here, or he would not have ended up with such an unnecessarily complicated mechanism for Mercury. An attentive analysis[20] shows that Ptolemy has no redundant observations here except possibly among the four symmetrical ones used to establish the apogee line, and three of these deviate from the error curve.

It is clear that Ptolemy's data are presented for specific geometrical configurations. His observations are surprisingly bad while his final parameters are amazingly good. Finally, we have the disconcerting fact that most of Ptolemy's reported observations, faulty as they are, agree almost perfectly with his theory. In the foregoing discussion I have mentioned three possible reasons for this remarkable state of affairs. In many instances Ptolemy, a redoubtable mathematician, could have derived his parameters with the minimum number of ob-

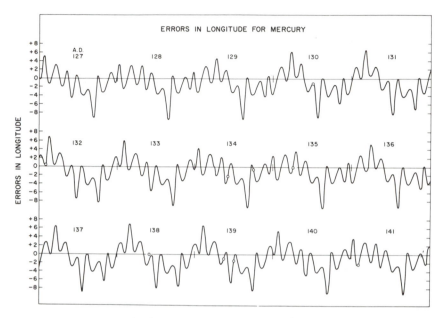

FIGURE 6. *Errors in the longitudes in degrees predicted from the* Almagest's *theory for Mercury, with the observations reported by Ptolemy shown as circles.*

servations, thereby forcing the theoretical positions to match the given data (hypothesis 3). One striking feature of Figure 3 suggests that, plausible as this hypothesis may be, it is incorrect. In principle, the observations for Mars ought to be as accurate as those for Jupiter or Saturn, as in no case does Mars move fast enough for the timing of the observation to be critical. Yet the four reported observations of Mars show the same large scatter that the theory for this recalcitrant planet exhibits. Jupiter and Saturn, with more accurate theories, also have less scatter in the observations. This strongly suggests that the recorded "observations" depend on the theory, and not *vice versa*.

Alternatively, Ptolemy could have selected from a much larger group those observations that happened to fit what he considered to be his best parameters (hypothesis 2). Finally, he could have "corrected" his observations in some unspecified manner (hypothesis 4). By strict contemporary standards either the selection or the hidden correction of data is considered reprehensible, though not necessarily fraudulent. In reality, weighting or rejecting some experimental data is probably the rule rather than the exception. It is instructive to look briefly at two well-documented non-astronomical examples.

The nineteenth-century American anthropologist, Samuel George Morton, a scholar of the highest reputation, measured the cranial capacities of some hundreds of human skulls. At that time virtually everyone assumed that brain size was directly correlated with brain power, a conclusion that is nowadays entirely discredited. Morton's results, published in 1839 (and more extensively in 1849), confirmed the expectations of the day: Caucasians had the biggest brains, American Indians the next, and Negroes the smallest. However, Morton, unlike Ptolemy, published not only his results but also his original data, and recently S. J. Gould reexamined the entire lot.[21] He found Morton's data to be "a patchwork of assumption and finagling controlled, probably unconsciously, by his *a priori* assumptions." For example, Morton, anxious not to distort his results by full inclusion of his overabundant samples of small Hindu skulls, deliberately excluded 14 specimens, a selection that increased the average size of the Caucasian group. Gould's reaveraging of all the data shows no significant differences in cranial capacity of the three races; nevertheless, he was able to find "no indication of fraud or conscious manipulation." He writes, "Unconscious finagling is probably the norm. . . . We measure greatness not by 'honesty,' but by insight," but he adds, "I do not condone or excuse finagling just because I regard much of it as intrinsic to scientific activity."

Another fascinating example of the selection of experimental data has been brought to light with Gerald Holton's inspection of Robert A. Millikan's laboratory notebooks, which record the measurements for his famous oil-drop experiment.[22] Despite Millikan's published statement that "*this is not a selected group of drops, but represents all of drops experimented on during 60 consecutive days*," the notebooks are peppered with such remarks as "*Publish* this surely, beautiful!" and "*Error high* will not use"; eventually 58 out of 140 drops were selected as "the drops experimented on." The others were considered abortive non-experiments.

At many stages in his research, Ptolemy must have been forced to choose and select between conflicting data. Yet not until the work of Kepler would an astronomer record the redundant and contradictory observations from which his theory was actually wrought. But can the observations actually found in the *Almagest* be adequately accounted for by a sophisticated selection from existing data? Although this is possible (and I certainly think that the evidence is strong for the existence of such a data pool), I am now inclined to believe that the most likely scenario for Ptolemy's procedures is neither the precise fitting of the given observations nor an exacting selection from existing data, but hypothesis 4, namely, that he has "corrected" his observations in some fashion which, regrettably, he has not bothered to state.

Because Ptolemy's parameters are generally pretty good, we must assume that he had a substantial data base beyond what is specifically preserved in his treatise. But a redundant data base with random observational errors must have yielded conflicting parameters that de pended on the particular minimal combination used. Ptolemy probably realized this, and quite likely he noticed that in using different combinations, certain results appeared more frequently, giving him the basis for a preferred set of parameters. At the same time, he might have realized that most, if not all, of his observations contained accidental observational errors. Although my suspicions remain highly speculative, I suspect that Ptolemy, convinced in the intrinsic soundness of his theory, simply replaced the only partially trustworthy observations by what he perceived to be the "correct" data.

Thus Ptolemy, like many of the brilliant theoreticians who followed him, was perfectly willing to believe that his theory represented nature better than the error-marred individual observations of the day. As one of America's Nobel laureates remarked to me, any good physicist would do the same today. Let me cite two parallel examples from more recent times.

Isaac Newton, thanking Flamsteed for a set of lunar observations wrote: "I am of opinion that for your Observations to come abroad with [my] Theory. . .would be much more for their advantage and your reputation then [sic] to keep them private till you dye or publish them without such a Theory to recommend them. For such a Theory will be a demonstration of their exactness and make you readily acknowledged the Exactest Observer that has hitherto appeared in the world."[23]

For a second example I turn to Einstein. One of his students has related an interesting incident that took place in 1919: "Once when I was with Einstein in order to read with him a work that contained many objections against his theory. . .he suddenly interrupted the discussion of the book, reached for a telegram that was lying on the windowsill, and handed it to me with the words, 'Here, perhaps this will interest you.' It was Eddington's cable with the results of measurement of the eclipse expedition. When I was giving expression to my joy that the results coincided with his calculations, he said quite unmoved, 'But I knew that the theory was correct'; and when I asked, what if there had been no confirmation of prediction, he countered: 'Then I would have been sorry for the dear Lord—the theory *is* correct.' "[24]

It is marvelous to find these foremost theoreticians so clearly voicing their belief in the primacy of theory over observations. As for Ptolemy, we can deplore his lack of explicit comment concerning his procedures; but the circumstances, rather than affording the occasion for moral judgments on his motivations, should challenge historians of astronomy to the task of reconstructing Ptolemy's pioneering trail to the most complete mathematical achievement in ancient astronomy. When Newton and Einstein are generally considered frauds, I shall have to include Ptolemy also. Meanwhile, I prefer to think of him as the greatest astronomer of antiquity.

Notes and References

[1] The authorship of the epigram is not completely established, but Franz Boll was inclined to attribute it to Ptolemy in "Das Epigramm des Claudius Ptolemaeus," *Sokrates* (*Jahresberichte des Philologischen Vereins*), vol. 9 (1921), pp. 2–12; Professor P. Kunitzsch has drawn my attention to this article.

[2] Ptolemaus, *Handbuch der Astronomie*, III,l, trans. by K. Manitius (Leipzig, 1912), p. 142. See also Ptolemy, *The Almagest*, trans. by R. C. Taliafero, *Great Books of the Western World*, vol. 16 (Chicago, 1952), p. 81.

[3] Mayer to Euler, 22 August 1753; adapted from the translation by Eric Forbes, *The Euler-Mayer Correspondence* (1751–1755), (New York, 1971), p. 75.

[4] "Did Ptolemy do any observing? Are not the observations he tells us he has made just calculations from his tables and some examples that serve for a better understanding of his theories?" J. B. J. Delambre, *Histoire de l'Astronomie Ancienne*, vol. 1 (Paris, 1817), p. xxv.

[5] *Almagest*, III,l, p. 79.

[6] John Britton, 1967. *"On the Quality of Solar and Lunar Parameters in Ptolemy's Almagest,"* doctoral dissertation, Yale University, 1967, p. 44.

[7] O. Neugebauer, *A History of Ancient Mathematical Astronomy* (New York, 1975). Apparently F. X. Kugler first noticed this relationship, as pointed out by Professor B. Goldstein; see Y. Maeyama, "On the Babylonian Lunar Theory," *Archives Internationales d'Histoire des Sciences*, vol. 28 (1978), pp. 21–35.

[8] The precession can be independently derived from observations of stellar declinations, and it has been noticed by Newton and earlier by A. Pannekoek, "Ptolemy's Precession," *Vistas in Astronomy*, vol. 1 (1955), pp. 60–66, that the 12 declinations recorded but not used by Ptolemy would lead to a correct value, whereas the six he selected confirmed the erroneous smaller value of 36" per annum. Raymond Mercier, *The British Journal for the History of Science*, vol. 12 (1979), p. 216, writes: "Neither, however, has realized that the value of 36" per annum would follow from

soundly observed occultations if they were reduced in the way indicated by Ptolemy. This reduction depends on the tropical longitude of the sun according to Hipparchus, and it was just this motion which entailed the erroneous rate of precession. . . . One can see that Ptolemy cannot be faulted really if he preferred the results based on occultations to those based on the more difficult measurements of stellar declinations."

[9] Robert R. Newton, *The Crime of Claudius Ptolemy* (Baltimore, 1977), p. 379.

[10] Robert R. Newton, *Ancient Planetary Observations and the Validity of Ephemeris Time* (Baltimore, 1976).

[11] Robert R. Newton, "On a Pedagogical Motivation for Ptolemy's Fabrication of Data," *Archaeoastronomy Bulletin*, vol. 2 (1978), pp. 7–9.

[12] O. Neugebauer, "Notes on Hipparchus," in *The Aegean and the Near East*, ed. by Saul S. Weinberg (Locust Valley, New York, 1956), pp. 292–96.

[13] Viggo M. Petersen, "The Three Lunar Models of Ptolemy," *Centaurus*, vol. 14 (1969), pp. 142–71.

[14] Victor E. Thoren, "Tycho Brahe's Discovery of the Variation," *Centaurus*, vol. 12 (1967), pp. 151–66.

[15] The lunar terms in Table 1 are taken from O. Neugebauer *op. cit.*, ref. 7, which in turn rests on Karl Stumpff, *Himmelsmechanik*, vol. 1 (Berlin, 1959), pp. 38ff, but neither author gives the specific form of Ptolemy's spurious term.

[16] These Copernican data are discussed in far greater detail in Owen Gingerich, 1978. "Early Copernican Ephemerides," *Studia Copernicana*, vol. 16 (1978), pp. 403–7 [reprinted as selection 12 in this anthology]. Subsequently it occurred to me that perhaps Copernicus had faked the two bad observations because he found it impossible to fit them satisfactorily with a reasonable variation in his parameters. An investigation has now shown that better observations on the dates in 1518 and 1523 could have been readily accommodated.

[17] Barbara Welther and I have written a FORTRAN program that produces ephemerides according to the *Almagest*, the Toledan and Alfonsine tables, *De revolutionibus*, etc., as well as modern values, the latter based on a computer code generously supplied by Professor Peter Huber formerly at the *Eidgenössiche Technische Hochschule* in Zurich and now at Harvard. I wish to thank Miss Welther for carrying out the computer runs and plotting for this paper.

[18] Robert R. Newton, *op. cit.*, note 11, states, "Then [J. P. Harrington] writes: 'Often Ptolemy gives only the minimum number of observations necessary to determine the parameters of his models.' If this were the case, the observations would necessarily agree exactly with his model, to the accuracy of the calculations. Harrington has apparently relied on a statement that Gingerich and several others have made. This is the statement that Ptolemy has five parameters in the model he uses for each of the outer planets, and that he determines these parameters by means of five observations. The first part of this statement is correct, but the second part is not. Ptolemy does not determine the mean motion in anomaly for any of the planets from the observations he quotes. . . ."

In my opinion R. R. Newton is insensitive to the astronomical situation with which Ptolemy was wrestling, and he therefore misidentifies the parameters Ptolemy is determining. Ptolemy realized that the solar motion is intimately connected with the epicyclic motion of the superior planets. Hence he *assumed* the motion in anomaly (i.e., in the epicycle) from the solar theory, and he employed a fifth, ancient observation (thereby improving the accuracy with a long time interval) to confirm his fifth independent parameter, the mean motion of each planet.

[19] See Owen Gingerich, "The Mercury Theory from Antiquity to Kepler," *Actes du XIIe Congrès International d'Historie des Sciences*, vol. 3A (1971), pp. 57–64 [reprinted as selection 23 in this anthology].

[20] From these six observations used initially to determine the apogee, Ptolemy also deduces the size of the epicycle and eccentricity, and from two of the remaining ones he establishes the size and position of a small crank device in the center of the orbit. Finally, he uses the ninth datum in conjunction with an observation from antiquity to establish the period of the epicycle.

[21] Stephen Jay Gould, "Morton's Ranking of Races by Cranial Capacity," *Science*, vol. 200 (1978), pp. 503–9.

[22] Gerald Holton, "Subelectrons, Presuppositions, and the Millikan-Ehrenhaft Dispute," in *The Scientific Imagination: Case Studies* (Cambridge, 1978), pp. 25–83 (see especially pp. 63–71).

[23] Newton to Flamsteed, 16 February 1694–5, no. 494 in *The Correspondence of Isaac Newton*, vol. 4 (Cambridge, 1967), p. 87.

[24] Quoted by Gerald Holton, *Thematic Origins of Scientific Thought* (Cambridge, 1973), pp. 236–7, from a manuscript by Ilse Rosenthal-Schneider. The telegram in question came from Lorentz rather than Eddington; it survives among the Lorentz papers in Leiden. Helen Dukas, Einstein's longtime personal secretary, tells me that she has confirmed this attitude of Einstein from another independent source. More recently, Klaus Hentschel has argued that this anecdote must be apocryphal ["Einstein's Attitude Towards Experiments," *Stud. Hist. Phil. Sci.*, vol. 23 (1992), pp. 593–624], although it does represent Einstein's later philosophical stance.

Ptolemy Revisited

R. Newton and I not only have a fundamental disagreement as to whether we have sufficient knowledge and insight into Hellenistic astronomy to evaluate Ptolemy's character and motivation, but we also differ in our appreciation of Ptolemy's contribution to the advancement of science. Newton weeps because Ptolemy's success "has probably caused us to lose almost all the vast body of accurate Hellenistic observations." In contrast, I have noted that Ptolemy was the first astronomer to show how to convert specific numerical data into the parameters of planetary models. In later centuries we find frequent changes of parameters in the planetary tables, but exceedingly rarely do we find the observational basis for making these changes. I am sorry that Islamic and medieval astronomers did not learn their lessons well enough from Ptolemy, although at least they had a framework from which to work.

Newton complains that I have not squarely addressed certain arguments that he has mentioned repeatedly. He fails to recognize that I was not attempting a review or rebuttal of his position, but instead I tried to focus on certain other quite interesting aspects that need to be considered in evaluating Ptolemy's success.

Neither time nor space nor inclination permits a sentence-by-sentence analysis of Newton's position as set forth in his paper.[1] Let me skip over the first part of it, where he once more recapitulates his

Selection 3 reprinted from *Quarterly Journal of the Royal Astronomical Society*, vol. 22 (1981), pp. 40–44.

own position, and go straight to the point where he attacks my statement that Ptolemy's parameters seem generally more accurate than his data base. Newton immediately provides the material to refute his own position. We all know that, at the initial level of sophistication, the motion of the Moon is the most recalcitrant and intractable case in the celestial mechanics of our system, and that Mars, because of its comparatively high eccentricity and close approaches, shows the most clearly the deficiencies of any planetary model. As Kepler later wrote, "This is why I consider it again an act of divine Providence that I arrived at Benatek at the time when [Longomontanus] was directed toward Mars; because for us to arrive at the secret knowledge of astronomy, it is absolutely necessary to use the motion of Mars; otherwise it would remain eternally hidden." Yet it is precisely for Mars and the Moon that Ptolemy comes closest to choosing the parameters that realize the potential accuracy of his models, as R. R. Newton states. The Moon is the more complicated case, and that is why I chose to examine it in some detail in my paper, and to show that Ptolemy's parameters were demonstrably better than the particular observations he cites. There is a curious parallel here to Kepler's results with Mars, for him the most difficult of the planets, yet the one for which his ephemerides gave the most reliable results. It seems to me that if Ptolemy concentrated his attention on Mars—indeed, it was the only case where he could have readily established his equant-eccentric model—then it is here (or alternatively, in the lunar case) where we must seek our insights into Ptolemy's procedures.

R. R. Newton states that Ptolemy does not use the observations he quotes to find his adopted mean motion of Mars, and that Ptolemy uses six observations to find only four parameters (eccentricity of the deferent orbit, its apsidal line, the epoch, and the size of the epicycle). I argue that Ptolemy uses five observations to find five parameters including the mean motion. In *Almagest* IX,3 Ptolemy says he will first set down the mean periods as calculated by Hipparchus, although these have been corrected by procedures that he will demonstrate in due course. He then says that Mars goes 42 revolutions plus 3 1/6° in 79 solar years plus 3.22 days; as O. Neugebauer has pointed out, the 42-79 combination is the one found in the Babylonian goal year texts. Because Ptolemy's mean motion tables (*Almagest* IX,4) are set up in increments of Egyptian years ($365\overset{d}{.}0000$) whereas the foregoing period is in terms of Ptolemy's solar year ($365\overset{d}{.}2467$), I shall tabulate Mars' period in both units:

	Ptolemy's tropical years	Egyptian years
Almagest IX,3	1.880 77	1.882 04
Almagest IX,4	1.880 77	1.882 04
Almagest X,9	1.880 77	1.882 04
Modern value[2]	1.880 86	1.882 14

In *Almagest* X,9 Ptolemy compares an observation from 272 B.C. with one of his own from A.D. 139. The early observation, an occultation of β Scorpii by Mars, seems quite plausible since recalculations show that the very close approach did take place; Ptolemy was rather astute to pick such an observation for the earlier epoch, as this sort of phenomenon guarantees a good positional accuracy. Unfortunately, he had problems converting the Greek calendar to his own, possibly because of an ambiguity with leap days, and hence he was one or two days wrong in the dating; the occultation occurred during the day on 16 January rather than early in the morning of 18 January as he thought. Because of the long temporal baseline, such an error of a day affects the period by only one part in the fifth decimal. Of course, Mars was not at the same place in its epicycle on 18 January -271 and 149,881.7 days later on 27 May $+139$, and therefore Ptolemy had to have the entire solar theory in hand to remove the effect of the epicyclic motion before he could establish the Martian mean motion. His procedure is rather foreign to contemporary tastes since he derives the synodic period rather than the sidereal period that is directly needed for setting up the mean motion table. Thus in *Almagest* X,9 he finds that the motion in the epicycle with respect to the center of deferent is 192 cycles plus 61°43′, leading to the synodic period of 779ᵈ938; given Ptolemy's solar period, the mean tropical period (686ᵈ944) is easily found as I have listed it above. Hence, I am convinced that Ptolemy has correctly used the early observation to derive the value of the mean motion he has claimed to find and which he used in the tables. R. R. Newton denies this, saying "the reader can easily verify the point himself by doing the required arithmetic." I have done this above with results contrary to his, but I must say that it was not easy to keep straight the periods in Julian years, sidereal years, tropical years, Ptolemaic solar years, Egyptian years, etc. Unfortunately, the *Crime* is not as lucid as the *Almagest*, so I cannot be sure if Newton has confused the units in tropical solar years of IX,3 with the units in Egyptian years of IX,4. In any event, it is far easier to verify a given period than to establish it in the first place, as Ptolemy must have done.

Since writing his *Crime*, R. R. Newton has decided that a sixth observation of Mars is involved; his Table 1 shows two observations on 30 May 139, but the table does not show that both of the observations give the same position, so that it would of course have been impossible for Ptolemy to derive a sixth parameter from the "additional" observation. Incidentally, it is quite remarkable to see Ptolemy deriving a highly accurate epicycle size from a pair of observations near opposition only three days apart. This seems to be another clear example where Ptolemy has introduced an observation for pedagogical purposes, but really used other quite different material to establish the actual parameter. I am prepared to believe that Ptolemy "laundered" his Mars observations to make them consistent with his determination of the epicycle size from the other observations that gave greater leverage on the solution.[3]

R. R. Newton's "most serious objection" to my paper is that I have ignored two of his three most important "proofs" of fabrication. I did mention and agree that the equinoctial and solstitial data appear to be calculated rather than observed as Ptolemy implies, although I discussed this situation primarily to indicate not only that this has been known for some centuries, but that there are also alternative suggestions as to why it came about. I felt, however, that the star catalog or the apogee of Mercury lay beyond the scope of my previous paper.

I consider Newton's statistical demonstration concerning the distribution of fractions of degrees in the star catalog to be the single most convincing and clever contribution that he has made.[4] He has shown that the distribution for the longitudes closely approximate that for the latitudes, *provided a shift of 40 minutes of arc is made* for the longitude. This strongly suggests that the catalog was set up on one reference system, and then updated by the simple addition of $n°40'$ to all the longitudes. Suppose that Ptolemy used an existing reference catalog from Hipparchus in order to establish the relative places of additional stars, and then precessed the results to his own epoch by increasing the longitudes by $2°40'$. Had Ptolemy actually made his observations as stated, with an armillary, this would leave a telltale sinusoidal variation in the latitudes, as Dennis Rawlins has pointed out. Such a variation is not found. Hence, it seems to me intrinsically more reasonable to suppose that Ptolemy appropriated an existing catalog, presumably derived from Hipparchus, rather than that he started from scratch as described in the *Almagest*. I don't doubt that Ptolemy wanted his opus to be the "complete" handbook, containing not only an extensive star catalog but also instructions for how to make a list from first principles. In a similar way theoretical astronomers today write general textbooks in which they describe how telescopes work and how observations are made even if they have never made observations themselves. It is unfortunate if Ptolemy failed sometimes to distinguish between the theoretical and the observational, but this scarcely makes him a criminal.

The case of the Mercury model is quite a curious one, but here Newton seems to ignore the fundamental observational constraint that plagued Ptolemy.[5] At sunset in September, when the ecliptic runs below the equator to the southwest, Mercury at eastern elongation will set before it is dark enough to measure, and similarly at sunrise in March with Mercury at western elongation, the sky will be too bright before Mercury is high enough above the horizon. "It is simply not true that older observers could not have located Mercury accurately at maximum elongation when the mean sun is at [the apogee or perigee points]," states Newton in his *Ancient Planetary Observations* (p. 464), but as counterexamples he goes on to cite an *evening* observation in April and a *morning* observation in October!

Poor Ptolemy! Because he couldn't get the symmetrical observations he needed, he blew the interpretation, coming up with an apsidal line 30° wrong. R. R. Newton has no such observational constraints, of course. He has picked from the almanac 51 longitudes at 80-day intervals, and has compared them with Ptolemy's predictions. Never mind that Mercury can be properly observed over only a small fraction of its trajectory, and not even at all of the maximum elongations. Newton's computer clearly shows him that Ptolemy's Mercury model would be improved if some of the parameters approached zero.

In fact, according to Newton's analysis, not only does Ptolemy continually cheat, but he is incompetent as well. For example, because he likes to keep things simple, Ptolemy places the so-called equant point opposite from the Earth an equal distance beyond the center, whereas the modern computer shows that he would have had better success (with the longitudes, that is) if the equant had not been equally spaced.

Why, then, does Ptolemy make such hard work of Mercury, with its extra wheel and very different organization? Why didn't he keep things simple by adopting the same model throughout? And if he loved forging observations, why didn't he just invent a September evening observation for Mercury and a March morning observation? At the level of accuracy to which Ptolemy is working, it is entirely the limited access of observations and not at all the eccentricity of Mercury's heliocentric orbit that has given the trouble. Clearly there are some very serious observational constraints involved that Newton has tacitly ignored. Ptolemy's Mercury model is so convoluted, in fact, that I can only believe that Ptolemy has taken mediocre observational data much too seriously, rather than that he has fudged perfectly good data to come up with something so different from all the other planetary cases.

R. R. Newton deserves credit for bringing so forcibly to our attention the inconsistencies and anomalies in Ptolemy's work. He has opened up some highly intriguing questions. Nevertheless, I believe, as J. D. Mulholland put it succinctly in another context,[6] "What is wanted in history is neither a measure of plausibility nor of ridiculousness. One wants to understand the evolution of ideas: *why* a particular idea was presented by a particular person at a particular time and *how* that idea related to the social, cultural, and intellectual milieu of the time and place. The question of truth or falsity is practically irrelevant." Although I consider the Mercury case to be a strong counterexample to Newton's argument, nevertheless I feel that it is far more interesting to find what aspects of the situation misled Ptolemy, rather than to brand him a criminal. As for R. R. Newton's litigation concerning Ptolemy's nefarious motives, I still have to adopt that peculiarly Scottish verdict: *not proven*.

Notes and References

[1] R. R. Newton, "Comments on 'Was Ptolemy a Fraud?'" by Owen Gingerich," *Quarterly Journal of the Royal Astronomical Society,* vol. 21 (1980), pp. 388–99.

[2] The mean motion of Mars per tropical century given by P. K. Seidelmann, L. E. Doggett, and M. R. De Luccia, *Astronomical Journal,* vol. 79 (1974), p. 58, has been converted from tropical years of $365^d 24220$ to units of Ptolemy's tropical solar years and Egyptian years.

[3] (Added for this anthology:) My comment that it was quite remarkable to see Ptolemy deriving a highly accurate epicycle cycle from a pair of oppositions only three days apart was based on my misunderstanding of the situation (which was shared by some of my colleagues). In the dynamical case of both a moving planet and a moving epicycle such a pair of observations in fact gives the strongest hold on the parameters, as Ptolemy no doubt knew. Hence my argument in the paper at this point was completely spurious.

[4] (Added for this anthology:) Subsequently I decided that Newton's evidence was not as strong as it first seemed; see Owen Gingerich and Barbara Welther, "Some Puzzles of Ptolemy's Star Catalogue," *Sky and Telescope,* vol. 67 (1984), pp. 421–23 [reprinted in Owen Gingerich, *The Great Copernicus Chase* (Cambridge, 1992)]. In the past few years there have been several important contributions concerning Ptoley's star catalog in the *Journal for the History of Astronomy* by M. Shevchenko (1990), Jaroslaw Wlodarczyk (1990), James Evans (1992, reviewing Gerd Grasshoff's *The History of Ptolemy's Star Catalog),* and Noel Swerdlow (1992).

[5] See my "The Mercury Theory from Antiquity to Kepler," *Actes du XIIe* Congrès International d'Histoire des Sciences, vol. 3A (Paris, 1971), pp. 57–64 [reprinted as selection 23 in this anthology].

[6] *Journal of the History of Astronomy*, vol. 11 (1980), p. 69.

Zoomorphic Astrolabes: Arabic Star Names Enter Europe

In the winter of 1966–67 the short-lived *Smithsonian Journal of History* featured on its cover a splendid fourteenth-century English astrolabe with an unusual rete filled with animal heads (Figure 1). According to the caption, the stars on this astrolabe were indicated by pointers "shaped in the form of dragon's teeth." Dragon's teeth they were not! Perhaps dog's tongues? In any event, my curiosity was piqued concerning zoomorphic astrolabes, and for a long time I have been on the lookout for them. Nevertheless, until recently I remained unaware of the ramifications of such a study, which leads directly into the fascinating problem of the transmission of Arabic astronomy into Europe.

The most important manuscript of Chaucer's *Treatise on the Astrolabe* (ca. 1391), in the Cambridge University Library, shows the star Alhabor (Sirius) with a dog's head (Figure 2).[1] Thus the zoomorphic figure makes a charming mnemonic for the Dog Star, the brightest fixed star of the nighttime sky.

Remarkably enough, the structural elements of the Smithsonian

Selection 4 reprinted from *From Deferent to Equant: A Volume of Studies in the History of Science in the Ancient and Medieval Near East in Honor of E. S. Kennedy*, ed. by D. King and G. Saliba, *Annals of New York Academy of Sciences*, vol. 500 (1987), pp. 89–104.

FIGURE 1. *The richly zoomorphic astrolabe CCA No. 2006 of the National Museum of American History in Washington.*

astrolabe rete are strikingly like this Chaucer manuscript diagram, and we must suppose that its maker either saw the diagram or the astrolabe from which the diagram was made (or another close relative). However, the superabundance of dog's heads is quite exceptional. No fewer

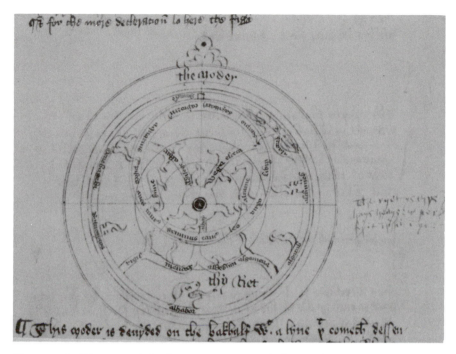

FIGURE 2. *The rete of the best manuscript of Chaucer's* Treatise on the Astrolabe, *Cambridge University Library Dd. 3.53, fol. 213v; Sirius or Alhabor,
the Dog Star, is represented with a dog's head, and the tropic of Capricorn is
shown as a dragon. Compare this with our Figure 3.*

than 18 dog's heads appear, and 18 dog's tongues pant toward their
designated stars. Furthermore, the immodest size of the heads give the
rete a bizarre, entirely idiosyncratic, appearance. Surely a significant
opportunity was lost by not figuring the pointers according to their
actual constellation figures: would it not have been more appropriate
to include a lion's head for Regulus in Leo or a bull's head for Alde-
baran in Taurus? Curiously, the maker has depicted Corvus as a raven,
so he cannot be accused of making a blind extrapolation from the Dog
Star alone.

Are there related astrolabes linking this one and the Chaucerian
diagram? In the course of this study I found two that bear such a close
resemblance to the manuscript that it is impossible to suppose that
they are independent (Figure 3). However, one must always be on
guard against latter-day forgeries inspired by the manuscript, which
became well known following the publication of Skeat in the last cen-

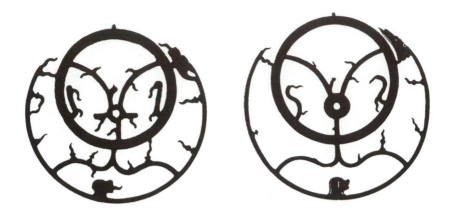

FIGURE 3. The retes of the "Chaucerian" astrolabes of Dr. Tomba (left) and the Oxford Museum of the History of Science (right).

tury, if not earlier.[2] One astrolabe, in a private collection (Tomba), is so similar to the diagram that it is tempting to believe that it is the very instrument from which the diagram was drawn. Measurements of the other, now in the Oxford History of Science Museum, show the star positions corresponding to A.D. 1600 (as opposed to before A.D. 1300 for the Tomba instrument), leading to the hypothesis that the Oxford astrolabe was a later, anachronistic example.

The Smithsonian astrolabe was once owned by the late Derek de Solla Price, who presumably supplied the fourteenth-century date noted in the *Smithsonian Journal of History*; no doubt he based his estimate on the precessional positions of the stars (corresponding to A.D. 1300) and on stylistic considerations. Sharon Gibbs, primarily on the basis of the writing style, suggests the fifteenth century in her catalog of the Smithsonian astrolabes.[3] The Tomba "Chaucerian astrolabe" could well have been made some decades before the Chaucerian manuscript was written in the early 1390s, so if the Smithsonian exemplar was derived from Tomba's, it could date from either the fourteenth or the fifteenth century. In any event, as we shall see, its 24 stars clearly fall into the same category as do the stars on the two dated instruments of the fourteenth century.

The dog's heads in both the Smithsonian astrolabe and the "Chaucerian" instruments are so conspicuous as to be deceiving—thus, in my initial search through standard sources including Gunther's *Astrolabes of the World*,[4] the Mensing catalog,[5] and the Billmeier catalog and its

FIGURE 4. *The oldest dated European astrolabe, British Museum 325, with many small zoomorphs. Reproduced by kind permission of the Trustees of the British Museum.*

supplement,[6] I found neither dog's heads nor other animals among the European instruments. However, a hint in Gunther's text led to a particularly interesting zoomorphic astrolabe, a fourteenth-century English instrument in the British Museum, and in fact the earliest *dated* European astrolabe (Figure 4).[7]

A subsequent examination of the earliest English astrolabes revealed that zoomorphic features, rather than being almost nonexistent, were in fact commonplace. The great majority of the fourteenth-century Western astrol?' ;s are English, and, in the century before Chaucer's *Treatise on the Astrolabe*, most of them apparently included zoomorphs as part of their Gothic heritage. A glance at the great Sloane astrolabe of the British Museum (Gunther 290), a magnificent specimen nearly half a meter in diameter and presumably made late in the thirteenth century, shows the standard star indicators. A more careful look, however, reveals no fewer than 34 fantastic animal pointers: small, subtle, and whimsical, like the gargoyles of medieval cathedrals or like the decorative sketches in the margins of medieval manuscripts.[8] Gunther describes the Sloane indicators as "tongues pointing from dog's heads," an initially suitable characterization for all but three, which more nearly approximate bird's heads. But then, on still closer inspection, a disquieting feature is visible: wings on at least eight of the creatures! Are they dragons? I am inclined to describe most of them that way.

Another but smaller astrolabe that exhibits the same structural pattern of the rete, and which has closely related star names, is Max Elskamp's Gothic astrolabe in the Musée de la Vie Wallonne in Liège.[9] The Elskamp instrument avoids dragons, but displays a score of bird's heads for pointers, plus a larger bird for Corvus.

Most of the fourteenth-century astrolabes described by Gunther also have zoomorphic features. "Mensing No. 26" (Gunther 295), now at the Adler Planetarium, has a dog's head not for the Dog Star, but for three in the summer evening sky, with the dog's ears pointing to *Altair* and *Delfin*, and its tongue indicating *Mu[cida] Eq^i* (ϵ Peg).[10] "Tsevi Herz's Astrolabe" (Gunther 293), in the Kensington Science Museum, has a hound's head for the Dog Star, a raven for Corvus, and the points of a satyr's beard (?) for Aquila, Delfin, and *Mu[cida] Eq^i*, as well as a flower (thistle?) for *Lanceator* (Arcturus or α Boo), *Lucida coro* (α CrB), and *Pal S Ser* (δ Oph). This astrolabe includes a serpent that borders the rete from *Cor* (Antares or α Sco) to its tail at *Cauda [Capricorni]* (δ Cap).[11]

The "Painswick Astrolabe" (Gunther 299), now at the Oxford History of Science Museum, has a similar serpent marked *Alacrab* (α Sco) at the head and *Denebalgedi* (δ Cap) at its tail. "Blakeney's London Astrolabe" (Gunther 292), at the British Museum, has a double-headed serpent bordering the rete, marked *Calbalacrab* (literally "heart of the scorpion") at one head and *Libideneb* ("goat's tail") at the other.[12]

The most charming of all these fourteenth-century zoomorphic English astrolabes is the one mentioned earlier and shown in Figure 4. With a date of 1326, it antedates Chaucer's work by 65 years. This British Museum astrolabe (BM 325) not only represents the Dog Star (named *Alhabor*) with a hound's head, but there are several other whimsical zoomorphs. Two human faces appear just inside the ecliptic perimeter, with the hat of one designated *Mucida Equi* ("nose of the horse," ϵ Peg) and the other *Alramech* (Arcturus or α Boo), while the collar of the same figure points to *Mirac* (β And). As we shall note later, the representation of Vega as a bird, "the soaring eagle," is occasionally found on Islamic astrolabes; here the beak of a bird marks *Wega*, while its tail points to *Alhae* (α Oph). Yet another bird does triple duty, its beak pointing to *Aldeboran*, its forward wing to *Elgeuse* (α Ori), and its lagging wing to *Rigil* (β Ori). At least half a dozen of the remaining pointers are also delicately sculpted bird's heads. The entire rete is bordered by a long serpent, its head pointing out *Alacrab* (α Sco), and its tail designating *Denebalgedi* (δ Cap).

In the course of searching for other zoomorphic candidates, it became clear that these instruments were among the earliest European astrolabes, and that a large proportion of all fourteenth-century astrolabes are English, in sharp contrast to the situation in the following century. Then followed the sobering realization that only two of these astrolabes are actually dated (BM 325, with a date of 1326, and the Blakene astrolabe of 1342),[13] and few have well-defined provenances. Are these really fourteenth-century English workmanship? As for dates, they all have star positions corresponding to 1300 or earlier (with the exception of the Oxford "Chaucerian" astrolabe noted above). Some of them include calendrical material specifically English, and their plates include latitudes relevant to Britain. Stylistically, the retes of these astrolabes fall into two categories, some with a quatrefoil or trefoil decoration, and others with a curved Y form; both seem to be characteristically English, although further comparison with stained-glass windows of this period would be necessary to settle this point. The evidence, though far from being decisive, confirms the traditional attribution to fourteenth-century England.

TABLE 1. Star Names

Kunitzsch or Skeat number and name		BM Sloane	Liège Elskamp	Adler Mensing 26	Whipple
1 Mirach	β And	Rigilalsabie	Rigilalsabie	Mirac	Mirak
1a	γ And	Mirac andro			
2 Batenkaytoz	ζ Cet	Batencaitoz	Batncaitoz	Betn Caitoz	Batuchaythos
2a	η Cet				
3 Pantenkaitoz	ζ Cet	—	—	—	—
4 Enif	α Ari	Caput ar'	—	Enif	Cenok
5 Finis fluxus	θ Eri	—	—	—	—
6 Menkar	α Cet	Menkar	Menkar	Menkhab	Menkar
7 Algenib	α Per	Algemb	Algemb	Algenib	Algeneb
7a	β Per				
8 Algetenar	γ Eri	Augetenar	Augetauar	Augetenar	Augetenar
9 Aldebaran	α Tau	Aldebran	Aldebran	Aldebarn	Aldeboram
10 Alhaioh	α Aur	Alhaioc	Alhaioc	Alhaioc	Alhaok
11 Rigil	β Ori	Rigil	Rigil	Rigel	Rigil
12 Algeuze	α Ori	Bedelgeuze	Algeuze	Elgeuse	Elgeuze
13 Alhabor	α CMa	Alhabor	Alhabor	Alhabor	Alhabor
14 Razalgeuze	α Gem	Caput gemino	—	Raz	—
15 Algomeyza	α CMi	Algomeiza	Algomeiza	Algomerza	Algomeiza

Comments by column: The first seven instruments are of the quatrefoil type, the others of the Y type, with the three most closely resembling the "Chaucer type" at the end. None of the astrolabes distinguish u from v, and the transcriptions here do not distinguish the rather arbitrary capitalizations of the retes.

Kunitzsch or Skeat: The transcription from Cambridge University Library Ii 3.3, fol. 70, originally given and numbered by Skeat has been repeated by Kunitzsch as his Type VIII. Because the Kunitzsch list is only being used as a convenient framework and not as a definitive system for these instruments, the modern identifications have been modified to conform with the actual usages on the astrolabes. The original manuscript apparently introduced two names and positions for β Peg, but the astrolabes commonly designate both β Peg and α And in this region of the sky, so the stars are here distinguished (but note the ambiguity and confusion as to which star is the horse, Alferaz). The designations of 17 and 21 as μ UMa and θ UMa have been interchanged. None of these astrolabes name both 40 Libedeneb and 44 Denebalgedi (δ Cet), but two that appear to point to θ Cet with the name Libedeneb have been flagged. All but one of the astrolabes that show Deneb Caitoz clearly designate β Cet rather than ι Cet.

BM Sloane = British Museum 324 = Gunther 290. Except for Sceac, the names have been taken from Morley's list in Gunther (1932), vol. 1, p. 44, which is generally much superior to the transcription in vol. 2, p. 463 (see ref. 4); I have checked my list against the instrument.

Liège, Musée de la Vie Wallonne, Max *Elskamp* collection = CCA 457. (*CCA = Computerized Checklist of Astrolabes*—see ref. 4). The names have been read from the very clear color plate XXVII in Michel (1966) (see ref. 9).

TABLE 1. (continued).

Kunitzach or Skeat Number and Name		Paris Kugel	Sothebys	Oxford Bailly's Rete	Blakene A.D. 1342
1 Mirach	β And	Rigil	Mirac	Mirac	—
1ª	γ And				
2 Batenkaytoz	ζ Cet	Batab	Betukaytoz	Batenkaitoz	
2ª	η Cet				Batencaytoz
3 Pantenkaitoz	ζ Cet	—	—	—	Pantacaytoz
4 Enif	α Ari	Cedec	[not named]	—	Enif
5 Finis fluxus	θ Eri	—	—	—	—
6 Menkar	α Cet	Menkar	Narisce	Menkhar	Menkar
7 Algenib	α Per	Algemb	Algenib	Algē	
7ª	β Per				?Algon
8 Algetenar	γ Eri	Augetenar[b]	Algetenar	Augethanar	Angethenar
9 Aldebaran	α Tau	Aldebra	Aldbaran	Aldebaran	Aldebaran
10 Alhaioh	α Aur	Alhaioc	Alayhioc [sic]	Alhaioc	Alhayot
11 Rigil	β Ori	—	Rigil	Rigil	Rigil
12 Algeuze	α Ori	Algeuze	Elgeuze	Elgeuze	Algeuze
13 Alhabor	α CMa	Alhabor	Alhabor	Alhabor	Alhabor
14 Razalgeuze	α Gem	Razel	—	—	—
15 Algomeyza	α CMi	Algomeiza	Algomeiza	Algomeiza	Algomeyza

[b]Mislabeled Rigil, corrected on rim.

Adler Planetarium *Mensing* 26 = Gunther 295 (= Gunther 200). I had the advantage of a preliminary transcription by S. Lubar.

Whipple Museum, Cambridge, England (diameter approximately 30 cm; not in the *CCA*). There are indeed two stars clearly labeled Cenok (4 and 45).

Paris, Alexis *Kugel* Collection = *CCA* 546. This rete is almost indistinguishable in design and size (118 mm) from the one in a Belgium private collection illustrated in *La mesure du temps dans les collections belges* (Bruxelles, 1984), and both contain the same marginal correction for the star Augetenar.

Sotheby's sale of 18 June, 1986, item 125, described by Francis Maddison as "circa 1300." Now in a European private collection.

Oxford Museum of the History of Science, *Bailly's rete* = Gunther 294, dated by the museum as 1320 ± 20.

Blakene = British Museum 326 = Gunther 292. One of the most difficult to read unambiguously (for example, star 49 may be Scecor rather than Stetor), and some pointers are apparently bent.

TABLE 1. (continued).

Kunitzsch or Skeat Number and Name		Oriel College	Merton College	Oxford Painswick	BM 325 A.D. 1326
1 Mirach	β And	—	Mirac	[not named]	Mirac
1ª	γ And				
2 Batenkaytoz	ζ Cet	Batonk~	Batonkaitoz	Batncaytos	Batnchaythos
2ª	η Cet				
3 Pantenkaitoz	ζ Cet	—	—	—	—
4 Enif	α Ari	—	—	—	—
5 Finis fluxus	θ Eri	—	Finis flux9	—	—
6 Menkar	α Cet	—	Menkar	Mencar	Menkar
7 Algenib	α Per	—	Alge~	—	—
7ª	β Per			Algon	
8 Algetenar	γ Eri	—	Augethanar	—	Angetenar
9 Aldebaran	α Tau	—	Al deb~	Aldeborā	Aldeboran
10 Alhaioh	α Aur	Ahok-	Alhaios	—	Alhayok
11 Rigil	β Ori	—	Rigil	Rigil	Rigil
12 Algeuze	α Ori	Elgeuze	Elgeuze	—	Elgeuse
13 Alhabor	α CMa	Alhab~	Alhabor	Alabor	Alhabor
14 Razalgeuze	α Gem	Esv~	Raz	—	—
15 Algomeyza	α CMi	Algom̄	Algomeiza	Algomeyza	Algomeyza

Oriel College = Gunther 296; in the History of Science Museum, Oxford.

Merton College, Oxford. The names have been corrected from Gunther's rather inadequate list (1923) (see ref. 14) by John North, who kindly inspected the astrolabe for me, and from my own readings of an excellent color transparency supplied by *Scientific American* and published with my article "Islamic Astronomy" in the April, 1986 issue.

Oxford Museum of the History of Science, *Painswick* = Gunther 299.

BM 325 = British Museum 325 = Gunther 291.

TABLE 1. (continued).

Kunitzsch or Skeat Number and Name		Smithsonian	Tomba	Oxford
1 Mirach	β And	—	—	—
1ᵃ	γ And			
2 Batenkaytoz	ζ Cet	—	Batukaytos	Batncatos
2ᵃ	η Cet			
3 Pantenkaitoz	ζ Cet	Pentecaitos	—	—
4 Enif	α Ari	—	—	—
5 Finis fluxus	θ Eri	—	—	—
6 Menkar	α Cet	Menchar	Menkar	—
7 Algenib	α Per	Algnib		—
7ᵃ	β Per		Algol	
8 Algetenar	γ Eri	—	—	—
9 Aldebaran	α Tau	Aldebaran	Aldeboran	Aldeboran
10 Alhaioh	α Aur	—	Alhayok	Alhayoc
11 Rigil	β Ori	Rigil	Rigil	Rigil
12 Algeuze	α Ori	Algeber	—	—
13 Alhabor	α CMa	Alabor	Alhabor	Alabor
14 Razalgeuze	α Gem	—	—	—
15 Algomeyza	α CMi	Algomeisa	Algomeyza	Algomeo

Smithsonian 2006, National Museum of American History, Washington, D.C. Here I had the advantage of the list in Gibbs and Saliba (1984). I am grateful to Dr. Gibbs for a prepublication copy of the relevant section of this work, which enabled me to correct various errors in my own transcription and to complete the part of the table relating to stars hidden on the photograph by the brass index. By oversight Gibbs's Table D omits *denebalgida*, and *zed* was accidentally read upside down to make *raz* (given as *rax alawe*).

Tullio Tomba (private collection), examined at the National Maritime Museum, Greenwich. Star 3 might be Batnkaytos rather than Batukaytos.

Oxford Museum of the History of Science (diameter approximately 14 cm, and almost a twin of the preceding rete in size and style; not in the *CCA*). This instrument is unique among this set in giving precessional positions for around 1600 instead of 1300 or earlier.

TABLE 1. (continued).

Kunitzsch or Skeat number and name		BM Sloane	Liège Elskamp	Adler Mensing 26	Whipple
16 Markep	ρ Pup	Markeb	Markeb	Markeb	Markeb
17 Egregez	θ UMa	—	—	Subpede	—
18 Aldiran	Hya	—	—	—	—
19 Alfart	α Hya	Alfard	Alfard	Alfard	Alfard
20 Calbalezed	α Leo	Cor leonis	Cor leonis	Cor	Cor
21 Alrucaba	μ UMa	Coniuncte	Coniunte	—	[not named]
22 Corvus	Crv	—	—	De corvo	[bird]
23 Dubhe	α UMa	Edub	Alruca (!)	Dubhe	Edub
24 Denebalezed	β Leo	Cauda leonis	—	Cauda	Cauda
25 Algorab	γ Crv	Algorab	Algorab	Algorab	Algorab
26 Alchimec	α Vir	Achimec	Achimec	Assimech	Alchimek
27 Bennenaz	η UMa	Benenaz	Benenaz	Bencenaz	Benenaz
28 Alramech	α Boo	Alramec	—	Alramec	Alramek
29 Alfeca	α CrB	Alfeca	Alfeca	Elfeca	Elfeca
31 Yed	δ Oph	Ied	Ied	Yed	Yed
32 Calbalacrab	α Sco	Cor scorpionis	Cor scorpionis	Cor us∼	Alacrab
33 Alhaue	α Oph	Alhaue	Alhaue	Alhawe	Alhawe

Kunitzsch or Skeat Number and Name		Oriel College	Merton College	Oxford Painswick	BM 325 A.D. 1326
16 Markep	ρ Pup	—	Markeb	—	Markeb
17 Egregez	θ UMa	—	Egregez	—	—
18 Aldiran	Hya	—			
19 Alfart	α Hya	—	Alfard	Alfard	Alfard
20 Calbalezad	α Leo	—	Cor leonis	—	Cor leon
21 Alrucaba	μ UMa		Ursa	—	—
22 Corvus	Crv	Corv9	Corv9	Corvus	—
23 Dubhe	α UMa	—	—	—	Dub
24 Denebalezed	β Leo	Cau le	Cau le	—	—
25 Algorab	γ Crv	—	—	—	Algorab
26 Alchimec	α Vir	—	[broken off]	—	Alchimek
27 Bennenaz	η UMa	—	Benetnaz	[not named]	Benetnace
28 Alramech	α Boo	Alram̄	Alram∼	[not named]	Alramek
29 Alfeca	α CrB	—	Elfeca	Elfeca	Elfeca
31 Yed	δ Oph	Yed	Yed	—	Yed
32 Calbalacrab	α Sco	Cor	Cor	Alacrab	Alacrab
33 Alhaue	α Oph	Alawe	Alawe	Alauue	Alhae

TABLE 1. (continued).

Kunitzach or Skeat Number and Name		Paris Kugel	Sothebys	Oxford Bailly's Rete	Blakene A.D. 1342
16 Markep	ρ Pup	Markeb	Mareb	Markeb	—
17 Egregez	θ UMa	—	—	—	Egregez
18 Aldiran	Hya	—	—	—	Alturan
19 Alfart	α Hya	Alfard	[not named]	Alfard	Equs
20 Calbalezad	α Leo	Cor	Cor L	Cor	Cor leonis
21 Alrucaba	μ UMa	Alruca	9iuncte	Ursa	Alrica
22 Corvus	Crv	—	—	Corvus	Corvus
23 Dubhe	α UMa	—	Edu	Dubh	—
24 Denebalezed	β Leo	Denebel	Deneb	—	Deneber
25 Algorab	γ Crv	Algorab	Algor	Algorab	Algorab
26 Alchimec	α Vir	Alchimec	Aschimec	Assimech	Alchimek
27 Bennenaz	η UMa	Benenaz	Benenaz	Benetnaz	Benenaz
28 Alramech	α Boo	Alramec	Arame	Aramec	Alramek
29 Alfeca	α CrB	Alaeca [sic]	Elfeca	Elfeca	Elfeca
31 Yed	δ Oph	Ied	Ied	Yed	Yed
32 Calbalacrab	α Sco	Cor Scorpionis	[not named]	Cor	Calbalacrab
33 Alhaue	α Oph	Alhue	Raza[c]	Alhawe	Alhawe

Kunitzsch or Skeat Number and Name		Smith-sonian	Tomba	Oxford
16 Markep	ρ Pup	Markeb	—	—
17 Egregez	θ UMa	—	—	—
18 Aldiran	Hya		—	—
19 Alfart	α Hya	Alfard	—	—
20 Calbalezad	α Leo	Cor leonis	Cor leonis	Cor leonis
21 Alrucaba	μ UMa	—	Ursa	—
22 Corvus	Crv	Corvus	Corvus	Corvus
23 Dubhe	α UMa	—	—	—
24 Denebalezed	β Leo	—	Cauda	Cauda leonis
25 Algorab	γ Crv	—	—	—
26 Alchimec	α Vir	Alcimech	Alchmek	Alchimek
27 Bennenaz	η UMa	—	—	—
28 Alramech	α Boo	Aramech	Alramek	Alramek
29 Alfeca	α CrB	Alfeta	Elfeca	Elfeca
31 Yed	δ Oph	Zed	—	—
32 Calbalacrab	α Sco	Alacrab	Alacrab	Alarcab
33 Alhaue	α Oph	Alawe	Alhawe	Alhawe

[c]Order clearly erroneous.

TABLE 1. (continued).

Kunitzsch or Skeat number and name		BM Sloane	Liège Elskamp	Adler Mensing 26	Whipple
34 Rahtaben	γ Dra	Raztaben	Raztaben	Taben	Thaben
35 Wega	α Lyr	Alvaca	Alvaca	Wega	Wega
36 Altair	α Aql	Altair	Altair	Altair	Althayr
37 Delfin	ε Del	Delfin	Delfin	Delfin	Delfin
38 Alrif	α Cyg	Alref	Alredf		
39 Addigege	α Cyg			Addigege	Aldigege
	α Cap				
40 Libedeneb	δ Cap	—	—	—	—
42 Aldiran	α Cep	Alderaim		Dherat	—
43 Enifelferaz	ε Peg	Menkeb	Alref (!)	Mu eq[i]	Mucida Equi
44 Denebalgedi	δ Cap	Denebalgedi	Denebalge	Dheneb algedi	Denebalgedi
45 Sceach	δ Aqr	Sceac	Sceac	Schcack	Cenok
46 Alferaz	β Peg	Bedalferaz	Alferaz	[Hu equi][a]	Humer9 Equi
46a	γ Peg	—	—	—	—
47 Mentichel	α And	Alferaz	—	Alfraz	Alferaz
48 Denebkaitoz	β Cet	Denebcaitoz	Denebcaitoz	Dheneb Caitoz	Denebchaytos
49 Sceder	α Cas	—	Sceder	Sceder	Skeder

Kunitzsch or Skeat Number and Name		Oriel College	Merton College	Oxford Painswick	BM 325 A.D. 1326
34 Rahtaben	γ Dra	—	Taben	—	Taben
35 Wega	α Lyr	Wega	Wega	Wega	Wega
36 Altair	α Aql	Alth∼	Altahir	Altair	Altair
37 Delfin	ε Del	Delfin	Delfin	Delfin	Delfin
38 Alrif	α Cyg				
39 Addigege	α Cyg	—	—	—	Aldigege
	α Cap				Cor corni
40 Libedeneb	δ Cap	Libed ′ᶜ	Libed∼ᶜ	—	—
42 Aldiran	α Cep	—	—	—	Aldera
43 Enifelferaz	ε Peg	—	Mu equi	—	Mucida equi
44 Denebalgedi	δ Cap	—	—	Denebalgedi	
45 Sceach	δ Aqr	—	Sceach	—	Cenok
46 Alferaz	β Peg	Hu9 equi	Hu Equi	[not named]	Hum9 equi
46a	γ Peg	Ala	—	—	—
47 Mentichel	α And	—	Alferaz	—	—
48 Denebkaitoz	β Cet	—	—	—	—
49 Sceder	α Cas	—	Sceder	—	Sheder

[a]Erased.   ᶜθ Cet.

TABLE 1. (continued).

Kunitzach or Skeat Number and Name		Paris Kugel	Sothebys	Oxford Bailly's Rete	Blakene A.D. 1342
34 Rahtaben	γ Dra	—	—	Taben	—
35 Wega	α Lyr	Alvaca	Razta[c]	Wega	Wega
36 Altair	α Aql	Altair	Altair	Altair	Altair
37 Delfin	ε Del	Delfin	Delfin	Del	Delfin
38 Alrif	α Cyg	Elref	Alwega[c]		Alrif
39 Addigege	α Cyg			A?digege	
	α Cap				
40 Libedeneb	δ Cap	—	Deneb	—	Libideneb
42 Aldiran	α Cep	—	Alder	Dhera	Aldiraz
43 Enifelferaz	ε Peg	Elferaz	Enif	Mu E	—
44 Denebalgedi	δ Cap	Denebalsedi	—	Denebalgedi	—
45 Sceach	δ Aqr	Sceac	Scehae	Sceach	—
46 Alferaz	β Peg	Alferaz	Bezalf	Hu-Eq[u]i	Alpheraz
46a	γ Peg				—
47 Mentichel	α And	—	Alfer	Alferaz	Menchef
48 Denebkaitoz	β Cet	Denebcaitos	Denebca	Denebkaitoz	Denebcaytoz[d]
49 Sceder	α Cas	—	—	Sced	Stetor

Kunitzsch or Skeat Number and Name		Smith-sonian	Tomba	Oxford
34 Rahtaben	γ Dra	Rasaden	—	—
35 Wega	α Lyr	Wega	Wega	Wega
36 Altair	α Aql	Altaire	Altayr	Altayer
37 Delfin	ε Del	Delphin	Deyfin	—
38 Alrif	α Cyg			
39 Addigege	α Cyg	—	—	—
	α Cap			
40 Libedeneb	δ Cap	—	—	—
42 Aldiran	α Cep	—	—	—
43 Enifelferaz	ε Peg	—	—	—
44 Denebalgedi	δ Cap	Denebalgida	Denebalgedi	Denebalgedi
45 Sceach	δ Aqr	—	—	—
46 Alferaz	β Peg	Humer9 eq	Humer9 eq[i]	Humer9 Equi[f]
46a	γ Peg	—	Ala	—
47 Mentichel	α And	—	—	Atera
48 Denebkaitoz	β Cet	—	Denebcaytos	—
49 Sceder	α Cas	—	—	—

[c]Order clearly erroneous.   [d]ι Cet.   [f]Pointer bent?

FIGURE 5. *The exuberantly zoomorphic astrolabe of as-Sahl al-Asturlabi an-Nisaburi, reproduced courtesy of the Germanisches Nationalmuseum, Nuremberg.*

Another stylistic similarity of these instruments is the particular choice of star names, which differs from those of the continent or of the fifteenth century. (In the absence here of alternative lists, this point will have to be granted on faith, but skeptics can consult the comparative list given by Gunther,[14] for example.) The names are generally direct transcriptions from the Arabic, with four almost invariable exceptions: *Cor leonis* and *Cauda leonis* for α and β Leonis, and *Humerus equi* and *Mucida equi* for β and ϵ Pegasi. The Arabic star names follow (with differing orthography) the Arabic designations of the "Chaucer-type" list VIII in Kunitzsch's *Typen von Sternverzeichnissen*,[15] (which comes from Skeat's edition of Chaucer's *Treatise on the Astrolabe*[16]). Kunitzsch has shown that this list is a melange of an old astrolabe star list that appears in Spain at the end of tenth century and a later list compiled in Paris by one John of London in 1246. Clearly these instruments document a key period in the transmission of Arabic star names into common English usage. Excepting for the Latin or Greek names of a few first-magnitude stars (Sirius, Regulus, Spica, Procyon, Antares, Arcturus), these Arabic names have stuck, though with a few variants such as Betelgeuse for *algeuze* and Deneb for *[deneb] ad-digege*.

Because of the potential significance of these names in documenting the flow of Islamic science into Europe, I have carefully recorded the names from a number of early English astrolabes in Table 1. It was my hope that such a list might help distinguish key differences between subgroups, but instead, it shows the remarkable homogeneity of these as a group, with the possible exception of two of the earliest examples, the Great Sloane Astrolabe and Elskamp's Gothic astrolabe, which as a subgroup exhibit the most differences of nomenclature compared to the others.[17] However, the star list shows no systematic differences between the quatrefoil and Y types (except that the nature of the quatrefoil rete provides more opportunities for star pointers), nor does it distinguish the more conspicuously "Chaucerian" type of astrolabe found in the last three columns of the table.

In the 1400s the center of Western astrolabe making moved to the Paris workshop of the artisan Jean Fusoris, and by the early 1500s the leading maker was Georg Hartmann of Nuremberg. Both were craftsmen of clean lines and serious purpose, and playful beasts or zoomorphs had no role in their workmanship.

For further zoomorphic touches we must move to the Islamic world. In general, Islamic orthodoxy worked against representational art, although by the thirteenth and fourteenth centuries the contact with Western art generated such an interest that some Islamic artists incorporated living forms into their work. As Richard Ettinghausen has pointed out, however, representational artists were of the lowest social classes, along with usurers, tatooers, and buyers of common (nonhunting) dogs.[18] It is thus not surprising that the hound's head makes essentially no appearance on Eastern astrolabes!

Eventually Islamic art expanded to some extent into depiction of man and beast, and this affected the production of astrolabes (and globes) as well. Nevertheless, one rarely finds a comparable display to the Smithsonian astrolabe or BM 325. The traditional soaring and swooping eagles of the Islamic nomenclature—Vega and Altair—were occasionally figured on Eastern astrolabes and on some of the earlier preserved specimens. Vega appears as a bird already in 1062 A.D. on a Byzantine astrolabe, Gunther 2. Both Vega and Altair are depicted as birds on the unusual geared thirteenth-century astrolabe of Muhammad ibn Abi Bakr of Isfahan; in addition, the head of a horse appears in the region of Pegasus.[19] Two astrolabes made around the same time by the Cairo maker Abd al-Karim show Vega as a bird, and the second of these is filled with delightful animal heads, fish, birds, and even Ophiuchus (who looks more like a dancer than a serpent bearer).[20] To a large extent his creatures are mnemonic—a horse, a goat, a scorpion, a sea serpent—although he clearly includes an elephant (but no camel!) and he avoids dogs.

Somewhat later in the thirteenth century the relatively quiet ingenuity of Abd al-Karim's astrolabe gives way to the thorough-going exuberance of as-Sahl al-Asturlabi an-Nisaburi (Figure 5).[21] The instrument bears three birds (pointing to Vega, Altair, and Corvus), a lion, a horse, a serpent, and seven men (Ophiuchus, who holds the serpent; Hercules kneeling; Bootes with the lance—he is sometimes called "Lanceator" and the Arabic equivalent of Arcturus is *simak ramih*, the "simak armed with a lance"; Cassiopeia/Cepheus; Orion; *shamiyya*, the northern [Sirius] = Procyon; and an anomalous figure who points to the eye of the bull. The figures do not necessarily occupy the space allotted to their constellations, but merely fall in the general area so they can point properly with various (and sometimes unmentionable) parts of their anatomy. These figures (with a few exceptions) are reminiscent of the drawings of as-Sufi and their representations on Islamic celestial globes.

A much later Indian astrolabe from around 1650, the so-called Jaipur A instrument, shows many playful elements, almost resembling a child's puzzle in the subtle and intricate way in which the animals have been incorporated into the tracery.[22] Not only are Vega and Altair shown as birds, but Deneb (of Cygnus) as well. Procyon and Sirius appear as dog's heads, possibly uniquely in the Islamic world. There also is a bull, a horse, and a river, and probably a whale and a bear among several other figures awaiting decipherment!

A similarly configured astrolabe was purchased by the Kensington Science Museum at the Christie's auction of 31 October 1985. Made by Jamal ad-Din in A.H. 1077 (A.D. 1666), the 25-cm diameter instrument shows at least 16 pictorial representations.

While containing a splendid variety of zoomorphs, these Islamic creations are too rare and dissimilar to be called a school, or even to reveal influences from one to another. The only Islamic "school" of astrolabists using zoomorphic elements involves a group of nineteenth- or twentieth-century forgers. One of the characteristics of these late forgeries is the depiction of Vega as a bird even though the makers being faked never represented "the soaring eagle" zoomorphically.[23]

In conclusion, this study has found a relatively homogeneous group of fourteenth-century zoomorphic astrolabes from England, and a much more diverse group of Islamic zoomorphic astrolabes from the thirteenth century (with scattered later examples). The greater interest seems to lie with the English group because of their testimony to an important period in the transmission of Arabic star names into the West. Undoubtedly more questions have been raised than solved by this study: Did astrolabe making enter Europe primarily through England? Were the astrolabes that have traditionally been called English really made in England, and did most of them precede the Black Death of 1349? Are the zoomorphs simply part of a larger tradition that includes manuscripts and other art forms? I hope that such questions will challenge art historians and paleographers as well as historians of science to look further at these puzzles.

Notes and References

[1] Cambridge University Library Dd.3.53 was used as the basis of Skeat's edition of Chaucer's astrolabe treatise (see next note); our Figure 2 comes from it, and it is also reproduced i.a. in Robert T. Gunther, *Early Science in Oxford*, vol. 5 (Oxford, 1929), p. 6. A similar rete appears in Bodleian Ms Rawl D.913, f. 41. The rete in Cambridge University Library Ii.3.3, f. 66v does not show a dog's head, but the tropic of Capricorn is represented as a dragon; it is quite similar to the fourteenth-century Bodleian Ms Ash 1522 #8, fol. 87v.

[2] Walter W. Skeat, ed., *A Treatise on the Astrolabe; addressed to his son Lowys by Geoffrey Chaucer, A.D. 1391* (London, 1872; reprinted New York, 1967).

[3] Sharon Gibbs with George Saliba, *Planispheric Astrolabes from the National Museum of American History*, Smithsonian Studies in History and Technology No. 45 (Washington, D. C., 1984).

[4] Robert T. Gunther, *Astrolabes of the World* (Oxford, 1932; reprinted London, 1976). These two volumes provide the standard numbering for the 332 examples described. This system is used and extended in Sharon L. Gibbs, Janice A. Henderson, and Derek de Solla Price, *A Computerized Checklist of Astrolabes* [*CCA*] (New Haven, 1973).

[5] Max Engelmann (cataloger), *Collection Ant. W. M. Mensing, Amsterdam, Old Scientific Instruments (1479–1800)*, vol. 2, Plates (Amsterdam, 1924).

[6] C. H. Josten, *A Catalogue of Scientific Instruments (13th–19th century), The Collection of J. A. Billmeier, C.B.E.* (Oxford, 1954), and Francis R. Maddison, *A Supplement to a Catalogue of Scientific Instruments in the Collection of J. A. Billmeier, Esq., C.B.E.* (Oxford and London, 1957).

[7] F. A. B. Ward, *A Catalogue of European Scientific Instruments in the Department of Medieval and Later Antiquities of the British Museum* (London, 1981), pp. 112–13, Plate LI.

[8] A good example of a whimsical and grotesque marginal figure is found in a thirteenth-century English astrolabe manuscript now in the Herzog August Bibliothek in Wolfenbüttel, 51.9 Aug 4°, f. 92.

[9] *CCA* 457; see Henri Michel, *Traité de l'astrolabe* (Paris, 1947), Plate III, and even better, Henri Michel, *Les cadrans solaires de Max Elskamp* (Liège, 1966). Michel's date of ca. 1200 is surely much too early.

[10] Among the astrolabes not mentioned by Gunther is a large, early Gothic astrolabe at the Whipple Museum, somewhat reminiscent of the Adler-Mensing astrolabe, which has a dog's head for the Dog Star as well as a raven for Corvus. This astrolabe is apparently not in *CCA* (ref. 4).

[11] Only a few of the original Gothic-script names remain on this instrument; most have been erased and replaced with later Latin names, so this astrolabe has not been included in our Table 1.

[12] Ward, (ref. 7), p. 113, Plate LI; with a few exceptions, Gunther did not illustrate the British Museum astrolabes because the museum declined to help subsidize his work, but a photograph can now be examined in Ward's book.

[13] Both of the dated astrolabes fall in the first half of the fourteenth century. John North has pointed out to me a dramatic decline in the production of English scientific writings following the Black Death of 1349. It is likely that the production of astrolabes followed a similar decline, but it would probably require the work of a paleographer to establish the dates more precisely on the other early English astrolabes.

[14] Robert T. Gunther, *Early Science in Oxford*, vol. 2 (Oxford, 1923), pp. 222–25.

[15] Paul Kunitzsch, *Typen von Sternverzeichnissen in astronomischen Handschriften des zehnten bis vierzehnten Jahrhunderts* (Wiesbaden, 1966). Professor Kunitzsch tells me that there are about 200 manuscripts of the parent text of the Type VIII list, not counting its secondary branches.

[16] Skeat, (Ref. 2), pp. xxxvii–xxxix.

[17] Kunitzsch has pointed out that many of the variant readings on these two astrolabes (Algemb, Coniuncte, Raztaben, Alvaca, Bedalferaz) agree closely with his Type VI list.

[18] Richard Ettinghausen, "The Man-Made Setting," in *The World of Islam*, ed. by Bernard Lewis (London, 1976), p. 62.

[19] Gunther 5, A.H. 618 = A.D. 1223–24, the only known instrument by this maker.

[20] Gunther 103 and 104. The latter, more richly zoomorphic instrument, is dated A.H. 633 = 1235–36. For a discussion of the dating of this British Museum astrolabe, see L. A. Mayer, *Islamic Astrolabists and their Works* (Geneva, 1956), p. 30.

[21] Gunther 137, but a very confused entry; see Paul Kunitzsch, *Arabische Sternnamen in Europa* (Wiesbaden, 1959), p. 47. This silver-and-brass astrolabe, presumably acquired by Regiomontanus in the 1460s, appears in color on the cover and as entry 2 of the Germanisches Nationalmuseum exhibition catalogue, *Treasures of Astronomy*, ed. by Gerhard Bott (Nuremberg, 1983).

[22] Gunther 74, illustrated in silhouette on 203 and in gold as the cover design. A better illustration is found in G. R. Kaye, *The Astronomical Observatories of Jai Singh* (Calcutta, 1918), Figure 5 of Plate II.

[23] See Owen Gingerich, David King, and George Saliba, "The 'Abd al-A'imma Astrolabe forgeries," *Journal for the History of Astronomy*, vol. 3 (1972), pp. 188–98 [reprinted as selection 5 in this anthology].

The 'Abd al-A'imma
Astrolabe Forgeries

with D. King and G. Saliba

'**A**bd al-A'imma is probably the best known and certainly the most prolific of the Persian astrolabists. The three-dozen known examples of his workmanship are characterized by elaborate but pleasing decorations and a high standard of calligraphy. Yet not a single detail of his life is known. From the few instruments that are dated and from dedicatory inscriptions, we know that he flourished in the Safavid period early in the eighteenth century. He not only made complete instruments himself but also decorated others for Khalīl Muḥammad, for Muḥammad Ṭāhir, and for Muḥammad Amīn b. Muḥammad Ṭāhir. Many of Khalīl Muḥammad's other astrolabes were decorated by Muḥammad Bāqir, who in turn also decorated some instruments for 'Abd al-'Alī. Hence, from these interconnections we can conclude that 'Abd al-A'imma was a leading member of a school of astrolabists probably centered in Isfahan and active from about 1678 (A.H. 1089) until the downfall of the Safavid dynasty in 1722 (A.H. 1135).[1]

Our initial studies disclosed that the astrolabes carrying the name of 'Abd al-A'imma fall into two distinct categories. Although both are distinguished by rich calligraphy and elaborate metalworking, the instruments in the larger group are accurately designed from an astro-

Selection 5 reprinted from *Journal for the History of Astronomy,* vol. 3 (1972), pp. 188–99.

nomical point of view (see Figure 1), whereas those of the second group can at best be called degenerate. Curiously, virtually every example of the degenerate group carries a date below 'Abd al-A'imma's name in a cartouche on the reverse side, but few of the good group are dated, and rarely within the signature cartouche. Consequently, we systematically obtained photographs or examined directly all the dated astrolabes of 'Abd al-A'imma the Younger listed in Mayer's *Islamic Astrolabists* (except number I, whose present location is unknown), as well as the first he attributed to 'Abd al-A'imma the Elder.[2] This investigation reinforced our initial working hypothesis: because the degenerate astrolabes are generally dated, we suppose they are deliberate forgeries rather than innocent but poor imitations. In addition, we hypothesize that "'Abd al-A'imma the Elder" is a spurious entity arising from a grotesque misdating of two of the forgeries.

Group I: Astronomically Correct Astrolabes

Instruments in this group are accurately engraved and fully decorated with graceful foliated spiders ('ankabūts or retes) not too "busy" in their patterns. In common with other astrolabes of the Isfahan school, they have an almost stereotyped reverse side that includes in the upper right quadrant two sets of arcs, one for the solar altitude at midday and the other for the altitude of the Sun when in the azimuth of the qibla (the direction of Mecca) (see Figures 2 and 3). The astrolabes either made or decorated by 'Abd al-A'imma that we have examined are listed in Table 1.

By means of photographs, we have been able as a group to examine over a dozen correct 'Abd al-A'imma astrolabes. These include three very similar instruments that were formerly in the Hoffman Collection and are now at the Smithsonian Institution's Museum of History and Technology in Washington [now National Museum of American History]; the large and particularly fine example made by Muḥammad Amīn and decorated by 'Abd al-A'imma, in the City Art Museum of St. Louis, and almost a twin to one in the Indian Museum in Calcutta, which is illustrated by Gunther[6]; an astrolabe in the National Maritime Museum, Greenwich; and three examples in the Adler Planetarium in Chicago. The photographs enabled us to note the close similarity between two of the Adler astrolabes and the Smithsonian ones, and between the third Adler astrolabe and the Greenwich example. Another astrolabe similar to these last two was examined by the first author in the Victoria and Albert Museum. Perhaps the rather stan-

(a)

(b)

FIGURE 1. *Astronomically correct 'Abd al-A'imma astrolabes. The instrument in (a) is typical of the smaller variety, with a diameter of 95 mm, whereas the spider in (b) represents the larger instruments, with a diameter of 160 mm. Left: A88, courtesy of the Adler Planetarium, Chicago; right: XVIII, courtesy of the trustees of the National Maritime Museum, Greenwich.*

dardized sizes and patterns of 'Abd al-A'imma's astrolabes made possible his unusual productivity.

In the course of this examination, we became increasingly aware of the remarkable similarity in the tracery of the spiders of these instruments. Although no two are identical, the astrolabes made by 'Abd al-A'imma when he was not collaborating with another craftsman exhibit the same flowing pattern outside the ecliptic, as well as one of two general patterns within the ecliptic, depending on the size of the instrument (Figure 1). Only on the smaller astrolabes is there a straight bar through the pole to the ecliptic; it never extends beyond. Furthermore, a very similar pattern is used by contemporary members of the Isfahan school such as 'Abd al-'Alī (compare Gunther plates XXVII and XXIX). The spiders of the astrolabes containing the name of Khalīl Muḥammed are distinctively different, however, with a bar extending across the entire spider and a circular concentric arc in the region south of Cancer. It is curious to note that an astrolabe located at the Observatoire de Paris and containing only the name of 'Abd al-A'imma and not Khalīl Muḥammad is also of this same type.[7]

The conclusions drawn from the photographs were reinforced by the opportunity of the first author to examine directly 15 additional 'Abd al-A'imma astrolabes in Oxford, London, and Leningrad. These instruments, also listed in Table 1, are, with the exceptions noted below the table, examples of excellent workmanship and technical accuracy.[8]

Group II: Degenerate Astrolabes

The second group of instruments attributed to 'Abd al-A'imma also exhibit a rich decorative style of metalworking, but the astronomical details are by a craftsman who can at best be called incompetent. These astrolabes are listed in Table 2.[9] Typically, the network of the spider has complete bilateral symmetry (an obvious scientific impossibility, since the spider is a star chart, and its leaves serve as pointers for the brighter stars). In all these astrolabes, both the ecliptic on the spider and the grid on the plates are badly distorted, as shown by the ecliptic projection ratios in Table 2.[14] This is revealed most obviously by the failure of the equator, horizon, and east-west line to intersect at common points on the plates. Furthermore, the plates are not notched, so that it is difficult to hold them in a fixed position. On the back of the astrolabe (see Figures 2 and 3), the three most typical errors are that (1) the upper left quadrant is not graduated with the required 60×60

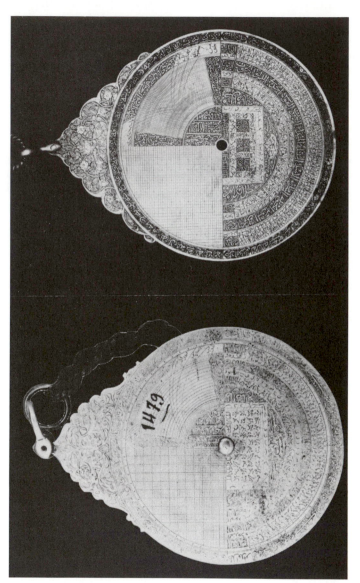

FIGURE 2. *Verso of a correct 'Abd al-A'imma astrolabe (left) exhibits the proper 60×60 grid in the upper left trigonometrical quadrant, in contrast to the nonsensical 47×39 divisions on the degenerate instrument (right). Compare also the divisions around the circumference. Although astronomically useless, the astrolabe at right displays first-class metalworking; note the Hegira date of 1127 below the signature in the cartouche centered under the shadow square. (Left: A74, courtesy of the Adler Planetarium; right: XXIX, courtesy of the Freer Gallery of Art, Smithsonian Institution, Washington, D.C.)*

FIGURE 3. *Verso of a correct 'Abd al-A'imma astrolabe (left) shows the graduation in spacing in the concentric arcs in the upper right quadrant, in contrast to the degenerate instrument (right). Note the radial lines that edge the shadow square in the correct astrolabe, compared to the forgery. (Left: XXVIII, Smithsonian Institution; right: XVII, Museum of Fine Arts, Boston.)*

TABLE 1. Some Astronomically Correct Astrolabes Made or Decorated by 'Abd al-A'imma

Maker	Mayer	Location	ICA (Ref. 3)	Collection	Ref.
'Abd al-A'imma	XV	Washington, Smithsonian	39	Hoffman 16	
'Abd al-A'imma	XXI	Washington, Smithsonian	40	Hoffman 18	
'Abd al-A'imma	XXVIII	Washington, Smithsonian	37	Hoffman 20	
M. Amin	III	St. Louis City Art Museum	1171		4
'Abd al-A'imma	XVIII	National Maritime Museum	1039	A14-36 1	
'Abd al-A'imma	—	Chicago, Adler Planetarium		A74	
'Abd al-A'imma	—	Chicago, Adler Planetarium		A88	
'Abd al-A'imma	—[a]	Chicago, Adler Planetarium		A90	
'Abd al-A'imma	—	Paris, Landau Collection			
'Abd al-A'imma	V[b]	Oxford, Mus.Hist.Sci.	38		
'Abd al-A'imma	VII[c]	Oxford, Mus.Hist.Sci.			
'Abd al-A'imma	XI[d]	Oxford, Mus.Hist.Sci.	35		
'Abd al-A'imma	XII	Oxford, Mus.Hist.Sci.	36		
'Abd al-A'imma	XIII = XVI	Oxford, Mus.Hist.Sci.	1087		
'Abd al-A'imma	XXII	Oxford, Mus.Hist.Sci.	11		
'Abd al-A'imma	XXVII	Oxford, Mus.Hist.Sci.	1001		
Khalil M.	IX	Oxford, Mus.Hist.Sci.	1017	Billmeir 7	
Khalil M.	X	Oxford, Mus.Hist.Sci.	1018	Billmeir 8	
Khalil M.	XVI[e]	Oxford, Mus.Hist.Sci.	1019	Billmeir 9	
Khalil M.	XVIII[f]	Oxford, Mus.Hist.Sci.			
M. Ṭāhir	I	Oxford, Mus.Hist.Sci.	21		
'Abd al-A'imma	VI[g]	London, Victoria & Albert	34		
'Abd al-A'imma	XIX	Leningrad, Hermitage		VC 941	5
Khalil M.	XIX	Leningrad, Hermitage		VC 939	

[a]The third Adler astrolabe has an outrageously wretched spider, apparently a later replacement; it is dated 1127 on the signature cartouche, and is the sole example known to us of a good 'Abd al-A'imma instrument dated this way.

[b]Dated 124 = A.H. 1124 in the dedication on the throne.

[c]Dated A.H. 1134 in the dedication to Prince 'Alīqūlī, a royal Safavid.

[d]Dated A.H. 1115 around the rim.

[e]Dated 119 = A.H. 1119 on the alidade.

[f]This instrument includes a replacement spider in the degenerate style of those in Table 2.

[g]Contrary to Mayer, this instrument is not dated.

TABLE 2. Degenerate astrolabes attributed to 'Abd al-A'imma

| | | | | Ecliptic proj. ratio | | |
Mayer	Location	Collection	Date	Spider	Plate	Ref.
II	Paris	G. Charliat	A.H. 1117	1:3.2	1:3.0	
III	Oberlin	Allen Memorial Art Museum	1121	1:2.8	1:2.8	10
IV	Detroit	Institute of Arts	1121	1:2.6	1:2.6	11
XVII	Boston	Museum of Fine Arts	—	1:3.2	1:3.4	12
XXIX	Washington	Freer Gallery of Art	1127	1:3.0	1:2.6	
XXXII	Paris	G. Charliat	1125	1:2.0	1:2.0	
—	Chicago	Anonymous	1097	1:3.6	1:3.0	
—	Zurich	E. Mannheimer	1125	1:3.1	1:4.0	13
(...the Elder I)	London	M. Dineley	986	1:3.2		
				(correct ratio 1:2.33)		

grid; (2) the arcs concentric to the center in the upper right quadrant are equispaced; and (3) the calibration lines in the shadow square are not radial to the center.

The symmetrical spiders of this degenerate group not only are entirely different stylistically from the preceding group but, unexpectedly, bear distinctive relations to each other in style (Figure 4). The similarity of the traceries is especially striking in the first five cases of Table 2. With the exception of the Charliat XXXII and Chicago astrolabes, they all have a bar through the pole across to the edge of the spider, and in most instances this bar carries an inscription. Another curious point of distinction is the treatment of the star pointer for Vega (Nasr wāqi', "falling eagle"). In the first five examples, this star is represented as a bird's head, a device never found in the astronomically correct astrolabes of 'Abd al-A'imma. (Nevertheless, on the correct instruments he invariably adds an extra dot, like the eye of a bird, to the Vega leaf, giving an almost subliminal impression of an eagle's head.)

Of this degenerate group, only the Freer XXIX astrolabe has any star names on the pointers, and these are often more wildly displaced than even the symmetry of the design would enforce; the more than 30 leaf pointers of the Mannheimer astrolabe contain the names of the Sun, Moon, Mercury, Mars, and Saturn, used repeatedly. On the Charliat II, Freer XXIX, Chicago, and Mannheimer astrolabes, the

FIGURE 4. *Two 'Abd al-A'imma forgeries with similar symmetrical spiders and with jumbled inscriptions both on the thrones (or handles) and around the edge of the spiders. Note the birds' heads above the central pivots. (Left: XVII, courtesy of the Museum of Fine Arts, Boston; right: III, courtesy of the Allen Memorial Art Museum, Oberlin College, Oberlin, Ohio.)*

zodiacal signs are reversed around the ecliptic; on the Charliat II and Oberlin III examples, they are shifted by 180°; and on 'Abd al-A'imma the Elder I, there are 24 zodiacal signs! The Charliat XXXII astrolabe has no names at all on the ecliptic, but an irregular assortment of numbers. Around the outside edge, the 5° intervals are subdivided rather randomly into 7, 8, 9, 10, or 11 parts on these astrolabes, or else they have no subdivisions at all.

An unusual and characteristic feature of most of the degenerate astrolabes is the presence of an inscribed border around the spider, filled with a rather peculiar jumble of star names and astronomical words. It is interesting to note that no similar astrolabes are illustrated in the standard works by Gunther,[15] Michel,[16] or Maddison,[17] presumably because of their disreputable quality; however, other quite comparable spiders exist. Among the examples that have come to our attention are one by Muḥammad Zamān in the Freer Gallery of Art in Washington, which is stylistically close to the Chicago astrolabe and which shares the same defects on the reverse side as does this group; a silver astrolabe inscribed with the name Muḥammad b. Khiḍr (?), which was formerly in the Toledo Museum of Art in Ohio and which has a very similar spider but a crude earlier style kufic verso; another very similar silver astrolabe, unsigned, in the Smithsonian Institution collection; an unsigned example in the Yale Medical Historical Library in New Haven; and one by Shaykh Muḥammad 'Irāqi in the Fogg Art Museum (Harvard),[18] which has no graphs at all on the verso. Another example, in the Oxford Museum of the History of Science, is particularly interesting because it is attributed to another of 'Abd al-A'imma's Isfahan contemporaries, 'Abd al-'Alī (Mayer VI); it shares the characteristic spider, including the bird and bar, as well as the same defects on the verso.

The term "astrological astrolabes" has sometimes been euphemistically applied to such degenerate devices, but they could hardly serve for astrological calculations any better than for astronomical ones. In addition, on several of these examples the correspondence (on the verso) between the zodiacal signs and the lunar mansions or other astrological lists has been reversed.

A crowning incompetence, calligraphic rather than astronomical, appears on the throne (kursī) or handle of the Charliat XXXII, Oberlin III, and Boston XVII astrolabes. In place of the often used Quranic verse 255 of Sura II,

wasi'a kursiyyuhu 'al-samawāt wa-'al-'arḍ'

("His throne extends over heaven and earth"), there is written on the Charliat XXXII instrument

khuṭūṭ samawāt wa-'al-'arḍ fī 'al-bilād 'al-marqūma 'alā 'aṭrāfihā'

("The lines of the heavens and latitude in the cities marked at their ends"). This curious phrase evidently arises from a garbling of the throne verse with the inscription ordinarily found on the verso in the upper right quadrant,

khuṭūṭ sumū 'al-qibla fī 'al-bilād 'al-marqūma 'ala 'aṭrāfihā
bi-'al-irtifā' 'al-gharbī

("The azimuth lines of the qibla for the cities marked at their ends by the western altitude scale"), and in particular from a confusion between the words samawāt (heavens) and sumūt (azimuths), which differ only by an aleph, and between the words arḍ (earth) and 'arḍ (latitude). On the Oberlin III instrument appears the following corrupt but truncated version:

khuṭūṭ samawāt wa-'al-'arḍ fī 'al-bilād 'al-marqū[ma].'

On the Boston XVII astrolabe, the following jumble is found:[19]

    khūṭ samawāt wa-'al-'a[rd].'

The engravers were probably Persians ignorant of Arabic; this conclusion is strengthened by the orthography *mīdhān*, rather than *mīzān*, used for the constellation Libra throughout this set of degenerate astrolabes. These two words sound the same in Persian, but the latter is the correct spelling.

This evidence leads us to postulate the existence at an unknown date, but perhaps as early as the eighteenth century, of a decadent school of incompetent astrolabe makers, who included "copies" of 'Abd al-A'imma instruments in their production. A principal reason for supposing these astrolabes to be deliberate forgeries rather than innocent but degenerate imitations lies in the fact that they are dated. Eight of the nine degenerate astrolabes are dated in the cartouche below the maker's signature, with dates in the early eighteenth century, in contrast to the good 'Abd al-A'imma astrolabes, which are rarely dated in that position. Hence, the dates of the degenerate astrolabes could not have been copied; they must have been deliberately used to identify the instruments with the most glorious period of the Isfahan school.

It seems to us inherently unlikely that two 'Abd al-A'immas with decorative styles superficially so similar would have been working as contemporaries and using the same signature. On the other hand, it does seem likely that the reputation of the best-known Persian astrolabist would attract forgers even at a relatively early date, especially when we read an account written in Teheran in 1875 that "some ancient astrolabes bearing the names of renowned makers, such as Abdul Ameh, still exist in Persia and are valued at the most extravagant prices"[20] and that around 1850 "the few remaining examples were sought by the greatest in the kingdom and were easily sold for 50 ducats, so much did they love to have one in their sight, although many could not understand one iota of it."[21]

In any event, the astrolabes of the degenerate group could not have been made or used by a competent astronomer and surely were not made by the same skilled craftsman who inscribed the splendid instruments in the Oxford, Smithsonian, and Adler Planetarium collections.

Notes and References

[1] We reject the hypothesis that 'Abd al-A'imma was the name of a workshop rather than of a specific person. Why would there be such a complex pattern of collaborations between the named astrolabists of the Isfahan school and 'Abd al-A'imma if

the latter were simply the name of a workshop? As we shall show, two distinct styles are associated with 'Abd al-A'imma's name. The workshop hypothesis fails to explain why two, and only two, such distinctive styles emerged.

The dates for the Isfahan school are deduced from dated astrolabes mentioned in L. A. Mayer, *Islamic Astrolabists and Their Works* (Geneva, 1956). In this paper, we follow Mayer's numeral notation. Note that for 'Abd al-A'imma, VIII = X, XIII = XVI; and XXIX–XXXI are included in the additions near the end. XXXII is described in Mayer's supplement in Richard Ettinghausen (ed.), *Aus der Welt der Islamischen Kunst* (Berlin, 1959), pp. 293–296.

[2] We wish to thank Dr. Francis Maddison, Curator of the Museum of the History of Science in Oxford, for access to the Museum's incomparable collection of astrolabe photographs; in this way we could examine, for example, the first instrument of 'Abd al-A'imma the Elder, the present location of which was unknown to us. We also wish to thank Alain Brieux, who supplied the photographs of the Landau and Georges Charliat astrolabes, and Roderick and Marjorie Webster for photographs of the Chicago astrolabes. Unfortunately we have not yet had an opportunity to examine the second astrolabe of 'Abd al-A'imma the Elder, which is in the estate of the late E. S. David in Forest Hills, New York.

[3] D. J. Price, "An International Checklist of Astrolabes," *Archives internationales d'histoire des sciences*, vol. 8 (1955), pp. 243–63, 363–81. Price's numbers below 1000 correspond to the Gunther numbers (see ref. 6).

[4] C. P. D[avis], "Muhammedan Metal Work," *Bulletin of the City Art Museum of St. Louis*, (October, 1926), p. 53.

[5] L. E. Maistrov, *Nautnye Pribori* [Scientific Instruments], (Moscow, 1968), p. 44, Plates 85–88.

[6] R. T. Gunther, *The Astrolabes of the World*, vol. 1 (Oxford, 1932), pp. 146–47.

[7] Described but unfortunately not illustrated in L. A. Sédillot, "Description d'un astrolabe construit par Abd-ul-Aima, ingénieur et astronome persan," *Annales de l'Observatoire Impérial de Paris*, Mémoires, vol. 9 (1868), pp. 164–71.

[8] Besides examining the actual instruments in the Museum of the History of Science in Oxford, the first author inspected the photographs of some additional 'Abd al-A'imma astrolabes in other collections. Although most of these examples fell unambiguously into the stylistic patterns described in the text, two exceptions can be noted. 'Abd al-A'imma XXIII, in the collection of Claudius Côte, Lyon, is stylistically similar to the good astrolabes in broad outline, but the details are worrisome and are in several instances wrong, or at least crude by 'Abd al-A'imma's standards. This example can perhaps be called a copy, but not a forgery. Dr. Maddison concurs that this work is anomalous and suggests as one possibility that 'Abd al-A'imma's name was added to another maker's astrolabe.

'Abd al-A'imma X, now in the Musée Alaoui but formerly in the Meyerhof collection (= VIII), has not been made with the sure hand characteristic of 'Abd al-A'imma; there are many false lines and a peculiar spill of a legend into the trigonometric grid in the upper left quadrant on the verso. Concerning this instrument, we must suspend judgment.

[9] We exclude from this list 'Abd al-A'imma XXX in the collection of the Buffalo Society of National Sciences because it is clearly a crude imitation, not even worthy of consideration in this curious group of decorative forgeries.

[10] L. A. Mayer, "An Astrolabe by 'Abd al-A'imma," *Allen Memorial Art Museum Bulletin*, vol. 14, no. 1 (1956), pp. 2-6. Mayer did not realize that the astrolabe was a forgery and apparently did not even notice any of its deficiencies.

[11] Adele Coulin Weibel, "A Persian Astrolabe," *Bulletin of the Detroit Institute of Arts*, vol. 25 (1946), pp. 59–61.

[12] Mehmet Aga-Oğlu, "Two Astrolabes of the Late Safawid Period," *Bulletin of the [Boston] Museum of Fine Arts*, vol. 45 (1947), pp. 79–84. Aga-Oğlu passed over the principal deficiencies in the astrolabe.

[13] Illustrated in the Sotheby's sale catalog for 19 December 1966, item 74, sold to E. Mannheimer for £350.

[14] On a stereographic projection, the ratio of the distances from the central pivot to the nearest and farthest points of the ecliptic is necessarily 1:2.33—in Table 2 we designate this as the ecliptic projection ratio.

[15] R. T. Gunther, *op. cit.*

[16] Henri Michel, *Traité de l'astrolabe* (Paris, 1947).

[17] [F. R. Maddison], *A Supplement to a Catalogue of Scientific Instruments in the Collection of J. A. Billmeir* (Oxford, 1957).

[18] Illustrated by Owen Gingerich, "Rara Astronomica," *Harvard Library Bulletin*, vol. 19 (1971), Plate IV, 117–39.

[19] Aga-Oğlu, *op. cit.*, noticed that the throne verse was corrupt on the Boston astrolabe, but did not recognize its source.

[20] R. Murdoch Smith, *Persian Art* (London, 1876), p. 36; also quoted by Aga-Oğlu, *op. cit.*

[21] Translated from A. Kržiž, "Das persisch-arabische Astrolabium des Ab-dul Aimeh," *Das Weltall*, vol. 5 (1905), pp. 121–30, 144–52.

Alfonso X as a Patron of Astronomy

W hatever may be the reputation of Alfonso X now—as a poet, as a patron of music, or as a great lawgiver whose image is emblazoned on a medallion in the U.S. House of Representatives—surely in the late Middle Ages and Renaissance he was known primarily as the author of a set of astronomical tables. Beginning in the 1320s, manuscript copies of the *Alfonsine Tables* rapidly displaced the earlier *Toledan Tables*, and in 1483 the printed editio princeps appeared, followed in the next six decades by four more published editions.[1]

Today we know that Alfonso was not actually the author of these tables; intensely interested as he may have been in their production, his primary role was one as patron both for the tables and for a much larger astronomical corpus, which was for the most part a Castilian translation of a series of Arabic works.

Unlike the *Alfonsine Tables*, the astronomical corpus did not receive a wide distribution in the Middle Ages, nor was it printed until the last century. Then, in a splendid burst of Castilian nationalism, Manuel Rico y Sinobas published it in five unwieldy folio volumes under the title *Libros del Saber de Astronomía*.[2] Anthony Cárdenas has recently shown quite convincingly that the title would have been correctly

Selection 6 reprinted from *Alfonso X of Castile, the Learned King (1221–1284)*, ed. by F. Márquez-Villanueva and C. A. Vega, Harvard Studies in Romance Language, vol. 43 (1990), pp. 30–45.

given as *Libro del Saber de Astrologia* (or simply *Del Saber de Astro-logia*),[3] but Rico, like his enlightened nineteenth-century colleagues, eschewed astrology to the extent of trying to launder Alfonso. He quoted approvingly from the French historian Bailly that the Alfonsine canons were "infectés de l'erreur commune (la astrología) à tous ces siècles."[4]

Rico's attempt to suppress the common error of all those centuries was anachronistic because, in Alfonso's day, a thirteenth-century ruler under strong Arabic influences would very likely have had an interest in astrology. Nevertheless, the word *astrologia* was used almost interchangeably with *astronomia*, and most of what appears in this corpus would by modern standards be called astronomy, so Rico's title probably gave his 1860s audience a better idea of its contents than the more accurate label "Book of the Knowledge of Astrology."

What, in fact, do these sixteen books contain? First, there is a large four-part section on the eighth sphere. This is the standard medieval term for the sphere of the starry heavens lying beyond the seven spheres for the planets (that is, for the Moon, Mercury, Venus, the Sun, Mars, Jupiter, and Saturn, in that order from the fixed, central Earth). The book on the constellations of the eighth sphere begins with *Ossa Menor*, the little bear, and, as is also the case with the 45 constellations that follow, it includes a diagram with sectors delineating astrological influences. To that extent, this is perhaps the most astrological of all the sections.

The text describing the stars in each constellation is quite interesting. For example, it says that in the constellation *Aguila* the brightest star is called in Castilian *bueytre volante* or in Arabic *alnacr altayr*.[5] The Castilian is a direct translation of the the Arabic's "flying eagle," but it is the Arabic name Altair that today designates the star. Along the ecliptic, we find *Escorpion*, the Latin name, but in Castilian it is *alacrán*, very similar to its Arabic etymon, *alacrab*, which is also given.[6] The bright star is in Castilian *coraçon*, very close to the Latin *Cor*, and in Arabic *calb*. Today we have gone back to the Greek for *Antares*, the name of the star on the ecliptic that is so red that it can be mistaken for the ruddy planet Mars.

As I indicated before, most of the works in this astronomical corpus are translations from the Arabic, and this is no exception. Steinschneider in the last century showed that this work derives from as-Sufi, a tenth-century Persian astronomer.[7] The preface to this Alfonsine section states that the translation was carried out in 1276 by Yehuda ben Moses Cohen, a Jewish scholar who figures in a number of these

FIGURE 1. *Alfonso X medallion in the U.S. House of Representatives, great lawgivers series.*

astronomical projects, and Guillen Arremon Daspa, apparently a Christian collaborator.[8] From the examples just given, it is clear that these scholars reworked the material, and they undoubtedly added information from other sources.

None of the fifteen sections that follow are quite as long as the combined parts on the constellations. Each of them deal with astronomical instruments or devices. The first is a treatise on the use of the celestial globe. The translation by Yehuda ben Moses and Daspa was begun in 1259, and 18 years later was cast in its final form.[9] There follows a work on the spherical astrolabe, a rare instrument that derives directly from an astronomical globe. The preface to this treatise is by Alfonso himself, who says that no suitable book could be found for translation, and therefore Rabiçag or Rabbi Sag—otherwise known as Isaac Ibn Sid—had been commissioned to create a work specially for the occasion.[10] This statement gives us a special insight into Alfonso's role in the undertaking. Clearly he wanted an encyclopedic corpus, and he took an active role in formulating the requirements for his series. As we shall see from the dates, some of these astronomical translations must have begun soon after his coronation in 1252, but the idea of an integrated corpus achieved its final form near the end of his troubled reign, in the late 1270s.

In contrast to the spherical astrolabe, the plane astrolabe is a more convenient, more sophisticated, but also more common instrument. In a certain sense it can be considered as a logical derivative from the spherical astrolabe, and so it appears next in the sequence. Not only was the plane astrolabe relatively common, but so were treatises about it. Thus, around a century later, when Geoffrey Chaucer set about to compose a treatise on the astrolabe in Middle English, he had no difficulty modifying a Latin work from Spain that was popularly attributed to Messahalla, an eighth-century Jewish astronomer from Basra.[11] Never mind that it was probably written by someone else and probably in Spain[12]—the diagrams are necessarily similar to Alfonso's version, as are the rules for use. What treatise was actually used and who was the translator or editor is not stated,[13] so perhaps an insoluable mystery remains. In any event, we can be confident that there were plenty of sources for this part of the corpus.

The universal astrolabe is the next logical generalization of the plane astrolabe, and thus it is no surprise to find it treated next in the corpus. In fact, two separate versions of it appear in consecutive books, the first being translated by Ibn Sid in Toledo and the second originally in the fourth year of Alfonso's reign and then again more completely by Bernaldo the Arab and Abraham the doctor in Burgos in 1277.[14] Both are attributed to the eleventh-century Moorish astronomer al-Zarqali, known in Spain as Azarquiel. However, the first form of the universal astrolabe seems actually to come from Ibn Khalaf, as is stated at the beginning of part two of the manuscript[15]; his treatise is preserved only through this Castilian translation. The second form of the instrument, called the "Açafeha," really does come from Azarquiel as advertised, and there is in the Escorial Library an Arabic manuscript with the same 100 chapters translated into Castilian in the Alfonsine corpus.[16]

The astrolabe in its various forms allows the astronomer to set the sky for a particular time and place, to know exactly which stars are rising or setting, which are on the meridian or zenith, and so on. However, the astrolabe works only for the fixed stars; to place the wandering planets, another complementary device is required. We might expect, therefore, that this device, the planetary equatorium, would follow next in the corpus, and so it does in Rico's edition.

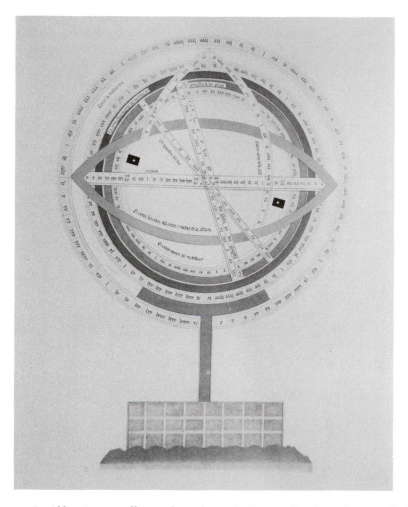

FIGURE 2. *Alfonsine armillary sphere from the Rico y Sinobas edition, vol. 2, plate between pp. 24 and 25.*

However, Professor Cárdenas points out that Alfonso himself preceded the equatorium with the armillary sphere,[17] and this has a strong internal logic because the armillary, which shows the three-dimensional celestial path of the planets, is to the equatorium as the celestial globe is to the astrolabe.

Not only is Alfonso's organizing hand seen in this arrangement, but he contributed the introduction. There he states that a work of Azarquiel serves as the basis for the first part, but that for the second part

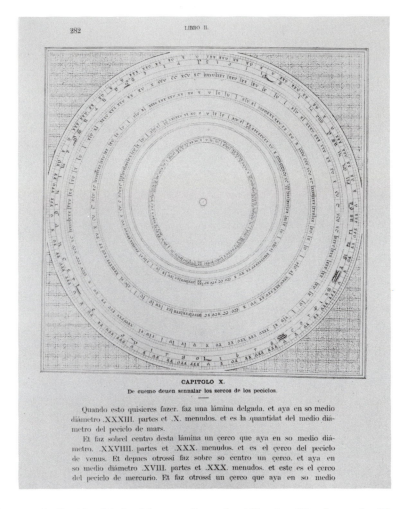

FIGURE 3. *Ovoid orbit for Mercury from the Alfonsine "Books on the Plates of the Seven Planets"; Rico y Sinobas, vol. 3, plate on p. 282.*

on the *use* of the armillary sphere, no suitable book could be found, and therefore it was specially written at his command by Ibn Sid of Toledo.[18] As for the two "Libros de las Láminas de las siete Planetas," as the two sections on the equatorium are called, these are attributed to Albucacim Abnaçamh and Azarquiel, respectively.[19] The first name is Ibn Sam'h, an Islamic astronomer of Seville in the first half of the eleventh century, on whose work al-Zarqali built.[20]

Perhaps the most fascinating thing about the Alfonsine equatorium concerns the particularly intricate path for Mercury. Let me pause to explain that in the Ptolemaic system the motion of each planet is basically represented by the combination of two circles. The planet itself is placed on the rim of an epicycle, which rides on the carrying circle or deferent. Elusive Mercury, always seen close to the Sun and therefore not well observed, posed an unusually difficult situation for Ptolemy. Because of faulty observations, Ptolemy found it necessary in his model to include an additional inner circle that displaced the deferent itself. Thus, while the model was composed of perfect circles, the combination motions created in effect an elliptically shaped deferent. As I indicated earlier, the Alfonsine corpus was seldom copied, and thus it had relatively little effect on the course of European astronomy. Nevertheless, knowledge of the odd, effective shape of Mercury's deferent did penetrate into Europe through Spain, and eventually became known to Kepler.[21] I hasten to point out that this particular ellipse has absolutely nothing to do with the real elliptical shape of Mercury's orbit, despite what Rico may have thought, but nonetheless it was a suggestive image and certainly part of Kepler's subconscious toolbox.[22]

Next in Alfonso's astronomical corpus comes a work on the quadrant. The quadrant is an observing instrument related to the astrolabe, and precisely why it follows the equatorium is not obvious. In any event, this is yet another case in which Alfonso in 1277 commissioned a new work from Isaac Ibn Sid to fill this gap in the series.[23] The next five books deal with timekeeping, and they likewise are written or edited by Ibn Sid. They deal with sundials, waterclocks, a mechanical device using quicksilver, standard candles, and the "palace of the hours," which seems to be a sundial building.

Finally, at the tail end of the compilation, comes the short book of the atacir, which tells how to make an astrolabic plate divided for the astrological houses. Thus the astronomical corpus is mostly a systematic book of instruments, which seems to have found its final form sometime after 1277.

Of the sixteen surviving volumes from the royal Alfonsine scriptorium, one comprises the text of this astronomical corpus, the *Libro del Saber de Astrologia*, Codex Complutensis 156 in the Biblioteca de la Universidad de Madrid.[24] It was used by Rico as the primary document for his edition[25]; since then it has suffered serious damage, particularly when the library of Madrid University became a battleground in the Spanish Civil War. For some years, a third of it was nearly illegible, but today it has been almost completely restored, thanks to

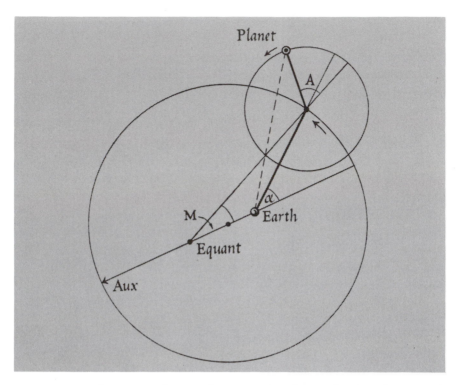

FIGURE 4. *Ptolemaic mechanism for a superior planet, showing the eccentric deferent, equant, and epicycle. From Derek Price,* The Equatorie of the Planetis (*Cambridge, 1955*).

the Centro Nacional de Restauración de Libros y Documentos.[26]

Another important collection from the royal Alfonsine scriptorium is today found in the Bibliothèque Arsenal in Paris, Codex 8.322, whose 142 folios contain Castilian versions of the tables of Albatani (together with a fragment of their canons) and those of Azarquiel, plus a work on the "quadrante sennero" (or "single quadrant"); the latter work was again a translation of Isaac Ibn Sid's, and the Albatani canons have been presumed as Ibn Sid's work.[27]

But even in Rico's day there was an embarrassing lack: neither of these key manuscripts included the work that made Alfonso famous—namely, the *Alfonsine Tables.*

In another, incomplete, manuscript of the astronomical corpus, however, Madrid National Library L97 (now MS 3306), Rico found a Castilian version of the instructions for using the tables, one that mentioned Yehuda ben Moses and Ibn Sid.[28] Unfortunately, this manuscript lacked the tables themselves, and, since then, no copy of the tables in a form that matches these Castilian instructions has ever been found.

Nevertheless, we know a great deal about these tables because, in another form, they spread throughout Europe, emanating from Paris in the late 1320s.[29] In order to appreciate these tables, we must again consider the Ptolemaic theory, as well as one of Alfonso's most famous and possibly apocryphal sayings.

First, let us review the traditional epicyclic theory of that second-century Alexandrian, Claudius Ptolemy. If we wish to calculate actual planetary positions from the epicyclic model, we must know the parameters that specify the actual geometry. We must know, for example, the size of the epicycle with respect to the deferent, the period of motion of the planet in the epicycle, and of the epicycle along the deferent, and so on. Altogether there are seven such parameters, of which five are independent for each planet.[30] Given the parameters and the geometry of the model, we could find the longitudinal direction of the planet for some specified time, but in the days before pocket calculators this would have been rather hard work. Even *with* a small calculator the procedure is tedious enough to ruin a morning. So this is where astronomical tables come in. By precalculating certain fundamental parts of the motion, the whole operation becomes much more manageable.

Basically, the procedure works like this: Suppose we know where Mars was on 1 January 1252. (That was the year of Alfonso's coronation, and, among others, precisely this quantity is tabulated.[31]) Since then to the Alfonsine anniversary, 732 years have elapsed. If we know how far Mars moves on the average in 732 years, we could add it onto this base position, and then we would know approximately where the center of the epicycle of Mars was on New Year's Day in 1984. If we want it for 18 November, we would need to know how much it would move in ten completed months, and how much more in seventeen completed days. The *Alfonsine Tables* provide for this, too, but in their common form, as issued from Paris, they basically provide the motion in days rather than in years and months, but they make all this very easy by providing tables that show how many days there are in 700 years, in 20 years, in 12 years, in 10 months, and so on. Furthermore, not only does this table give the daily motion of the center of the epicycle, but a similar one gives the motion within the epicycle. If you know where the epicycle is, then all you need is some correction back

Tabula Equationū Martis ♂						
Linee nu. merico munes	Equa tio centri	diuersitas proportionalis		Logi tudo longior	Equa tio argumenti Adde	Logi tudo prior
g / g	g m m	m	m	g m m	g m m	g m m
31 / 29	11 23	0	4 / 1	2 30 / 3	33 40 / 18	1 52 / 3
32 / 28	11 24	0	5 / 1	2 32 / 2	33 58 / 18	2 55 / 3
33 / 27	11 24	0	6 / 1	2 35 / 3	34 15 / 17	2 57 / 2
34 / 26	11 24	0	7 / 1	2 37 / 2	34 32 / 17	3 0 / 3
35 / 25	11 24	0	8 / 1	2 39 / 2	34 49 / 17	3 3 / 3
36 / 24	11 24	0	9 / 1	2 42 / 3	35 6 / 17	3 6 / 3
37 / 23	11 23	I	10 / 1	2 44 / 2	35 23 / 17	3 9 / 3
38 / 22	11 23	0	11 / 1	2 47 / 3	35 40 / 17	3 12 / 3
39 / 21	11 22	I	12 / 1	2 49 / 2	35 56 / 16	3 15 / 3
40 / 20	11 21	I	13 / 1	2 51 / 2	36 12 / 16	3 19 / 4
41 / 19	11 20	I	14 / 1	2 54 / 3	36 28 / 16	3 22 / 3
42 / 18	11 19	I	15 / 1	2 56 / 3	36 43 / 15	3 25 / 3
43 / 17	11 17	2	16 / 1	2 59 / 3	36 58 / 15	3 28 / 3
44 / 16	11 15	2	16 / 0	3 1 / 2	37 13 / 15	3 32 / 4
45 / 15	11 13	2	17 / 1	3 4 / 3	37 27 / 14	3 36 / 4
46 / 14	11 11	2	18 / 1	3 7 / 3	37 41 / 14	3 39 / 3
47 / 13	11 9	2	19 / 1	3 10 / 3	37 55 / 14	3 43 / 4
48 / 12	11 6	3	20 / 1	3 13 / 3	38 9 / 14	3 47 / 4
49 / 11	11 3	3	21 / 1	3 16 / 3	38 23 / 14	3 50 / 3
50 / 10	11 0	3	22 / 0	3 19 / 3	38 36 / 13	3 54 / 4
51 / 9	10 57	3	22 / 1	3 22 / 3	38 49 / 13	3 58 / 4
52 / 8	10 53	4	23 / 1	3 25 / 3	39 1 / 12	4 1 / 3
53 / 7	10 49	4	24 / 1	3 28 / 3	39 13 / 12	4 5 / 4
54 / 6	10 45	4	25 / 1	3 32 / 4	39 24 / 11	4 9 / 4
55 / 5	10 41	4	26 / 1	3 35 / 3	39 35 / 11	4 13 / 4
56 / 4	10 37	4	27 / 0	3 39 / 4	39 45 / 10	4 17 / 4
57 / 3	10 33	4	27 / 1	3 43 / 4	39 56 / 11	4 21 / 5
58 / 2	10 29	4	28 / 1	3 46 / 3	40 5 / 9	4 26 / 4
59 / 1	10 25	4	29 / 1	3 50 / 4	40 14 / 9	4 30 / 5
60 / 0	10 21	4	30 / 1	3 54 / 4	40 23 / 7	4 35 / 5

FIGURE 5. *Table of equations of Mars from the* Alfonsine Tables, *shown here in the first edition, Venice, 1483.*

or forth to account for where the planet is within the epicycle, and this is furnished by the so-called table of equations. This table contains not just one column of corrections but five, so the matter is clearly a little more complicated than I have stated. If we think about the geometry again, we will see why this is so. The Earth is not located at the center of the deferent, so even if the epicycle is moving uniformly around on the deferent, it will not appear that way from the Earth since sometimes the epicycle will appear closer and thus seem to be moving faster. Another column of corrections will take care of that detail. But if the epicycle varies its distance from the Earth, then a given motion within the epicycle will have a larger effect when it is closer than when it is farther. This is just a little more subtle, but Ptolemy, in one of his greatest mathematical tricks, found how to cover all the possible cases by multiplying together two numbers from two different columns[32]; he slightly complicated matters by using sometimes one pair of columns and sometimes a different pairing column, so that three more columns were required.

Two centuries after Alfonso, a Polish student at the University of Cracow bought a newly printed set of the *Alfonsine Tables*, and he used them very hard, as the stains indicate.[33] His name was Nicholas Copernicus, and two decades later he came up with a radical new Sun-centered arrangement for the planets. At the end of his first small tract proposing this heliocentric arrangement, he exclaimed, "Behold, the entire ballet of the planets in only 34 circles."[34] To nineteenth-century ears, this sounded as if he had enormously simplified things, and so the rumor began that Ptolemy had required 80 circles in his system, and that Alfonso had used even more, piling epicycles upon epicycles.[35] And then that famous snide, almost blasphemous, remark attributed to Alfonso apparently gained great currency, that if he had been present at creation, he could have given the Good Lord some hints![36] Matters got so bad that the 14th edition of the *Encyclopaedia Britannica,* published in the 1960s, stated that, by the time of Alfonso, there were 40 epicycles for each planet![37]

If we pause to think rationally about the structure of the *Alfonsine Tables*, we must realize that the addition of a single epicycle on an epicycle would add many more columns to the tables. Not only would there be a column to correct for the new epicycle, but the tables would have to take into account whether the smaller epicycle was on the near side or the far side of the bigger epicycle, and in turn whether the bigger epicycle is near or far from the Earth. Even Yehuda ben Moses and Ibn Sid would have had difficulty calculating all these extra columns, not to mention what would have happened with an epicycle on an epicycle on an epicycle.

In fact, what they produced was even purer Ptolemy than the eleventh-century *Toledan Tables* on which they were based—that is, the models stayed exactly the same and the parameters were closer to Ptolemy's original numbers than those in the *Toledan Tables*.[38] Alfonso's astronomers had done a good, if not highly original, job. The tables were the basis of all astronomical almanacs and ephemerides until the middle of the sixteenth century, and Copernicus's revisions were not all that much better with respect to the mundane task of predicting planetary positions.[39] Alfonso's name was surely known to every university student through the sixteenth century, since astronomy was a required subject. I have my doubts that Alfonso ever said that, if he had been present at creation, he could have given a few suggestions. But I have no doubts at all that he got his money's worth as a royal patron of astronomy!

Notes and References

[1] Venice, 1483; Venice, 1492; Venice, 1518 (1521 on colophon); Venice, 1524 (date on f. Q4); Paris, 1545 (reissued 1553). The occasionally reported issue of 1488 is a ghost.

[2] Five volumes, Madrid, 1863–67, hereafter cited as Rico y Sinobas. Volume 5, part 2, was never published.

[3] Anthony J. Cárdenas, "A New Title for the Alfonsine Omnibus on Astronomical Instruments," *La Coronica, Spanish Medieval Language and Literature Newsletter,* vol. 8 (1980), pp. 172–78; see also his "Toward an Understanding of the Astronomy of Alfonso X, El Sabio," *Indiana Social Studies Quarterly,* vol. 31 (1978–79), pp. 80–90.

[4] Rico y Sinobas, vol. 1, p. L.

[5] Rico y Sinobas, vol. 1, p. 45.

[6] Rico y Sinobas, vol. 1, p. 75.

[7] For references to Steinschneider, see p. 143 in Alfred Wegener, "Die astronomischen Werke Alfons X," *Bibliotheca Mathematica,* ser. 3, vol. 6 (1905), pp. 129–85; Wegener, who is better known for the theory of continental drift, has here in his doctoral research produced an invaluable analysis of Alfonso's astronomical corpus.

[8] Rico y Sinobas, vol. 1, p. 7.

[9] Rico y Sinobas, vol. 1, p. 153.

[10] Rico y Sinobas, vol. 2, p. 113. Moritz Steinschneider seems to have been the first to identify Rabiçag with Isaac Ibn Sid, an observer of eclipses in 1263–66; see his *Die Hebraeischen Ubersetzungen des Mittelalters und die Juden als Dolmetscher* (Berlin, 1893, reprint Graz, 1956). He has been universally followed in this by everyone including George Sarton, *Introduction to the History of Science,* vol. 2, part 2 (Baltimore, 1931), but excepting John Esten Keller, who lists the two as separate individuals in his *Alfonso X, El Sabio* (New York, 1967), p. 136.

[11] Robert T. Gunther, *Chaucer and Messahalla on the Astrolabe,* vol. 5 of *Early Science in Oxford* (Oxford, 1929).

[12] Paul Kunitzsch, "On the Authenticity of the Treatise on the Composition and Use of the Astrolabe Ascribed to Messahalla," *Archives Internationales d'Histoire des Sciences,* vol. 31 (1981), pp. 42–62.

[13] Rico y Sinobas, vol. 2, p. 225.

[14] Rico y Sinobas, vol. 3, pp. 3 and 135. The expression "Abrahem su alfaqui" can be translated as "Abraham the lawyer"; Cassell's *Spanish Dictionary* gives "alfaqui" as "Mohammedan doctor of laws." However, if the word "alfaquin" is intended, the meaning is "sage" or "doctor [of medicine]"; Evelyn S. Procter, *Alfonso X of Castile: Patron of Literature and Learning* (Oxford, 1951), pp. 124–25, identifies Abraham as a physician.

[15] Rico y Sinobas, vol. 3, p. 11.

[16] David A. King, "On the Early History of the Universal Astrolabe in Islamic Astronomy," *Journal for the History of Islamic Science*, vol. 3 (1979), pp. 244–57, esp. p. 248.

[17] Anthony J. Cárdenas, "A Medieval Spanish Collectanea of Astronomical Instruments: An Integrated Collection," *Journal of the Rocky Mountain Medieval and Renaissance Association*, vol. 1 (1980), pp. 21–28.

[18] Rico y Sinobas, vol. 2, p. 1.

[19] Rico y Sinobas, vol. 3, p. 241; Professor Cárdenas has supplied the spelling Abnaçamh from MS 156.

[20] Emmanuel Poulle, *Équatoires et Horlogerie Planétaire du XIIIᵉ Siècle* (Paris, 1980), pp. 195–200.

[21] Willy Hartner, "The Mercury Horoscope of Marcantonio Michiel of Venice," *Vistas in Astronomy*, vol. 1 (1955), esp. pp. 118–22; reprinted in *Oriens-Occidens* (Hildesheim, 1968).

[22] See Owen Gingerich, "Kepler's Place in Astronomy," *Vistas in Astronomy*, vol. 18 (1975), esp. pp. 272–74 [reprinted as selection 19 in this anthology.]

[23] Rico y Sinobas, vol. 3, p. 287.

[24] Anthony J. Cárdenas, "The Complete *Libro del Saber de Astrologia* and Cod. Vat. Lat. 8174," *Manuscripta*, vol. 25 (1981), pp. 14–22.

[25] Rico y Sinobas, vol. 5, pp. 6–10 ("Codice Num. 2") and pp. 103–8.

[26] Cárdenas, *op. cit.*, pp. 15–16, plus a private communication.

[27] Georg Bossong, *Probleme der Übersetzung wissenschaftlicher Werke aus dem Arabischen in das Altspanische zur Zeit Alfons des Weisen* (Tübingen, 1979), p. 81; see also his *Los Canones de Albateni* (Tübingen, 1978), esp. pp. 5–10.

[28] Rico y Sinobas, vol. 5, pp. 12–14 ("Codice Num. 4") and pp. 109–13. His transcription is found in vol. 4, pp. 111–83.

[29] See Emmanuel Poulle, "Jean de Murs et les tables Alphonsines," *Archives d'histoire doctrinale et litttéraire du moyen âge*, vol. 47 (1980), pp. 241–71 and also his articles "John of Lignères," "John of Murs," and "John of Saxony" in *Dictionary of Scientific Biography*, vol. 7 (New York, 1973); John North, "The Alfonsine Tables in England," in *Prismata, Festschrift für Willy Hartner* (Wiesbaden, 1977), pp. 269–301, which supersedes J. L. E. Dreyer, "On the Original Form of the Alfonsine Tables," *Monthly Notices of the Royal Astronomical Society*, vol. 80 (1920), pp. 243–62.

[30] Owen Gingerich, "Ptolemy, Copernicus, and Kepler," *Great Ideas Today 1983* (Chicago, 1983), esp. pp. 149–51 [reprinted as selection 1 in this anthology].

[31] For a description of the calculating procedure, see Emmanuel Poulle and Owen

Gingerich, "Les positions des planètes au moyen âge: application du calcul électronique aux tables Alphonsines," *Académie des inscriptions et belles-lettres, comptes rendus des séances*, (1967), pp. 531–48; for all the numbers, see Emmanuel Poulle, *Les Tables Alphonsines avec les canons de Jean de Saxe* (Paris, 1984).

[32] See Olaf Pedersen, "Ptolemy's Method of Interpolation," in his *A Survey of the Almagest* (Odense, 1974), pp. 85–89.

[33] This volume of the 1492 *Alfonsine Tables* is now at the Uppsala University Library, Copernicana 4.

[34] Paraphrased from Copernicus's *Commentariolus*, trans. by Edward Rosen, *Three Copernican Treatises* (New York, 1939, 1959, 1971), p. 90.

[35] Owen Gingerich, " 'Crisis' versus Aesthetic in the Copernican Revolution," *Vistas in Astronomy*, vol. 17 (1975), pp. 85–93 [reprinted as selection 11 in this anthology]; Robert Palter, "An Approach to the History of Early Astronomy," *History and Philosophy of Science*, vol. 1 (1970), pp. 93–133.

[36] Cárdenas (private communication) reports that Jeronimo de Zurita (1512–80) mentions but rejects the statement; José Soriano Viguera, *Contribución al conocimiento de los trabajos astronónomico desarrollados en escuela de Alfonso X el Sabio* (Madrid, 1926) states that nothing of the sort can be found in Alfonso's writings. I would like to take this opportunity to thank Professor Cárdenas for his many helpful remarks concerning this paper.

[37] *Encyclopedia Britannica* (Chicago, 1969), vol. 2, p. 645.

[38]

Table of Parameters

	Sun	Moon	Mercury	Venus	Mars	Jupiter	Saturn
			Eccentricity				
Ptolemaic	0.041 67	0.828 06	0.050 00	0.020 83	0.100 00	0.045 83	0.056 94
Alfonsine	0.037 82	0.827 88	0.050 00	0.018 91	0.100 00	0.052 04	0.056 94
Toledan	0.034 72	0.828 06	0.050 00	0.017 22	0.100 00	0.045 83	0.056 94
			Direction of apsidal line at A.D. 1				
Ptolemaic	65.50 00	—	188.63 33	53.63 33	114.13 33	159.61 67	231.63 33
Alfonsine	71.42 31	—	190.65 92	71.42 31	115.20 36	153.61 67	233.39 51
Toledan	77.83 33	—	197.50 00	77.83 33	121.83 33	164.50 00	240.08 33
			Mean motion				
Ptolemaic	0.9856 35	13.1763 82	3.1066 99	0.6165 09	0.5240 60	0.0831 22	0.0334 89
Alfonsine	0.9856 46	13.1763 95	3.1067 02	0.6165 16	0.5240 68	0.0831 27	0.0334 97
Toledan	0.9856 09	13.1763 56	3.1067 02	0.6165 25	0.5240 38	0.0830 91	0.0334 56

[39] Owen Gingerich, "The Accuracy of Ephemerides 1500–1800," *Vistas in Astronomy* vol. 28 (1985), pp. 339–42.

The 1582 "Theorica Orbium" of Hieronymus Vulparius

The Florentine craftsman Girolamo Della Volpaia or Hierony-mus Vulparius (ca. 1530–1614) constructed a variety of astro-nomical instruments[1] including nocturnals,[2] armillary spheres,[3] and prismatic sundials.[4] Among them is a rare device for demonstrat-ing the detailed arrangement of crystalline spheres for Mercury in the Ptolemaic system, a device which in the sixteenth century was called a "theorica orbium," literally a "model of the spheres." The example illustrated here is M4 in the Adler Planetarium collection and visible on Plate II of the scarce Mensing catalog.[5] The maker's name is stamped on the underside of a brass ring that is part of a substantial stand (primarily of wrought iron) for the instrument: the inscription reads HIERONIMVS VULPARIA FACIEBAT FLORENTIAE FACIEBAT ANNO DOMINI MILLE CINQVECENTO LXXXII DIE JUNI.

The basic purpose of the theorica orbium is to demonstrate how an eccentric planetary orbit can be arranged with auxiliary spheres so that the inner and outer surfaces are still concentric with the Earth, this being an essential requirement for a plenum system of homocentric crystalline spheres. Such a scheme is illustrated in two dimensions in Islamic manuscripts of the fourteenth century (and perhaps earlier) and it gained wide currency in the West through its publication in the many editions of Peurbach's *Theoricae novae planetarum* beginning in 1473.

Selection 7 reprinted from *Journal for the History of Astronomy*, vol. 8 (1977), pp. 38–43.

THEORIA MERCVRII.

THEORICA ORBIVM MERCVRII.

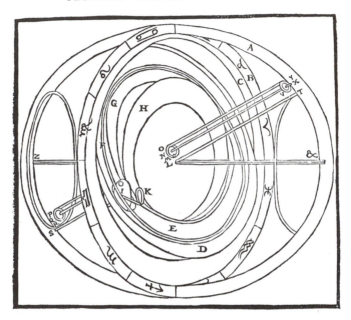

FIGURE 1. *In Schöner's* Opera mathematica *(1551) the spheres that change Mercury's apsidal line rotate about the equatorial axis, whereas those that change the eccentricity rotate about the ecliptic axis. (Uppsala Observatory Library.)*

The three-dimensional form of the theorica orbium appears in Johann Schöner's *Opera mathematica* (Nuremberg, 1551) in connection with his aequatorium astronomicum (Figure 1). E. O. Schreckenfuchs printed a similar device in his folio *Commentaria in Novas theoricas planetarum G. Purbachii* (Basel, 1556), and the same blocks were used by Christian Urstitius in his octavo edition of the Peurbach (Basel, 1558). In addition, Urstitius included diagrams of a more clearly mechanical nature; our Figure 2 illustrates a particularly interesting copy of the work, annotated in 1599 by Laurentius Paulinus Gothus,[6] first professor of astronomy at Uppsala University.

Whether Vulparius or the other makers of brass theoricae orbium were directly influenced by the mechanical form shown in these books

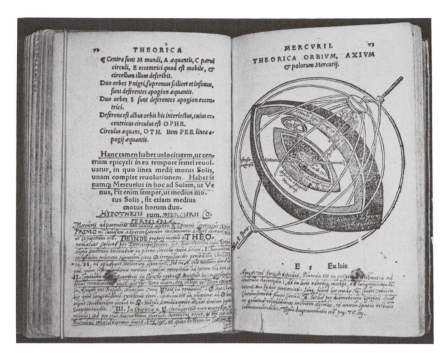

FIGURE 2. *Laurentius Paulinus Gothus, first astronomy professor at Uppsala, has added parallel Copernican explanations to his copy of the Urstitius edition,* Questiones novae in Theoricas novas planetarum *(Basel, 1573). (Uppsala Observatory Library.)*

is a moot point. In any event, they could have worked just from the 1473 Peurbach.

The spheres for Mercury are particularly complicated because, according to the Ptolemaic model, the effective eccentricity of the Mercurian orbit (that is, of the deferent carrying its epicycle) oscillates between two extreme values. The spheres are clearly shown in Figure 3, reproduced from Regiomontanus's original publication of Peurbach's treatise. If we sequentially number the five Mercury spheres shown by Peurbach, then the middle sphere, number 3, carries the epicycle. Spheres 2 and 4 are coupled together, and their rotation causes the oscillation in eccentricity. Spheres 1 and 5 are also coupled, and their slow rotation produces the precessional change in the apsidal line of Mercury's orbit.[7]

In constructing his mechanical model (Figure 4), Vulparius indeed coupled spheres 2 and 4 together as required by the theory, but he had

FIGURE 3. *The Mercury model, hand-colored, in the original Regiomontanus's edition of Peurbach's* Theoricae novae planetarum. *(Used by permission of Yale Medical Library.)*

no simple way to fix sphere 1 to sphere 5. Thus, our illustration shows them somewhat askew. Like the circles in Peurbach's book, Vulparius's spheres are only approximately drawn to scale. Curiously, Vulparius has represented Mercury's epicycle by a bead with an axis in the plane of the deferent rather than perpendicular to it.

A similar metal model, but for the Moon and Sun rather than Mercury, is found in the National Technical Museum in Prague.[8] Because the stand is missing, the maker's name is lacking, but it is not impossible that Vulparius also made the Prague instrument. Another theorica orbium, but for a simpler use than Mercury, and stylistically

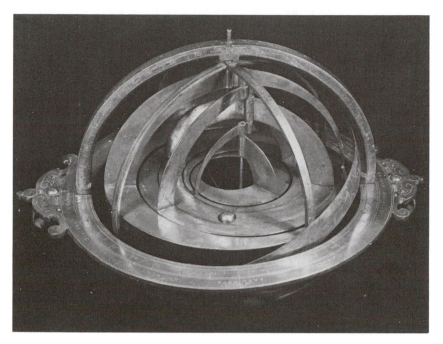

FIGURE 4. *The 1582 "theorica orbium" or "orbarium" of Hieronymus Vulparius, now in the Adler Planetarium, Chicago.*

somewhat different, was made in 1557 by Hieronymus's father, Camillus, and is in the History of Science Museum in Florence.[9] Yet another, by an unknown maker, is also in the Florence museum.[10]

Most of these instruments have been previously categorized as "armillary spheres," but clearly they serve a quite different purpose. An armillary sphere shows the two-dimensional mapping of the ecliptic, equator, tropics, and poles on the sky. The theoricae orbium show the three-dimensional nesting of the heavenly spheres. Alternatively, the name planetarium has been assigned to these devices. I find this term objectionable for two reasons: first, because it is sometimes used for orrery-like constructions,[11] and more importantly because it fails to emphasize the real teaching purpose, namely the arrangement of the spheres rather than of the planets. An analogous neologism would be "orbarium," which I suggest as a modern translation of "theorica orbium."

The name "planetarium" or "spherical planetarium" might well be adopted for another category of spherical models that is at first glance similar to the theorica orbium or orbarium. This device displays a planetary orbit with an epicycle, generally inclined to exhibit the motion in latitude. These do *not* show the eccentrically nested spheres, and hence are not orbaria. A beautiful set is found in the Vatican Museum, and the Deutsches Museum contains other, less elegant, examples.[12]

The existence of the brass theoricae orbium reveals the interest and belief of sixteenth-century astronomers in the reality of the celestial crystalline spheres. The classification of Renaissance instruments will be more exact if these few rare theoricae orbium (or orbaria) are distinguished from planetaria and armillary spheres.

Notes and References

[1] Carlo Maccagni, "The Florentine Clock- and Instrument-Makers of the Della Volpaia Family," *Des Globusfreund*, no. 18–20 (1970) pp. 92–98, gives a checklist of 14 instruments, 11 signed by Vulparius, 2 unfinished ones attributed to him, and an armillary sphere presumably in Perugia but known only through a 1929 exhibition. One of these, a solar quadrant dating from 1577 in the Museo Archeologico Nazionale dell' Umbria in Perugia, is listed as an astrolabe, apparently erroneously, in M. L. Bonelli, *Catalogo con Aggiornamenti* (Florence, 1952) and so copied in S. L. Gibbs, J. A. Henderson, and D. de Solla Price, *A Computerized Checklist of Astrolabes* (New Haven, 1973). An earlier version of Maccagni's paper appears in *Actes du XII*e *Congrès International d'Histoire des Sciences, Paris 1968*, vol. 10 A (1971), pp. 65–73.

[2] Two nocturnals are illustrated in color in Figure 62 of M. L. Righini-Bonelli, *Il Museo di Storia della Scienza a Firenze* (Florence, 1968). The first, from 1568, has been issued in a cardboard facsimile by Marcello Felli, *L'Orologio Notturno* (Florence, 1974).

[3] A real armillary sphere from 1564 is shown in Figure 35 of Righini-Bonelli, *op. cit.*, and described as Item 74 on p. 161. Another is in the Kensington Science Museum, illustrated on their colored postcard 441.

[4] A sundial signed by Vulparius and another attributed to him are listed as Items 188 and 239 (pp. 175 and 178) of Righini-Bonelli, *op. cit.*

[5] *Collection Ant. W. M. Mensing, Amsterdam, Old Scientific Instruments* (1479–1800), cataloged by Max Engelmann (Amsterdam, 1924). The present instrument is rather roughly described as a "grosse Armillarsphäre."

[6] Laurentius Paulinus Gothus added both Copernican and Tychonic schemes to this copy of *Questiones novae in Theoricas novas planetarum G. Purbachii* (Basel, 1573) by Christianus Urstitius or Wursteisen [J. C. Houzeau and A. Lancaster, *Bibliographie Générale de l'Astronomie* (reprint London, 1964), item 2661]. For a transcription and Swedish translation, see N. V. E. Nordenmark, "Laurentius Paulinus Gothus föreläsningar vid Uppsala universitet 1599 över Copernicus hypotes," in *Arkiv för Astronomi*, vol. 1 no. 24 (1951), pp. 261–300.

[7] The Mercury model, including a sculptured representation from 1527 and an extensive discussion of Peurbach and his precursors, is found in Willy Hartner's "The Mercury Horoscope of Marcantonio Michel of Venice," in *Vistas in Astronomy*, ed. by A. Beer, vol. 1 (1953), pp. 84–138, and reprinted on pp. 440–95 of his *Oriens-Occidens* (Hildesheim, 1968).

[8] See Zdeněk Horský and Otilic Škopova, *Astronomy Gnomics, A Catalogue of Instruments in the National Technical Museum, Prague* (Prague, 1968), pp. 138–39 and Plate XXXIX.

[9] Righini-Bonelli, *op. cit.*, Figure 33 and Item 65 on p. 160.

[10] *Ibid.*, Figure 32 and Item 77 on p. 161.

[11] See, for example, the planetaria from 1751 and 1653 in Ernst Zinner, *Deutsche und Niederländische Astronomische Instrumente des 11.–18. Jahrhunderts*, Tafeln 4 and 5.

[12] Two instruments in the Deutsches Museum are illustrated in Ernst Zinner, *Entstehung und Ausbreitung der Coppernicanischen Lehre* (Erlangen, 1943), Abh. 51–52.

The Search for a Plenum Universe

L uminiferous aether will prove to be superfluous," wrote Albert Einstein in his celebrated 1905 paper on special relativity. With this, he apparently sounded the death knell for one of the most intriguing ideas in the history of thought—the concept of a plenum universe completely full of something fundamental but invisible.

For the ancient Greek philosophers, that mysterious something was the weightless, transparent, permanent substance of the heavens. Lying above the terrestrial earth, water, air, and fire, it was literally the quintessence, a fifth element. Philosophers of the Middle Ages conceptually crystallized the aether into smooth, glassy, solid spheres, still weightless and invisible. Copernicus was ambivalent about their role, but before the end of the sixteenth century Tycho Brahe had shattered the crystal spheres with his geo-heliocentric cosmology. Scarcely had the celestial aether been exorcised when Descartes reintroduced it in the form of pervasive material vortices that drove the planetary rhythms. In turn, the Cartesian plenum gave way to the luminiferous aether of Newton and Huygens, a "subtile fluid" that bore the vibrations of light across space. For two centuries this luminiferous aether played an increasingly significant role in physics until Einstein with a stroke rendered it irrelevant.

In the Greek view, the celestial aether occupied the region of the

Selection 8 reprinted from *Great Ideas Today 1979,* ed. by M. J. Adler and J. Van Doren (Chicago, 1979), pp. 68–86.

stars and planets, filling every conceivable space beyond the zone of fire. Carved in concentric spheres, it provided the carriers for individual planets, which were envisioned as luminous nodules within the aethereal plenum. Because the heavens were eternal and because circular motion was seen as unending, the Greek philosophers associated uniform circular motions (or combinations thereof) with these aethereal spheres.

"Eudoxus was the first of the Greeks to concern himself with hypotheses of this sort," wrote Simplicius in his commentary to Aristotle's *De caelo*, "Plato having set it as a problem to all earnest students of this subject to find the uniform movements by which the motions of the planets can be explained."[1] Eudoxus, a brilliant mathematician who discovered much of what has become Euclid's *Elements*, proposed a nested series of spheres pegged together with different axes so that the individual motions of various spheres combined to produce the apparently irregular planetary motion. Eudoxus's theory was more to be admired than used, because it failed to provide reliable numerical predictions. Nevertheless, the homocentric spheres of Eudoxus became, in the next generation, a fundamental feature of Aristotle's cosmology.

Aristotle's medieval followers paraphrased his view of the plenum in an epigram: *Nature abhors a vacuum*. There is no evidence that Aristotle himself said this, but he would have known what it meant. To preserve the plenum and to ensure a smooth mechanical operation of the Eudoxian spheres within the high heavens, Aristotle added enough new spheres to counteract the motion of each planet, an inside-out set to reduce the motion back to the zero level so that one planet's spheres could be neatly linked to the next. This cost Aristotle about 55 spheres, not unduly many considering the apparent complexity of the planetary motions. Perhaps more disappointing was the failure of the scheme to account for the clearly different distances of the planets from the Earth. "This is indeed obvious," remarked Simplicius, "for the star called after Aphrodite [Venus] and also the star of Ares [Mars] seem, in the middle of their retrogressions, to be many times as large, so that the star of Aphrodite actually makes bodies cast shadows on moonless nights."[2]

An attempt to model the celestial motions accurately enough for future predictions came two centuries after Aristotle with the astronomer Hipparchus (fl. between 147 and 127 B.C.). Borrowing numerical parameters from the Babylonians and forging them into a geometric model, Hipparchus succeeded well enough with the Sun and Moon to predict the times of eclipses, but evidently he failed completely with the planets. The line of mathematical philosophers was so thin in those ancient times that not until A.D. 140, nearly three centuries after Hipparchus, did a sufficiently gifted theoretician emerge to establish a satisfactory system that could predict planetary positions. Ptolemy (Claudius Ptolemaeus), an astronomer of Alexandria, linked major circles to account for the motions of each planet. On the larger circle, called the deferent, moved a subsidiary circle, called the epicycle, and their combined movements roughly reproduced both the direct and the so-called retrograde motions of the planets.

To bring his model into closer agreement with the observed planetary motions, Ptolemy introduced several important auxiliary devices. Rather than center the deferent circle on the Earth itself, he placed it somewhat eccentric to the Earth, thereby generating the varying angular speed of the Sun and epicycles as seen from the Earth. Second, he proposed an off-centered axis of uniform angular motion, which brought the differing lengths of the retrogressions into agreement with observations. This latter device, known as the equant, violated the precept of uniform motion around the circumference of the circle and, as we shall see, provoked increasing philosophical criticism of Ptolemy in later centuries.

Ptolemy described his planetary models in a great handbook of mathematical astronomy originally entitled the *Syntaxis*, but generally known as the *Almagest* or "the Greatest," as it was designated by Islamic admirers. The *Almagest* thus provided the basis for what is often called the Ptolemaic system; in fact, the book describes the individual parts with no attempt to fit the pieces together into a grand cosmological scheme, as Aristotle had done in his *De caelo*. However, in the late 1960s a long overlooked section of another Ptolemaic treatise, the *Planetary Hypotheses*, was discovered in an Arabic translation.[3] This section, lacking in the surviving Greek text, showed that Ptolemy had indeed created the Ptolemaic system.

The heretofore missing text, although essentially mathematical in nature, fits beautifully into the Aristotelian scheme with its plenum of nested spheres. In it Ptolemy determines the ratio of the nearest and farthest approach of each planet, taking into account not only the epicycle but also the eccentric position of the deferent circle. In this scheme he arranged the nearest approach of Mercury to fall immediately beyond the farthest excursion of the Moon, the nearest approach of Venus immediately beyond the farthest excursion of Mercury, and so on through the spheres for the Sun, Mars, Jupiter, and Saturn. Since

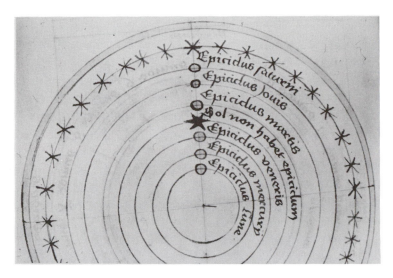

FIGURE 1. *Drawn by a medieval scribe, this highly schematic Ptolemaic system shows the planetary circles and epicycles arranged around the Earth. From a manuscript of Alcabitius (1474) in the Crawford Collection, Royal Observatory, Edinburgh, Permission of the Astronomer Royal of Scotland.*

Ptolemy had already established the distance of the Moon in terrestrial units, he could then specify the dimensions of the remaining mechanisms in absolute terms. Furthermore, the arrangement satisfied the Aristotelian requirements, provided that invisible aether existed to fill all the gaps around the epicycles.

Nevertheless, Ptolemy's geometrical devices were compromised by three features that made it difficult to construct a mechanical model of pure aether. First, the motion of the Moon was sufficiently complicated so that he had been obliged to add a further circle, a kind of interior crank that brought the Moon closer to the Earth at first and third quarter. Second, Mercury required an extra circle to account for certain observed positions. (The data had caused Ptolemy to suppose that this elusive planet had a closest approach to the Earth in two different places, an idiosyncrasy challenged neither by the Islamic astronomers nor by Copernicus, although it is now known to rest on faulty observations.) The interior circles required by the lunar and Mercury models clearly got in each other's way whenever anyone tried to diagram a comprehensive mechanical model of the system. Finally, the tidy picture of nested spheres was disturbed by the equant, an objectionable "additional center," which, if represented mechanically, required a linkage from an interior point through the intervening spheres to the planet in question.

The Ptolemaic-Aristotelian cosmology clearly fascinated the Islamic astronomers, who adopted it as their world view. But criticism of Ptolemy on philosophical grounds emerged in the eleventh century in the work of Ibn al-Haytham (Alhazen, 965–ca. 1040). In his *Doubts on Ptolemy*, Ibn al-Haytham complained that the equant failed to satisfy the requirement of uniform circular motion. At the same time he objected to Ptolemy's lunar model, although he passed in silence over the complicated construction for Mercury. A more extreme criticism was launched in the following century by the Cordoban astronomer and philosopher, Averroës (Ibn Rushd, 1126–1198), who wrote:[4]

> In my youth I hoped it would be possible for me to bring this research to a successful conclusion. Now, in my old age, I have lost hope, for several obstacles have stood in my way. But what I say about it will perhaps attract the attention of future researchers. The astronomical science of our days surely offers nothing from which one can derive an existing reality. The model that has been developed in the times in which we live accords with the computations, not with existence.

Averroës was thoroughly critical of the Ptolemaic theory; he found its eccentric epicycles, and above all its equant, completely unacceptable. Nevertheless, he failed to propose any satisfactory alternative.

Perhaps in response to these criticisms, the Moorish astronomer al-Bitrūjī attempted late in the twelfth century to formulate a strictly concentric geocentric model, somewhat reminiscent of the Eudoxian scheme and equally disastrous. Although Bitrūjī's system spread throughout much of Europe in the thirteenth century, it attracted no continuing support.

Elsewhere, at the other end of the Islamic world, a fresh critique of the Ptolemaic mechanism was undertaken in the thirteenth century by Naṣir ad-Dīn at-Ṭūsī. In his *Tadhkira*, or "Memorandum," Ṭūsī launched a thorough exposition of the shortcomings of Ptolemaic astronomy. Ṭūsī found the equant particularly objectionable; he also found the Mercury and lunar models inadequate, presumably because of the difficulty of fitting their spheres together for a consistent plenum universe. In the *Tadhkira*, Ṭūsī proposed an ingenious linkage of two circles whose uniform motions together produced a straight-line reciprocating motion, and this device (now called the Ṭūsī couple) provided a replacement for the equant, a replacement that generated the same apparent angular motion by means of two additional epicyclets rather than by the interior off-center axis. Hence, Ṭūsī not only preserved the principle of uniform circular motion, but he achieved it with a mechanism that would fit neatly within each planet's own set of spheres. He also tried, apparently with less success, to eliminate the philosophically objectionable central mechanisms of the Moon and Mercury.

Mu'ayyad ad-Dīn al-'Urḍī and, later, Quṭb ad-Dīn ash-Shīrāzī, both workers at Ṭūsī's Marāgha observatory, offered an alternative arrangement of the Ṭūsī couple to provide another approach to the planetary models, but one that still retained the eccentric circles.[5] A completely concentric rearrangement of the planetary mechanisms was finally achieved by the Damascene astronomer, Ibn ash-Shāṭir, around 1350. He wrote:[6]

> I found that the most distinguished of the later astronomers had adduced indisputable doubts concerning the well-known astronomy of the spheres according to Ptolemy. I therefore asked Almighty God to give me inspiration and help me to invent models that would achieve what was required, and God—may He be praised and exalted, all praise and gratitude to Him—did enable me to devise universal models for the planetary motions in longitude and latitude and all other observable features of their motions, models that were free from the doubts surrounding previous ones.

By using the Ṭūsī couple, Ibn ash-Shāṭir succeeded in transferring not only the equant but the central circles in Ptolemy's constructions for the Moon and Mercury from the interior regions of the model to the outer parts, thereby clearing the way for a perfectly nested and mechanically acceptable set of celestial spheres. However, his solution remained generally unknown in medieval Europe, was eventually forgotten, and was ultimately rediscovered only in the late 1950s.

Meanwhile, although technical understanding of Ptolemy's astronomy had nearly vanished in Latin-speaking Europe, the concentric spheres of terrestrial elements and of crystalline celestial aether entered the artistic repertoire. Piero di Puccio's calm but majestic fresco on the wall of the Camposanto in Pisa adds spheres of angels and God himself beyond the crystal dome of stars—a typical Christianized ver-

FIGURE 2. *The nested aethereal spheres of the planets surround the terrestrial spheres of earth, water, air, and fire in Giovanni di Paolo's* Expulsion from the Garden of Eden, *a Christianized mid-fifteenth-century version of Aristotle's cosmos. Metropolitan Museum of Art, Robert Lehman Collection (1975.1.31).*

sion of the fourteenth century. Giovanni di Paolo's colorful nested spheres, now a prize of the Lehman Collection of the Metropolitan Museum of Art in New York City, depicts the same tradition in the fifteenth century; and by 1493 a lavishly illustrated book, the *Nuremberg Chronicle*, gave still greater currency to this tidy, compact view of the universe.

The *Sphaera* of the thirteenth-century cleric John of Holywood, or Sacrobosco, conveyed the principal knowledge of astronomy available in the Middle Ages, but it dealt with planetary theory only in its unsatisfactory final chapter. Hence a *Theorica planetarum* (anonymous, though often mistakenly attributed to Gerard of Cremona) was added to the standard "textbook" collection of Latin astronomical manuscripts. Brief and faulty as the *Theorica* was, it established the astronomical vocabulary, and it gave a preeminence to Ptolemy's geometric theory. Not until the mid-fifteenth century was the old *Theorica* successfully overhauled and replaced by a new and longer *Theoricae novae planetarum*, written by Georg Peurbach, professor of astronomy at Vienna. What was new in Peurbach's little treatise was principally a set of diagrams showing how crescent-shaped zones of crystalline aether could fill in the spaces and thus make geometric units from Ptolemy's eccentric circles and epicycles. Such diagrams had already been included in Ptolemy's *Planetary Hypotheses*, but precisely how Peurbach got them remains a mystery. In any event, the introduction of printing in the fifteenth century served to disseminate this arrangement, and the simple machinery of the nested spheres became the standard iconography of the astronomical Renaissance.

In 1473, the same year that Peurbach's "New Theories of the Planets" was first printed, Nicolaus Copernicus was born. When young Copernicus studied astronomy as an undergraduate at Cracow in the 1490s, he must have seen Peurbach's pictorial representation of the crystalline spheres. Whether his teachers expressed any doubt about the arrangement, as the Islamic astronomers had been doing for several centuries, we do not know. But it is highly suggestive that Copernicus attacked the Ptolemaic equant at the very beginning of his earliest extant tract on astronomy. Equally remarkable are the great pains Copernicus took to replace the interior circles of the lunar and Mercury models by alternative mechanisms at the perimeter of their orbits. Concerning these problems, he wrote:[7]

> Our predecessors assumed a large number of celestial spheres principally in order to account for the apparent motion of the planets through uniform motion, for it seemed highly unreasonable that the heavenly bodies should not always move uniformly in a perfectly circular figure. Callippus and Eudoxus, who endeavored to solve this problem by the use of concentric spheres, were unable to account for all the planetary movements. . . . Therefore it seemed better to employ eccentrics and epicycles, a system which most scholars finally accepted. Yet the planetary theories of Ptolemy and most other astronomers, although consistent with the numerical data, also seemed quite doubtful. For these theories were not adequate unless certain equants were also con-

ceived; it then appeared that a planet moved with uniform velocity neither on its deferent nor with respect to its proper center. Hence a system of this sort seemed neither perfect enough nor sufficiently pleasing to the mind.

Copernicus, of course, did far more than replace the mechanical blemishes of the Ptolemaic system. He almost literally turned cosmology inside-out by proclaiming that the Sun, not the Earth, was the fixed center of the universe. In the most profound philosophical sense, the ultimate and most revolutionary consequences of his work were the twin recognitions that the Earth was but one of the planets and that the Sun was but one of the stars. Apparently Copernicus himself only vaguely grasped the implications of the first of these concepts, and not at all the second. Nevertheless, as the greatest cosmographer since Ptolemy, Copernicus recognized the intrinsic elegance of the heliocentric system. Ptolemy had used two major circles for each planet; Copernicus saw that the great Earth-Sun circle could serve in common for every pair. Thus, instead of ten major circles for the five planets and another for the Sun as Ptolemy had required, Copernicus needed only five major circles plus the Earth-Sun circle, which furnished the common measure and hence the distance scale for the entire scheme. The intrinsic economy of this rearrangement offered to sixteenth-century astronomers the most persuasive argument for heliocentric cosmology, and once glimpsed, the idea was too compelling to be abandoned. Furthermore, this ordering placed Mercury, the swiftest planet, closest to the Sun and the lethargically moving Saturn the most remote; "In no other way," wrote Copernicus, "do we perceive the clear harmonious linkage between the motions of the planets and the sizes of their orbs."[8]

But did Copernicus still envision the solar system filled with solid crystalline spheres? Unfortunately, he is virtually silent on this subject. The Ptolemaic blemishes seem minor compared to the radical reordering that he finally adopted. Yet the equant, the inner lunar circle, and the construction for Mercury, which Copernicus worked so hard to eliminate, are objectionable principally in the context of a plenum universe and with the Aristotelian spheres. Could it be that Copernicus had merely stumbled onto the heliocentric arrangement as he responded to a current philosophical problem concerning the physical reality of the Ptolemaic mechanism?

How and why Copernicus adopted the heliocentric cosmology remains a mystery. We can, however, glean a few shreds of evidence to support the following speculative reconstruction. Early in his career the young Copernicus somehow became aware of the inadequacy of the Peurbachian models for a physically acceptable universe of crystalline spheres (just as Ibn al-Haytham had perceived these inadequacies four centuries earlier). In tinkering with the arrangement of circles, Copernicus must surely have seen how the Earth-Sun motion kept recurring in each planetary model—indeed, the organization of the numbers in his well-thumbed copy of the *Alfonsine Tables* made this quite obvious. It was an easy step to transpose the epicycles for Mars, Jupiter, and Saturn to a common circle coincident with the solar orbit, all within a geocentric framework. This had the effect of placing the planets in orbit around the Sun, while the Sun in turn orbited the Earth. In fact, precisely such notes still exist—on one page of the handful of his surviving manuscripts.[9] But straightaway a difficulty arose for Copernicus; the dimensions of the two primary circles for Mars were such that when the deferent and epicycle were interchanged, the circle carrying Mars around the Sun intersected the circle carrying the Sun around the Earth. How could the spheres be saved? By a still more radical transformation—choosing the Sun rather than the Earth as the fixed center! As Copernicus's disciple Rheticus remarked, Mars unquestionably comes closer to the Earth than the Sun does, "and therefore it seems impossible that the earth should occupy the center of the universe."[10] Rheticus's statement is a stark non sequitur unless seen in the context of the intersecting spheres.

At this point (as we may speculate) Copernicus revised his goal: instead of saving the physical reality of the spheres with a geocentric system, he would try a heliocentric system—not just heliostatic, but absolutely heliocentric, unblemished by eccentrics or equants. The program is entirely analogous to Ibn ash-Shāṭir's but converted to the heliocentric case, and in fact Copernicus's initial solution follows that of Ibn ash-Shāṭir so closely that many contemporary historians suppose that some currently missing link accounted for the westward transmission of the Arabic ideas. Such was the scheme outlined by Copernicus in his *Commentariolus*, a small tract closely modeled on Peurbach's *Theoricae novae planetarum* and written sometime prior to 1514.

But in 1515 the full text of Ptolemy's *Almagest* appeared in print for the first time, and perhaps for the first time Copernicus recognized the formidable task awaiting a cosmologist who sought to supersede the Alexandrian. The labor required nearly three more decades, and in the process Copernicus abandoned his attempt to set up a strictly helio-

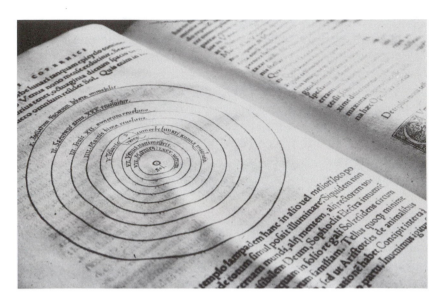

FIGURE 3. *The heliocentric planetary spheres in the first edition of Copernicus's* De revolutionibus *(1543). Besides the circles shown for Mercury, Venus, Earth, Mars, Jupiter, and Saturn, two extra circles show schematically the bounds of the Moon's circle, which is centered on the Earth.*

centric astronomy. In his research Copernicus discovered that the eccentric orientation of the planetary orbits slowly changed, and it seemed easier to him to represent this with circles not centered on the Sun rather than with an additional epicyclet at the perimeter of the planetary orbits. Thus, the order of circles in his magnum opus, the *De revolutionibus*, paralleled the original arrangement of al-'Urḍī rather than that of Ibn ash-Shāṭir.

If Copernicus had indeed been motivated originally by a fervent ambition to save the crystalline spheres, he must have cooled to this idea as his thinking matured. The idea of a moving Earth composed of ponderous terrestrial elements and a fixed Sun made of weightless aether destroyed the coherence of Aristotelian physics. As the distinguished Danish observer Tycho Brahe was to complain later in the century, "Copernicus's theory in no way offends the principles of mathematics. Yet it gives to the earth, this lazy, sluggish body unfit for motion, a movement as fast as that of the aethereal torches."[11]

Only a few vestiges of Copernicus's original commitment to cosmic spheres remain in his *De revolutionibus*, as when he says: "The space remaining between Venus' convex sphere and Mars' concave sphere must be set apart as also a sphere or spherical shell, both of whose surfaces are concentric with those spheres."[12] In contrast, when he discusses the Earth-Sun mechanism, he describes three alternative modes quite indifferently as to whether the auxiliary wheels are placed in the center or at the perimeter of the model. Perhaps by the time he wrote this later section he was taking the role of the mathematician rather than that of the cosmographer, having failed at his attempt to achieve a physically acceptable celestial cosmology consistent with physics of his day.

Tycho Brahe, the number two astronomer of the sixteenth century, would have loved to go down in history as one of the all-time-great cosmographers. Born in the generation after Copernicus's death, Tycho understood the beauty of the heliocentric system as well as its contradiction with the accepted Aristotelian physics, and for this he criticized Copernicus. But in electing to save physics by abandoning heliocentrism, he made the wrong choice and an apparent step backward in cosmology. History has awarded Tycho laurels as an all-time-great instrument builder and observer, but not as a cosmographer.

Nevertheless, Tycho's insistence on a physically real model of the universe moved astronomy a major step closer to modern science. For years Tycho toyed inconclusively with various planetary arrangements, but not until 1583 did he finally adopt the Tychonic system, precisely the geo-heliocentric model used briefly by Copernicus, in which the Sun, while circling the Earth, carried the other planets in orbit around itself.

From a twentieth-century vantage point, the distinction between Copernicus's heliocentric system and Tycho's geo-heliocentric plan may seem to be only a matter of relativity: two geometrically identical schemes with different reference points. But, as this discussion attempts to make clear, each model was embedded in a larger view of the physical world. Because of the intersection of the circles for Mars and the Sun in the geo-heliocentric plan, accepting this geometry necessarily implied rejecting the solid spheres, a consequence almost as radical as that for Copernicus in throwing the Earth into orbit.

With his analysis of the Great Comet of 1577, Tycho already had some evidence for doubting the existence of the crystal spheres, for the observations clearly placed the comet beyond the sphere of the Moon. Still, he considered the possibility that the comet could circle the Sun in the space left beyond the sphere of Venus. The sticking point, as it had been for Copernicus, was the "ridiculous penetration of orbs" of Mars and the Sun, as Tycho referred to it. "It happened that my own

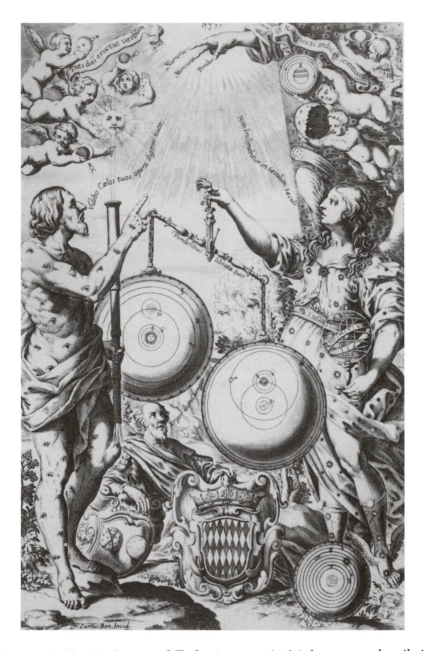

FIGURE 4. *The Earth-centered Tychonic system (right) hangs more heavily in Urania's balance than the Sun-centered Copernican system (left) in this frontispiece from J. B. Riccioli's* Almagestum novum *(1651). Note in the Tychonic scheme the intersection of the larger circle of Mars with the smaller circle that carries the Sun and planets about the Earth.*

discovery was suspect," he later wrote, "because I was so steeped in the opinion, approved and long accepted by nearly everyone, that the heavens were composed of solid spheres."[13]

A detailed study of Mars in its close approach of 1583 finally convinced Tycho that Mars indeed came nearer than the Sun, as the sphere-smashing Tychonic arrangement would require. The irony of all this, as Kepler later pointed out, was that in no way could Tycho's observations yield an accurate distance to Mars, and hence he made his decision on spurious grounds!

Faulty as Tycho's basis for accepting the geo-heliocentric system may have been, it nevertheless made a crucial and total impact on the young Kepler, who for ten months was Tycho's last but most brilliant understudy. History remembers Kepler for distilling from Tycho's data the elliptical shape of the planetary orbits, but he would probably have preferred to be known as the first astrophysicist. Like Tycho, Kepler noted the inconsistency between the heliocentric plan and the Aristotelian physics of his day, but, unlike Tycho, Kepler accepted the Sun-centered arrangement and sought to rework the physics. For Kepler, the rejection of the solid aethereal spheres left the planets with no mechanical means to maintain their motion. As a searcher after physical causes, he naturally turned to magnetism, for he knew of the lodestone's mysterious ability to attract iron across seemingly empty space. Such a force emanating from the Sun could, Kepler believed, push the planets in their orbits.[14] Furthermore, his theory could be linked quantitatively to the fact that the closer the planets were to the Sun, the faster they completed their circuits.

Kepler's theory of action-at-a-distance was, in the eyes of many contemporaries, no proper substitute for the aether. Even Galileo labeled Kepler's notion occult,[15] and in the next generation Descartes hammered out an alternative world view based on a fluid plenum. Descartes agreed with Aristotle that nature abhors a void; he regarded space and substance as identical except in our mode of conception of them. "Space without matter is a contradiction," he once said. As a starting point for his geometry and physics, he imagined extension, and since matter was the essence of extension, matter must fill all of space. The subtle matter, set by God into vortical motion at creation, drove the planets and satellites. It also gave an easy out with respect to the troubling question of the center of the universe. The Earth is at rest with respect to its own vortex, said Descartes sagely, and the Sun is at rest with respect to the solar vortex.

Like the Eudoxian spheres, Descartes's vortical theory was more to

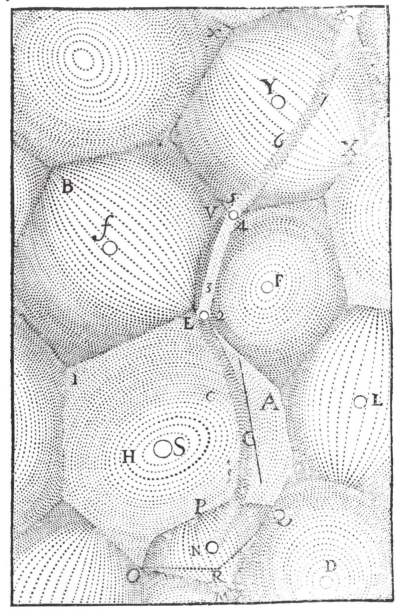

FIGURE 5. *The sinuous path of a comet through the celestial vortices as de-picted in Descartes's* Principia philosophiae *(1644).*

be admired than used. Several generations of continental scientists, soothed by the revival of the plenum, tried in vain to deduce mathematical consequences from it. Descartes had entitled his grand cosmological book *Principia philosophiae;* when Isaac Newton, who had for many years been attracted to the Cartesian philosopy, finally realized how dismally the vortices failed when dealing with elliptical orbits, he deliberately entitled his own work *Philosophiae Naturalis Principia Mathematica*—emphasizing that it was the *mathematical* principles of natural philosophy that concerned him.

Newton, of course, described his gravitational theory in the mathematics of geometry and of a nascent calculus. He described the consequences of gravity without saying what it was. Pricked by criticism that "action-at-a-distance" smacked of the occult, he added in later editions a scholium saying that as to the nature of gravity, he feigned no hypotheses. But privately, Newton wrote to Richard Bentley:

> That gravity should be innate, inherent and essential to matter so that one body may act upon another at a distance through a *vacuum*, without the mediation of anything else, by and through which their action and force may be conveyed from one to another, is to me so great an absurdity, that I believe no man who has in philosophical matters a competent faculty of thinking, can ever fall into it.[16]

Solid crystalline spheres were of course dead, and the material fluid of the Cartesian vortices had proved inadequate to explain the detailed phenomena of planetary motion. Nevertheless, like a phoenix rising from the ashes, an elusive aethereal plenum arose once more. In founding the wave theory of light, Newton's illustrious Dutch contemporary Christiaan Huygens introduced an aether in 1690 to explain the propagation of light. "One will see," Huygens wrote, "that it is not the same that serves for the propagation of sound. . . .It is not the same air, but another kind of matter in which light spreads; since if the air is removed from the vessel, the light does not cease to traverse it as before."[17] To have an aether to support the motion of light is necessary for the true philosophy, Huygens remarked elsewhere, "or else [we] renounce all hopes of ever comprehending anything in physics."[18]

Newton's *Optics* also carried a notice of the new aether:[19]

> . . .Is not the heat of the warm room conveyed through the vacuum by the vibrations of a much subtiler medium than air, which after the air was drawn out remained in the vac-

uum? And is not this medium the same with that medium by which light is refracted and reflected, and by whose vibrations light communicates heat to bodies?. . . .And is not this medium exceedingly more rare and subtile than the air, and exceedingly more elastic and active? And doth it not readily pervade all bodies? And is it not (by its elastic force) expanded through all the heavens? (Query 18).

And further:[20]

May not planets and comets, and all gross bodies, perform their motions more freely, and with less resistance in this aethereal medium than in any fluid, which fills all space adequately without leaving any pores, and by consequence is much denser than quick-silver or gold? And may not its resistance be so small, as to be inconsiderable? (Query 22).

Thus, in the age of Newton and Huygens, the luminiferous aether was born as the carrier of light. Remarking on the history of aether in the ninth edition of the *Encyclopaedia Britannica* (ca. 1860), James Clerk Maxwell wrote:[21]

Aethers were invented for the planets to swim in, to constitute electric atmospheres and magnetic effluvia, to convey sensations from one part of our bodies to another, and so on, till all space had been filled three or four times over with aethers. . . . The only aether which has survived is that which was invented by Huygens to explain the propagation of light. The evidence for the existence of the luminiferous aether has accumulated as additional phenomena of light and other radiations have been discovered; and the properties of this medium, as deduced from the phenomena of light, have been found to be precisely those required to explain electromagnetic phenomena.

Maxwell himself studied the interaction of electric and magnetic fields in space. He discovered their wavelike behavior and showed theoretically that these electromagnetic waves propagated at the velocity of light. His contribution, in the words of Joseph Larmor in the eleventh edition of the *Encyclopaedia Britannica*, "largely transformed theoretical physics into the science of aether."[22] Larmor's entry on "Aether" was long and technical. "Can we form a consistent notion of such a connecting medium?" he asks, almost in despair. "In carrying

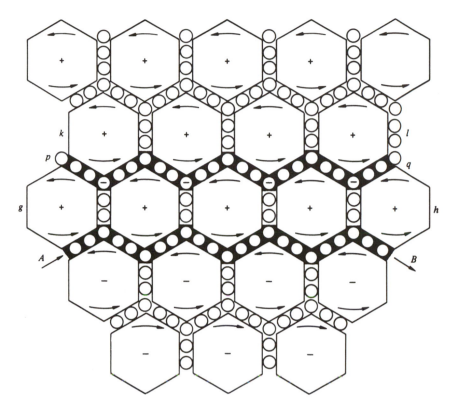

FIGURE 6. *James Clerk Maxwell's molecular vortices and electrical particles, part of the nineteenth-century elaboration of the luminiferous aether (1861).*

out this scientific procedure false steps will from time to time be made, which will have to be retraced, or rather amended," he adds; but, though published in 1911, the article carries no hint of Einstein's iconoclastic statement of 1905, that the luminiferous aether will prove to be superfluous.

Writing partly in jest about a decade later, Einstein remarked:[23]

> As regards the mechanical nature of Lorentz's ether, one might say of it. . .that immobility was the only mechanical property which Lorentz left it. It may be added that the whole difference which the special theory of relativity made in our conception of the ether lay in this, that it divested the ether of its last mechanical quality, namely immobility.

With special relativity, Einstein did not so much "solve" the problem of the aether as to make it irrelevant. His radically different choices of what were the pivotal points of physics essentially rendered aether a nonproblem. Indeed the luminiferous aether, stripped of its essential properties, withered on the vine. In our modern conceptions, space has become emptier and emptier. Light waves inhabit space quite independently of any material or aethereal substratum. But, in fact, the aether has simply undergone another metamorphosis. No longer needed as the bearer of light, an entirely new aether can symbolize the physical properties of space specified by Einstein's general theory of relativity in 1916. As Einstein himself wrote in 1920: "To deny the existence of the ether means, in the last analysis, denying all physical properties to empty space. But such a view is inconsistent with the fundamental facts of mechanics."[24]

Among Einstein's contemporaries were scientists like Sir Oliver Lodge, who had made their scientific reputations by describing the aether. By 1926, when the 13th edition of the *Encyclopaedia Britannica* appeared, not only had Einstein's theory been well established, but the idea of the nuclear atom as well; reluctant to abandon a concept once so firmly entrenched, Lodge could still write:[25]

> Meanwhile we must assume that the ether has a substantiality and wave-conveying structure beyond our present clear imaginings, with parts of it modified in an unknown way into electrons and protons. . . . There is very little doubt that matter is not an alien substance, but is essentially composed of it, being built up of the electrons and protons whose constitution has not yet been ascertained, but which must somehow be constituted of ether, perhaps in some sense analogous to that in which a knot in a piece of string is constructed of string.

Lodge went on to say:[26]

> The theory of Relativity has led some people—not many leaders of thought—to doubt if the ether can really exist. . . . Einstein was led by considerations of relativity to formulate a law of gravitation, not in terms of force or of action at a distance, but in terms of something in space, that is, in the ether, which results in a tendency of bodies to approach each other. It might be called a warp in space, or it might be called by other names: the names do not matter. . . .

The beauty of these results is overwhelming; but the idea that any mathematical scheme is more than a powerful method of exploration, and that a universe can be thus constructed in which physical explanations can be dispensed with, involves too simple and anthropomorphic a view of nature. The things calculated, and the things observed, cannot exhaust reality; an explanation is bound to be sought, and ultimately attained, in terms of the partially recognised but largely unexplored properties of the entity which fills space.

Because the word *aether* has been virtually banished from today's physics, Lodge's statement has a curiously conservative ring; yet Einstein himself would probably have agreed with much of it. As he grew older, Einstein long pondered the relation of electromagnetism, atoms, and gravitation. His unsatisfied search for a unified field theory was, in effect, an attempt to reconcile the metrical properties of space (the "gravitational ether" as he called it) and the electromagnetic properties of matter. "It would, of course, be a great step forward if we succeeded in combining the gravitational field and the electromagnetic field into a single structure," Einstein wrote in the article mentioned above. "The antithesis of ether and matter would then fade away, and the whole of physics would become a completely enclosed intellectual system, like geometry. . . ."[27]

Most of Einstein's contemporaries saw his search for a unified field theory as a frustrating quest after a will-o'-the-wisp. To Einstein it remained a compelling goal. To have two independent structures of space, one metric-gravitational and the other electromagnetic, was, to him, "intolerable to the theoretical spirit." Had Einstein lived to be a centenarian, he would undoubtedly have been gratified to learn that his elusive objective is once more a fashionable area of research. Whether gravitational, electrical, and nuclear interactions can be encompassed within a unified theoretical structure, and whether such a structure will be conceived as a plenary space with physical properties, remains to be seen. But if the history of the successive dynasties of aether is any guide, we can eventually proclaim:

The luminiferous aether is dead!

Long live the aether!

Notes and References

[1] Paraphrased from Thomas L. Heath, *Greek Astronomy* (London, 1932), p. 67.

[2] Sir Thomas Heath, *Aristarchus of Samos* (Oxford, 1913), p. 222.

[3] Bernard Goldstein, "The Arabic Version of Ptolemy's *Planetary Hypotheses*," *Transactions of the American Philosophical Society*, vol. 57, part 4 (Philadelphia, 1967).

[4] Roger Arnaldez and Albert Z. Iskandar, "Ibn Rushd," in *Dictionary of Scientific Biography*, vol. 12 (New York, 1975), p. 3.

[5] For this information I am indebted to Dr. George Saliba who has recently discovered the treatise of My'ayyad ad-Dīn al-'Urḍi.

[6] Adapted from the translation of Ibn ash-Shāṭir by David A. King, in *Dictionary of Scientific Biography*, vol. 12 (New York, 1975), p. 358.

[7] Revised from the translations of Edward Rosen, *Three Copernican Treatises* (New York, 1939; 1971), p. 57, and Noel Swerdlow, "The Commentariolus of Copernicus," *Proceedings of the American Philosophical Society*, vol. 117, no. 6 (1973), pp. 433–34.

[8] *Great Books of the Western World*, vol. 34, p. 528.

[9] See Swerdlow, *op. cit.,* pp. 427–29.

[10] In Rheticus's *Narratio prima*; see Rosen, *op. cit.,* p. 137.

[11] Tycho Brahe, *De mundi aetherei recentioribus phaenomenis*, in *Tychonis Brahe opera omnia*, ed. by J. L. E. Dreyer (Copenhagen, 1922), chap. 8, p. 156; for a translation of this chapter, see M. Boas and A. Rupert Hall, *Occasional Notes of the Royal Astronomical Society*, vol. 3 (1959), pp. 253–63.

[12] Nicholas Copernicus, *On the Revolutions*, trans. by Edward Rosen (Baltimore, 1978), p. 20; see also *Great Books of the Western World*, vol. 16, p. 525.

[13] Tycho Brahe to Caspar Peucer, 13 September 1588, in *Tychonis Brahe opera omina*, ed. by J. L. E. Dreyer, vol. 7 (Copenhagen, 1924), p. 130. Trans. by Christine Jones and quoted by Robert Westman in "Three Responses to the Copernican Theory," in *The Copernican Achievement*, ed. by R. Westman (Berkeley, 1975), p. 329.

[14] *Great Books of the Western World*, vol. 16, pp. 888–905.

[15] Galileo Galilei, *Dialogue Concerning the Two Chief World Systems*, trans. by Stillman Drake (Berkeley, 1953), p. 462.

[16] Isaac Newton to Richard Bentley, 25 February 1692/3, in *The Correspondence of Isaac Newton*, ed. by H. W. Turnbull, vol. 3 (Cambridge, 1961), p. 254; see also *Great Books of the Western World*, vol. 3, pp. 817–18.

[17] *Great Books of the Western World*, vol. 34, p. 558; see also *Great Books of the Western World*, vol. 3, p. 817.

[18] *Great Books of the Western World*, vol. 34, p. 554.

[19] *Great Books of the Western World*, vol. 34, p. 520.

[20] *Great Books of the Western World*, vol. 34, p. 521.

[21] Quoted in Joseph Larmor's article "Aether," in *Encyclopaedia Britannica*, 11th ed., vol. 1 (London and New York, 1911), p. 292.

[22] *Ibid.*

[23] Albert Einstein, "Relativity and the Ether," in *Essays in Science* (New York, 1920), p. 103.

[24] *Ibid.*, p. 106.

[25] Oliver Lodge, "Ether," in *Encyclopaedia Britannica*, 13th ed., vol. 29 (London and New York, 1926), pp. 1027–28.

[26] *Ibid.*, p. 1028.

[27] Einstein, *op. cit.*, pp. 110–11.

COPERNICUS AND THE HELIOCENTRIC UNIVERSE

The Astronomy and Cosmology of Copernicus

I t was close to the northernmost coast of Europe, in the city of
Toruń, that the King of Poland and the Teutonic Knights signed
and sealed the Peace of 1466, which made West Prussia part of
Polish territory. And it was in that city, just seven years later and
precisely 500 years ago, in 1473, that Nicholas Copernicus was born.
We know relatively few biographical facts about Copernicus and virtually nothing of his childhood. He grew up far from the centers of
Renaissance innovation, in a world still largely dominated by medieval
patterns of thought. But Copernicus and his contemporaries lived in an
age of exploration and of change, and in their lifetimes they put together a renewed picture of astronomy and geography, of mathematics
and perspective, of anatomy, and of theology.[1]

When Copernicus was ten years old, his father died, but fortunately
his maternal uncle stepped into the breach. Uncle Lucas Watzenrode
was then pursuing a successful career in ecclesiastical politics, and in
1489 he became Bishop of Varmia. Thus Uncle Lucas could easily
send Copernicus and his younger brother to the old and distinguished
University of Cracow. The Collegium Maius was then richly and unusually endowed with specialists in mathematics and astronomy; Hartmann Schedel, in his *Nuremberg Chronicle* of 1493, remarked that
"Next to St. Anne's church stands a university, which boasts many

Selection 9 reprinted from *Highlights in Astronomy of the International Astronomical Union*, ed.
by G. Contopoulos, vol. 3 (1974), pp. 67–85.

eminent and learned men, and where numerous arts are taught; the study of astronomy stands highest there. In all Germany there is no university more renowned in this, as I know from many reports." At the university the young Nicholas embraced the study of astronomy with a passion found only in the most exceptional of undergraduates. There he learned about the works of Sacrobosco, Regiomontanus, Ptolemy, and Euclid.

After leaving the Collegium Maius, Copernicus journeyed to the great university cities of Bologna, where he studied canon law, and Padua, where he studied medicine. Italy, then as now, bore the visible imprint of ancient Rome. It had become the recent home of Greek scholars, refugees from Byzantium, and in Italy Copernicus seized the opportunity to learn Greek. Italy was then in the high Renaissance, with Leonardo, Michelangelo, and Raphael creating their great masterpieces. But Copernicus, like many before him, had been drawn to Italy not for art but in search of a degree, and before he went home, he picked up a doctorate in canon law at the University of Ferrara. He thus became a lawyer by profession, with astronomy remaining an avid avocation.

In 1503, the 30-year-old Copernicus returned to Poland to take up a lifetime post as a canon of the Cathedral of Frombork, an appointment arranged through the benevolent nepotism of his uncle Lucas. Bishop Lucas was the head of the local government in Varmia, and the sixteen canons of the Cathedral Chapter constituted the next highest level of administration. In this northernmost diocese of Poland, Copernicus led an active and fruitful life for 40 years.

It was here that Copernicus served as an administrator of the Cathedral estates, collecting rents, resettling peasants, and writing an essay on currency reform. He served for a while as private secretary, personal physician, and diplomatic envoy for his uncle. And here in northern Poland, imbued with the spirit of Italian humanism, he made a Latin translation of a Greek work by Theophylactus Simocatta, a seventh-century Byzantine epistolographer, and perhaps he even painted his own self-portrait. Each of the Cathedral canons received an ample income derived from the peasants working the farmlands administered by the Chapter, and with such a tenured position Copernicus had the financial security to pursue his sideline of astronomical researches.

It was in Frombork that he wrote "For a long time I reflected on the confusion in the astronomical traditions concerning the derivation of the motion of the spheres of the Universe. I began to be annoyed that the philosophers had discovered no sure scheme for the movements of the machinery of the world, created for our sake by the best and most systematic Artist of all. Therefore, I began to consider the mobility of the Earth and even though the idea seemed absurd, nevertheless I knew that others before me had been granted the freedom to imagine any circles whatsoever for explaining the heavenly phenomena."

We do not know precisely when Copernicus began to meditate on the mobility of the Earth. He first announced his assumptions in an anonymous tract, today called the *Commentariolus*, that is, the Little Commentary. The *Commentariolus* was written before 1514, because in that year Matthew of Miechow, a Cracow University professor, cataloged his books and noted that he had "a manuscript of six leaves expounding the theory that the Earth moves while the Sun stands still." This brief document represents a first account of planetary motion, which was considerably extended and elaborated by Copernicus in later years. We do not know if the *Commentariolus* was widely distributed. In any event, it dropped completely out of sight until around 1880, when an example was found in Vienna and another in Stockholm. More recently a third copy has been found in Aberdeen, Scotland.

In Copernicus's day the sciences, and astronomy not least, were beginning to respond to the new opportunities offered by the printing press. It is interesting to notice that his lifetime of astronomical studies was to a large part made possible by his access to printed sources. During the Thirty Years' War, the Frombork Cathedral library was carried off to Sweden, and as a result most of his books are now found in the Uppsala University library. They include the beautiful Ptolemaic atlas printed in Ulm in 1486, Argellata's book on surgery, two editions of Pliny the Younger, plus works by Cicero, Herodotus, Hesiod, and Plato.

One of the earliest books he bought, presumably while he was still a student at the Collegium Maius, was the 1492 edition of the *Alfonsine Tables*. His personal copy is still preserved in its Cracow binding. These tables, originally constructed in 1273, represented the state of the art when Copernicus was a young man. They enabled him to calculate solar, lunar, and planetary positions for any date according to the Ptolemaic theory. Among the other scientific volumes remaining from Copernicus's personal library is the beautiful first edition of Euclid's *Elements*, printed by Ratdolt in 1483, and Stoeffler's *Calendarium Romanum magnum* of 1518. The annotations in this latter book show that Copernicus witnessed celestial phenomena on numerous occasions not mentioned in his published work.

A book that must have been enormously important during Copernicus's formative years was the Regiomontanus *Epitome of Ptolemy's Almagest*. His personal copy of this book is lost, but perhaps it is still waiting to be recognized by some sharp-eyed scholar. Our astronomer's principal access to the Ptolemaic theory must have come at first through the *Epitome*. It was not until after he had written the *Commentariolus* that the full text of Ptolemy's *Almagest* became available, in the edition printed in Venice in 1515. Copernicus studied the work carefully, as the manuscript notes and diagrams in the margins clearly show. Through this work he must have become more fully aware of the tremendous task facing any astronomer with the courage to construct a complete celestial mechanism.

During the 1520s, Copernicus worked extensively to elaborate his ideas, especially the planetary theory, if we are to judge by the scattered planetary observations recorded in his work. The *Commentariolus* had already hinted at a larger work, which Copernicus composed and continually revised during these years. By heroic good fortune, which we could scarcely have expected, his original manuscript has survived all these years. Perhaps the most priceless artifact of the entire scientific renaissance, it is now preserved in the Jagiellonian Library of Cracow University. The skilled draftsmanship, the precise hand, and, above all, the way in which he has elegantly written his text around the famous diagram of the heliocentric system (see Figure 1) convey the impression that this was a piece of calligraphy for its own sake, not a manuscript to be destroyed in the printing office, but an opus destined for the library shelf in the quiet cloisters of Frombork.

It is quite possible that his manuscript would have gathered dust, unpublished and virtually unknown, had it not been for the intervention of a young professor of astronomy from Wittenberg, Georg Joachim Rheticus. Exactly how Rheticus heard about Copernicus's work is still a mystery, although he may have seen a copy of the *Commentariolus*. In any event, he decided that only a personal visit to the source would satisfy his curiosity about the new heliocentric cosmology. Thus, in 1539, the 25-year-old Rheticus set out to that "most remote corner of the Earth," as Copernicus himself described it. Although he came from the central bastion of Lutheranism, the Catholic Copernicus received him with courage and cordiality.

Swept along by the enthusiasm of his young disciple, Copernicus allowed him to publish a first printed report about the heliocentric

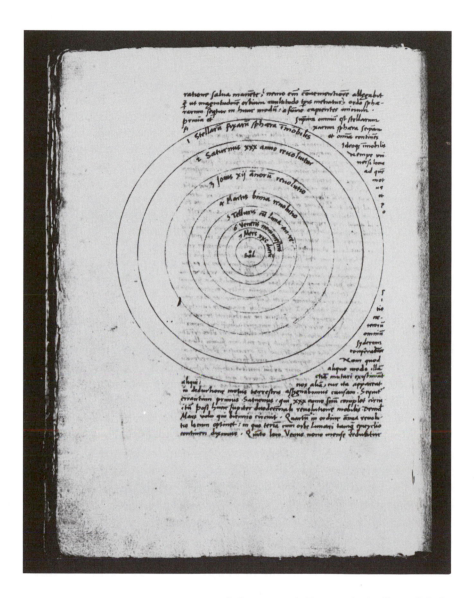

FIGURE 1. *Autograph manuscript of Copernicus's* De revolutionibus, *fol. 9v, showing the heliocentric system. Photograph by Charles Eames, courtesy of the Jagiellonian Library, Cracow.*

system. In a particularly beautiful passage of the *Narratio prima*, Rheticus wrote:

> With regard to the apparent motions of the Sun and Moon, it is perhaps possible to deny what is said about the motion of the Earth. . . . But if anyone desires to look either to the order and harmony of the system of the spheres, or to ease and elegance and a complete explanation of the causes of the phenomena, by no other hypotheses will he demonstrate more neatly and correctly the apparent motions of the remaining planets. For all these phenomena appear to be linked most nobly together, as by a golden chain; and each of the planets, by its position and order and very inequality of its motion, bears witness that the Earth moves.

Rheticus had not come to Polish Prussia empty-handed. He brought with him three volumes, the latest in scientific publishing, each handsomely bound in stamped pigskin. These he inscribed and presented to his distinguished teacher. Included were Greek texts of Euclid and Ptolemy, as well as three books published by Johannes Petreius, the leading printer of Nuremberg. By the time Rheticus returned to Wittenberg in September of 1541, he had persuaded Copernicus to send along a copy of his work, destined for Petreius's printing office.

Tantalizingly little information survives concerning the actual publishing of Copernicus's book. We do not know the time required for the printing, the size of the edition, the methods of distribution, or the price. A few things can be conjectured from the standard practices of the day. Thus we can deduce that if a single press was used for the folio sheets, the printing of the 404-page treatise would have taken about four months. It is likely that the type would have been redistributed and continually reused, so that a competent technical proofreader would have been required on the scene.

Wildly diverse guesses about the size of the first edition have appeared in the literature. At the present time, I have located approximately 200 copies; perhaps an additional hundred exist that I have not found, and I would appreciate help in locating other copies. These numbers suggest an edition of at least 400, and perhaps 500. If many more were sold, it seems improbable that a second edition of about the same size would have been required 23 years later. In any event, enough copies were issued so that its ideas could not easily be suppressed or forgotten.

By the time the printing had got under way, Rheticus had taken a professorship at Leipzig, too far from Nuremberg to assist directly with the proofreading. Thus the printer, Petreius, turned to a local scholar and theologian, Andreas Osiander, who had helped him on at least one previous occasion.

In order to disarm criticism of the unorthodox cosmology in the book, Osiander added an unsigned introduction on the nature of hypotheses. He wrote:

> It is the duty of an astronomer to record celestial motions through careful observation. Then, turning to the causes of these motions he must conceive and devise hypotheses about them, since he cannot in any way attain to the true cause. . . . The present author has performed both these duties excellently. For these hypotheses need not be true nor even probable; if they provide a calculus consistent with the observations, that alone is sufficient. . . . So far as hypotheses are concerned, let no one expect anything certain from astronomy, which cannot furnish it, lest he accept as true ideas conceived for another purpose, and depart from this study a greater fool than when he entered it.

I doubt that Osiander's anonymity stemmed from any malicious mischievousness, but rather simply from a Lutheran reluctance to be associated with a book dedicated to the Pope. In any event, Kepler and the other leading astronomers of that century were fully aware of the authorship; in Kepler's copy, preserved at the University of Leipzig, Osiander's name has been written above the introduction. There exists a presentation copy given by Rheticus to Andreas Aurifaber, who was then Dean of the University of Wittenberg. The inscription is dated 20 April 1543. Thus a copy of the book could have easily reached Copernicus before he died on 24 May 1543, but because he had been incapacitated by a stroke, he was probably unaware of Osiander's introduction.

Rheticus himself was so offended by the added introduction that he struck it out in the copies he distributed. He also deleted the last two words of the printed title *De revolutionibus orbium coelestium*. There is an old tradition that Osiander assisted the printer in changing the title from "Concerning the Revolutions" to "Concerning the Revolutions of the Heavenly Spheres." It is difficult to see precisely what Rheticus thought was offensive about the additional words except that, like the introduction, the expression "Heavenly Spheres" perhaps suggests too much the idea of model building. As I shall explain, the idea that astronomers were merely playing some kind of geometrical game had a widespread currency in the sixteenth century, and Osiander's preface simply served to reinforce what astronomers thought they saw in the major part of *De revolutionibus*. When we notice that Copernicus used an entirely different arrangement of circles for predicting latitudes than for predicting longitudes, we realize that any reader who studied the great bulk of the book carefully would necessarily have seen Copernicus as a builder of hypothetical geometrical models.

Despite the existence of the manuscript with its many layers of revisions, and even the *Commentariolus*, which provides a glimpse of an earlier formulation, we have no definite idea of the circumstances that caused Copernicus to adopt a Sun-centered cosmology. Attempting to answer this question is one of the intriguing problems that face Copernican scholars today.

If we, as twentieth-century astronomers, were to speculate freely, we might well invent some quite convincing causes. First, we might suppose that the *Alfonsine Tables* were no longer in accord with the actual observations. This is true, but mostly irrelevant. Second, we might imagine that successive generations of theory-patching had left the Ptolemaic system too cumbersome for practical use, so that a massive simplification was in order. This second supposition is entirely false.

Let us first consider the matter of predictions versus observations. Was Copernicus motivated to reform astronomy because the current almanacs were bad? Because we can compare fifteenth-century ephemerides with the far more accurate calculations carried out recently by Dr. Tuckerman at the IBM Corporation, we know nowadays that they often had errors of several degrees. But did Copernicus know this?

Soon after Copernicus had returned to Poland from Italy, the planets put on a particularly spectacular celestial show. Saturn and Jupiter, the slowest moving planets, moved into the constellation Cancer for one of their scarce conjunctions, once in 20 years. In addition, Mars, Venus, and Mercury, and eventually the Sun and the Moon, all congregated within this single astrological sign. In the winter of 1503–1504, Mars went into its retrograde motion, making repeated close approaches to Jupiter and Saturn.

My assistant, Barbara Welther, has charted for us the geocentric longitudes of the superior planets as a function of time (Figure 2). You can see how Mars bypasses Jupiter and Saturn in October 1503, and then, as all the planets go into retrograde, Mars backs up past Saturn and Jupiter, and then passes them directly once more in the winter of 1504. We have not shown the great conjunction of Jupiter

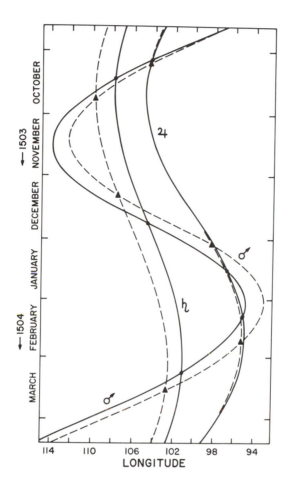

FIGURE 2. *Apparent motions of the superior planets just before the great conjunction of 1504. The solid lines and circles show the actual positions and conjunctions. The broken lines and triangles show the predicted positions and conjunctions.*

and Saturn at the end of May, because by that time they were too close to the Sun. We have marked with dashed lines the predicted positions of the planets according to the *Alfonsine Tables*. Notice particularly that in February and March the Mars predictions erred by 2° and Saturn by 1.5°, whereas Jupiter was predicted rather accurately. The

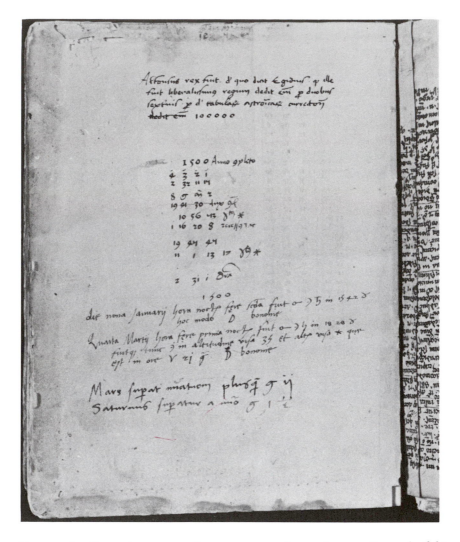

FIGURE 3. *Copernicus wrote these notes on observations at the end of his printed copy of the* Alfonsine Tables *(1492). By permission of the Uppsala University Library.*

predicted times of the conjunctions differ by about one or two weeks from the actual times shown by the intersections of the curves.

Anyone as interested in astronomy as Copernicus could scarcely have failed to observe these phenomena, but I was curious to know whether he had noticed these deficiencies in the *Alfonsine Tables*. Although there is no direct record that Copernicus made these observations, Dr. Jerzy Dobrzycki suggested to me a way whereby we can be certain that our astronomer followed the planetary motions in the year of the great conjunction. Bound in the back of his copy of the *Alfonsine Tables* are sixteen extra leaves on which Copernicus added carefully written tables and miscellaneous notes. Below the record of two observations made in Bologna in 1500, there is, in another ink, a cryptic undated remark in highly abbreviated Latin (Figure 3): "Mars surpasses the numbers by more than two degrees. / Saturn is surpassed by the numbers by one and a half degrees."

If we examine carefully the error pattern between the positions predicted for the superior planets by the *Alfonsine Tables* and the calculations made by Tuckerman, we find a virtually unique error pattern for February and March of 1504 corresponding to the note. Thus, our astronomer must have been fully aware of the discrepancies between the tables and the heavens.

Why, then, are such glaring inadequacies never mentioned by Copernicus as a reason for introducing a new astronomy? I believe the answer is quite simple. Copernicus knew very well that discrepancies of this sort could be corrected merely by changing the parameters of the old system. A new Sun-centered cosmology was hardly required for patching up these difficulties with the tables.

But furthermore, if we turn once more to the analysis made possible by modern computers, and if we examine the old ephemerides, we are shocked to discover that there is relatively little difference in the average errors before and after Copernicus. His work has scarcely improved the predictions.

Rather than condemn Copernicus, we should remember that he had no procedure for handling errors in a multiplicity of data. He had only a few score ancient observations, those recorded by Ptolemy in the *Almagest*. Since these were the minimum number required to establish the parameters, he was obliged to assume that they were perfect and to force his own parameters to fit them. From his own planetary observations he only slightly modified Ptolemy's eccentricities and apsidal lines, and he reset the mean longitude, somewhat akin to resetting the hands of a clock whose mechanism is still basically faulty. Copernicus himself must have realized that he had not achieved as much in this direction as he might have hoped, and perhaps this partly explains his reluctance to send his great work to the printer.

After *De revolutionibus* was published, Erasmus Reinhold reworked the planetary tables into a far handier form. His *Prutenic Tables* superseded the *Alfonsine Tables* remarkably quickly. This is actually very curious because, in the absence of systematic observations, nobody really knew how good or bad any of the tables were. In fact, it was not until Tycho Brahe that a regular series of observations established the inadequacies of all the tables.

Tycho himself was something of a child prodigy; when he saw an eclipse at age 13 it struck him as "something divine that men could know the motions of stars so accurately that they could long before foretell their places in relative positions." But three years later, at the great conjunction of Saturn and Jupiter in 1563, he was astonished and offended to discover that even Prutenic-based ephemerides foretold the event on the wrong day. From the time of that great conjunction onward, he kept regular observations of increasing precision that eventually became the basis for another sweeping reform of astronomy.

Let us now turn quickly to a second imagined defect in the ancient geocentric astronomy, which, if true, would give more than adequate grounds for introducing a new system. This is the story, widely repeated in the secondary literature, that by the Middle Ages the Ptolemaic theory had been hopelessly embroidered with epicycles-on-epicycles. I fear that we modern astronomers have been particularly fond of this legend because it reminds us of a Fourier series. In Ptolemy's original scheme, the Earth is placed near but not exactly at the center of a large orbital circle called the deferent. Each planet moves in a secondary circle of epicycle, which produces the retrograde motions of the sort that we have noted at the time of the conjunctions in 1504. From a modern heliocentric viewpoint we would say that the planetary epicycles are reflections of the Earth's own orbit.

About a century ago, the story began to propagate that Ptolemy's rather simple system had been overlaid with dozens of additional secondary circles. The seed for this mythology was planted by Copernicus himself when, at the end of his *Commentariolus*, he concluded: "All together, therefore, 34 circles suffice to explain the entire structure of the universe and the entire ballet of the planets." Nineteenth-century commentators used their imaginations to embellish Copernicus's simple claim. Without checking the facts, they created a fictitious pre-Copernican planetary theory hovering on the brink of collapse under the burden of incredibly complex wheels upon wheels.

I suspect that, at the end of the thirteenth century, Alfonso the Great may have contributed to the legend, because he supposedly told his astronomers that, if he had been present at creation, he could have given the Good Lord some hints. Again, modern electronic computers have helped us put this legend to rest. I have recomputed his planetary tables in their entirety to show that they are based on the classical and simple form of the Ptolemaic theory with only two or three minor changes of parameter in the whole set.

Next, I used these thirteenth-century tables to compute a daily ephemeris for 300 years, and this I compared with the best almanacs of the fifteenth and early sixteenth centuries. The comparison showed, without any question, that the leading almanac makers, such as Regiomontanus, were using the unembellished Ptolemaic theory as found in the *Alfonsine Tables*.

Is it possible that the epicycles-on-epicycles existed but simply did not get to the level of almanac making? The answer is both no and yes. From antiquity there were actually two competing cosmological views. First was the system of concentrically nested spheres, espoused by Aristotle because it made such a tidy, compact, mechanical universe. In contrast, the Ptolemaic system had large clumsy epicycles that were difficult to place in concentric nests.

Peurbach's *New Theory of the Planets*, the most important work on astronomy written in the generation immediately preceding the birth of Copernicus, added no new epicycles, but instead attempted to resolve the cosmological competition by incorporating large eccentric zones of crystalline aether. By providing something of an off-center tunnel for the epicycle, the mechanism for each planet could be contained within two concentric bounds. Thus in principle the entire planetary system of Ptolemy could be nested together within the homocentric aethereal spheres of Aristotle. Such was the *New Theory of the Planets*, and I hasten to say that this idea was not really new, as it had already been described by Islamic scientists, and proposed even earlier by Ptolemy himself.

In recent years, the historians of science have discovered that, interestingly enough, thirteenth- and fourteenth-century Islamic astronomers discussed one important case of an epicycle-on-epicycle, designed not to improve the fit to observations, but to satisfy a philosophical principle. Because this same philosophical point played a major role in the motivation of Copernicus, let me now return to his work and present the two major reasons that Copernicus himself gives as primary motivations for his astronomical work.

In the *Commentariolus*, our astronomer wrote concerning the planetary motions that "Eventually it came to me how this very difficult problem could be solved with fewer and much simpler instructions than were formerly used, if some assumptions were granted me." If we

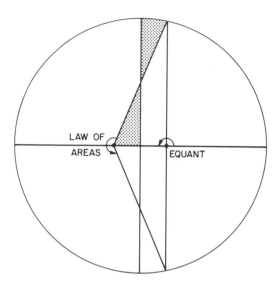

FIGURE 4. *The near-equivalence of the equant and the law of areas.*

put aside the spurious relevance of counting circles, the heliocentric
system does provide a profound simplification, and I must necessarily
return to this point before the end of the paper. However, Copernicus
awarded virtually equal weight to a second philosophical principle, the
Platonic-Pythagorean concept of uniform circular motion. Copernicus
opened his *Commentariolus* with an attack on the Ptolemaic equant,
which appeared to violate this principle of uniform circular motion.
The equant is a seat of uniform circular motion placed equal and
opposite to the Earth within the deferent circle; it drives the epicycle
around on the deferent more swiftly at the perigee than at the apogee.

Figure 4 illustrates the relation between Kepler's law of areas and
the equant; because the equant turns uniformly, the planet will move
in equal time in each of the four quadrants. The law of areas tells us
that the planet will move through these same arcs in equal times
provided that the areas swept out from the primary focus are equal.
Because the equant is at the empty focus, the shaded triangles are
virtually equal except that the upper one has a curved side; to this
extent, the equant is a good approximation to the true motion, espe-

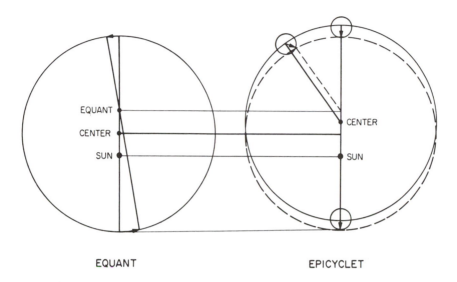

EQUANT

EPICYCLET

FIGURE 5. *Copernicus's replacement of the equant by a pair of uniform circular motions. The epicycle has a radius of* e/2 *and always moves to form the isosceles trapezoid shown above.*

cially at the quadratures. As Kepler was later to show, the major discrepancy occurs in the octants.

In any event, Copernicus despised the equant and he felt that Ptolemy had cheated by introducing it. Figure 5 shows how Copernicus replaced the equant with an eccentric circle and a small epicyclet. In the *Commentariolus* he preferred to use a concentric circle with a double epicyclet, which was precisely the same mechanism suggested two centuries earlier by Ibn ash-Shātir in Damascus; whether there was any transmission from those Islamic astronomers to Copernicus is still debatable. After Copernicus discovered the motion of the planetary apsidal lines, it became more convenient to use the eccentric circle and single epicyclet shown here. I shall not take the time here to explain the equivalence between this mechanism and the equant, but I shall simply say that the great bulk of the *De revolutionibus* involves the use of this mechanism.

Nowadays the epicyclet seems esoteric and forgettable. When we commemorate Copernicus, we praise his profound insight in seeing the philosophical and aesthetic simplicity of his system, but we try to ignore his infatuation with the second, very deceptive principle of uniform circular motion. I should now like to demonstrate that Copernicus's sixteenth-century successors, living in an age long before Newtonian dynamics, evaluated these philosophical principles in precisely the opposite way, rejecting the simplicity of the heliocentric cosmology but admiring the epicyclets.

About 1971, I had an interesting discussion with another Copernican scholar, Dr. Jerome Ravetz, and we asked ourselves if *De revolutionibus* actually had any careful readers. We speculated that there are probably more people alive today who have read this book carefully than in the entire sixteenth century. I have already introduced some of the candidates for that early era: Georg Rheticus, the Wittenberg scholar who persuaded Copernicus to publish his book; Erasmus Reinhold, the Wittenberg professor who stayed home but who later composed the *Prutenic Tables*; and Tycho Brahe, the great Danish observer. Others would include Johann Schöner, the Nuremberg scholar to whom Rheticus addressed the *Narratio prima*; Christopher Clavius, the Jesuit who engineered the Gregorian calendar reform; Michael Maestlin, Kepler's astronomy teacher; and Johannes Kepler himself.

At that time I was on a sabbatical leave from the Smithsonian Astrophysical Observatory, and two days after talking with Dr. Ravetz I happened to visit the remarkable Crawford collection of rare astronomical books at the Royal Observatory in Edinburgh. There I admired one of their prize possessions, a copy of the first edition of *De revolutionibus*, legibly annotated in inks of several colors. As I examined the book, I deduced that the intelligent and thorough notations were undoubtedly made before 1551, that is, within eight years after its publication. Our speculation from two days earlier seemed completely demolished, because it appeared that if intelligent readers were so rare, it would be unlikely that the very next copy of the book that I saw could be so carefully annotated. But then a second thought crossed my mind: Perhaps the Crawford copy had been annotated by one of the handful of astronomers we had mentioned. The list quickly narrowed to Rheticus, Reinhold, and Schöner, the only ones active before 1550. Internal evidence suggested Erasmus Reinhold, and although his name is not in the book, I soon found his initials stamped into the decorated original binding. Ultimately I was able to obtain additional specimens of Reinhold's distinctive handwriting, which settled the matter beyond all doubt.

One of the most interesting annotations in Reinhold's copy appears on the title page, where he has written in Latin "The axiom of astronomy: Celestial motion is uniform and circular or composed of circular and uniform parts." Reinhold was clearly fascinated by Copernicus's epicyclets and his adherence to the principle of circular motion. The paucity of annotations in the first twenty pages, which Copernicus devoted to the new cosmology, shows that Reinhold was not particularly interested in heliocentrism. Accepting Osiander's statement that astronomy was based on hypotheses, Reinhold was apparently intrigued by the model-building aspects. Whenever alternative mechanisms for expressing the motions appeared in the book, he made conspicuous enumerations with Roman numerals in the margins.

Because Reinhold published the *Prutenic Tables*, naming them in part for Copernicus, he is sometimes listed as an early adherent of the heliocentric cosmology. However, the nature of the tables makes them independent of any particular cosmological system, and although his introduction is full of praise for Copernicus, he nowhere mentions the heliocentric cosmology. With his great interest in hypothetical model building, there's reason to suspect that Reinhold was on the verge of an independent discovery of the Tychonic system; unfortunately, he died of the plague at an early age before he could consolidate any cosmological speculations of his own.

Flushed with the success of identifying Reinhold's copy, I resolved to examine as many other copies of the book as possible in order to establish patterns of readership and ownership, always hoping to find further interesting annotations. For three years I have systematically examined copies in such far-flung places as Budapest and Basel, Leningrad and Louisville, Copenhagen and Cambridge. In the process I saw and photographed several particularly interesting copies, including the *De revolutionibus* owned by Michael Maestlin, preserved in Schaffhausen, Switzerland; this is one of the most thoroughly annotated copies in existence. I also examined copies once owned by Rheticus, by Kepler, and by Tycho Brahe—the last being a heavily annotated second edition in Prague. In all, I had managed to see 101 copies by the spring of 1973. The investigation confirmed that the book had rather few perceptive readers, at least among those who read pen in hand. Despite this, however, the book seems to have had a fairly wide circle of casual readers, much larger than generally supposed.

In May of 1973, I had the opportunity to visit Rome, where there were seven copies of the first edition that I had not examined. My quest took me first to the Vatican Library, where I went armed with shelf mark numbers provided by Dr. Dobrzycki. Some of the books in the Vatican Library came there with the eccentric Queen Christina of

Sweden, who abdicated her throne in 1654, abandoning her Protestant kingdom for Rome. Her father, Gustavus Adolphus, had ransacked northern Europe during the Thirty Years' War and among other things had captured most of Copernicus's personal library. Dr. Dobrzycki had gone to Rome in search of Copernican materials that Queen Christina might have taken along. In the Vatican, he found an unlisted copy of Copernicus's book among the manuscripts, that is, a third copy beyond the two examples cataloged among their printed books. Fortunately, Dr. Dobrzycki gave me the number for the volume, which could not have been found in any of the regular Vatican catalogs.

When I examined this copy, I recognized that the extensive marginal annotations must have been made by a highly skilled astronomer. At the end were 30 interesting manuscript pages, full of diagrams made by someone working along the same lines as Tycho Brahe, and dated 1578. Although there was no name any place on the volume, I quickly conjectured that the annotations had been made by the Jesuit astronomer Christopher Clavius. In the first edition of his learned *Commentary on the Sphere of Sacrobosco*, published in 1570, he failed to mention Copernicus. But in the third edition, published in 1581—after the time these manuscript notes were written—he commented rather extensively and wrote "All that can be concluded from Copernicus's assumption is that it is not absolutely certain that the eccentrics and epicycles are arranged as Ptolemy thought, since a large number of phenomena can be defended by a different method."

In a state of considerable excitement, I contacted Dr. D. J. K. O'Connell, former Director of the Vatican Observatory, and with his help I obtained Xerox copies of two Clavius letters from the Jesuit Archives. I eagerly returned to the Vatican Library, only to have my hypothesis smashed within a few minutes. There was no possibility that the handwriting in the *De revolutionibus* could be that of Christopher Clavius.

I left Rome in a baffled and troubled state for a Copernicus conference in Paris. There, by a fantastic stroke of luck, I received the new Prague facsimile of the second-edition *De revolutionibus* with the annotations by Tycho Brahe. I think my heart must have skipped a beat when I saw the handwriting in the facsimile, because I then realized that the first edition in Rome was probably also in Tycho's hand. What I had discovered was the original working copy, probably the most important Tycho manuscript in existence. The example in Prague was a derivative copy, being annotated by Tycho for possible publication. I rebooked my flights, went back to Rome, and after I put the Prague facsimile side by side with the Vatican copy, it took only a few minutes to prove my conjecture. Afterward, the Vatican librarians traced the book to Queen Christina, who must have gained possession of it in 1648 when her troops captured the collections founded by Rudolf II in Prague.

Of many remarkable things about this copy, the first appears on the title page itself. We find the very same words that Reinhold had inscribed on the title page of his copy, "The axiom of astronomy: Celestial motion is uniform and circular, or composed of uniform and circular parts." I had already known that Tycho Brahe had visited Wittenberg on at least four occasions, and that, in 1575, three years before the dated annotations in this book, he had visited Reinhold's son and had seen Reinhold's manuscripts. In an article that I had written earlier for the Copernicus celebrations in Toruń, I had stated "We are tempted to imagine that Tycho's own cosmological views grew from seeds planted at Wittenberg by a tradition that honored Copernicus, but which followed Osiander's admonition that it is the duty of the astronomer to 'Conceive and devise hypotheses, since he cannot in any way attain the true causes'." [2] The newly found Tycho copy dramatically confirms this intellectual heritage, not only through this motto on the title, but within the book, where numerous annotations are copied word for word from Reinhold's copy. In particular, Tycho, like Reinhold, specifically numbered any alternative arrangements of circles indicated by Copernicus.

In the Tycho Brahe manuscript bound at the end of the Vatican *De revolutionibus*, the first opening is dated January 27, 1578, the day after the spectacular comet of 1577 had been seen for the last time. The diagrams on those two pages are heliocentric, and a note in the corner indicates that it was drawn according to the third hypothesis of Copernicus. In the next two weeks, Tycho explored additional heliocentric arrangements for the planets and geocentric models for the Moon. On February 14 and 15, he began to investigate *geocentric* constructions for Venus and Mercury, especially alternate positions of the single epicyclet for Venus and the pair of epicyclets for Mercury. He specifically noted that "This new idea occurred to me on February 13, 1578."

Three days later Tycho drew the most interesting diagram of the entire sequence, a proto-Tychonic system with the Earth at the center circled by the Moon and the Sun (Figure 6). Around the Sun are the orbits of Mercury and Venus. The three superior planets are still ar-

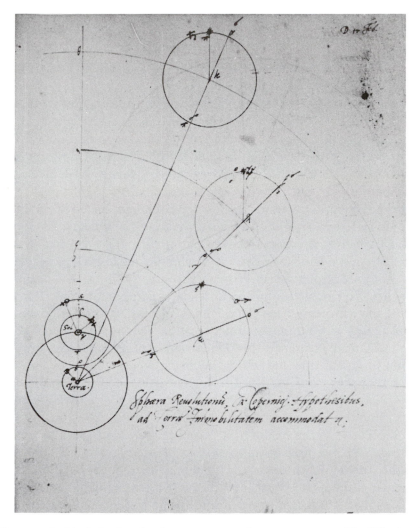

FIGURE 6. *Tycho's sketch of a geocentric planetary system, fol. 210v in the manuscript notes bound at the end of his annotated copy of* De revolutionibus *(1543), Vatican Library Ottob. 1901.*

ranged in circles about the Earth, but each epicycle has been drawn the same size as the Sun's orbit. To finish the construction of the Tychonic system, it is necessary only to complete the parallelograms for Mars, for Jupiter, and for Saturn. Tycho was now surely within grasp of his final system. But notice the caption: "The spheres of revolution ac-

commodated to an immobile Earth from the Copernican hypotheses."
Here we see Tycho playing the astronomical geometry game, greatly
under the influence of Copernicus, and somehow supposing that a
geocentric system is compatible with the teachings of the master.

It is very curious that Tycho did not publish his new system until a
decade later. Tycho was a dynamic young man of 31 when he wrote
this manuscript, already well established on the island of Hven, but
perhaps still uncertain where his observations for the reform of astronomy would lead him. A passage in his book implies that he did not
establish the Tychonic system until around 1583, five years after he
drew these diagrams. I can only suppose that these five years were an
important time of maturing. In that interval, Tycho must have speculated on the movement of the Great Comet of 1577, realizing that it
would have smashed the crystalline spheres of the ancient astronomy,
had they existed. Perhaps he began to look for greater certainty in
astronomy and to suppose that, after all, the observations made with
his giant instruments at his Uraniborg Observatory could lead beyond
hypothesis to physical reality. If so, like his contemporaries in that
pre-Newtonian, predynamical age, he must have viewed the physics of
the sluggish, heavy Earth as a most important phenomenon to be
preserved. Concerning the Copernican system, Tycho Brahe wrote:
"This innovation expertly and completely circumvents all that is superfluous or discordant in the system of Ptolemy. On no point does it
offend the principle of mathematics. Yet it ascribes to the earth, that
hulking, lazy body, unfit for motion, a motion as quick as that of the
aethereal torches, and a triple motion at that." I can well imagine that
Tycho believed he was making a great step forward toward understanding the physical reality of the universe when he adopted his own
geocentric system.

To us, the Tychonic system looks clumsy and wrong. To us, there is
something more neat and orderly about the heliocentric system. Indeed, it is precisely this elegant organization that Copernicus found
pleasing to the mind, and that led to his cosmology. In a powerful plea
for the heliocentric world view near the beginning of *De revolutionibus*,
Copernicus wrote:

> At rest in the middle of everything is the Sun. For in this
> most beautiful temple, who would place this lamp in another
> or better position? From here it can light up the whole thing
> at the same time. Thus as though seated on a royal throne,
> the Sun governs the family of planets revolving around it. In

> this arrangement, therefore, we discover a marvelous com-
> mensurability of the Universe, and an established harmoni-
> ous linkage between the motion of the spheres and their size,
> such as can be found in no other way. Thus we perceive why
> the direct and retrograde arcs appear greater in Jupiter than
> in Saturn and smaller than in Mars, and why this reversal in
> direction appears more frequently in Saturn than in Jupiter,
> and more rarely in Mars and Venus than in Mercury. All
> these phenomena proceed from the same cause, which is the
> Earth's motion. Yet none of these phenomena appears in the
> fixed stars. This proves their immense height, which makes
> the annual parallax vanish from before our eyes.

There is a whiff of reality here, especially in the resounding conclusion, "So vast, without any question, is this divine handiwork of the
Almighty Creator." Yet very few people in the sixteenth century
grasped the harmonious, aesthetic unity that Copernicus saw in the
cosmos. And that is why we must also salute another perceptive genius, born almost a century later than Copernicus. Like Copernicus,
Johannes Kepler saw the Sun seated upon its royal throne as the
governor of the planetary system, and he tried mathematically to find
the harmonious linkage between the motions of the spheres and their
sizes. To us Kepler's neo-Platonic attempts to find an archetypal geometrical structure in the planetary arrangement smack of mystical
numerology—yet this is hardly a criticism, considering that numerology has not been banished from modern cosmology. But more important, Kepler saw in the Copernican arrangement of the planets about
the Sun the real possibility of a celestial physics, and he made the first
groping steps toward a dynamics of the heavens—a dynamics that,
reshaped and powerfully formulated by Isaac Newton, ultimately
proved to be the primary justification for the heliocentric universe.

Although Copernicus is not celebrated for his observations, it was in
the Copernican tradition that Kepler and Galileo taught us to use our
senses to distinguish between the various hypothetical world views,
leaving only those consistent with the observations. In a way we are
still model builders, as Osiander suggested, but unlike Andreas Osiander and Erasmus Reinhold, we are no longer content to entertain
alternatives without trying to choose one as physically most acceptable. Modern science still plays its games, but in an entirely different
way than did the ancient and medieval astronomers. Certainly Copernicus, Tycho, Kepler, Galileo, and Newton are heroes in this epic

reformation in our understanding of what nature is and what learning and observation should be.

   Although I have said perhaps too much about the technical astronomy of Copernicus and rather little about his cosmology, I hope that within this broader context you have been able to appreciate all the more how unique was Copernicus's own intellectual adventure. Only in our own generation have we been able to break the terrestrial bonds; men flung out toward the Moon have seen the spinning Earth, a blue planet, sailing through space. Although rejected by the astronomers of his day, the Copernican idea became the point of departure for the law of universal gravitation. In reality, the Copernican quinquecentennial celebrates the origins of modern science and our contemporary understanding of the universe. In setting the Earth into motion, Copernicus was right: his daring idea still guides the unfinished journey of modern science.

Mathematical Postscript

   That the Copernican scheme of an eccentric deferent and a single epicyclet satisfies Keplerian motion to the first power of the eccentricity can be elegantly demonstrated as follows. If we neglect terms of e^2 and higher, the law of areas becomes

$$r^2\theta = na^2(1 - e^2) = na^2$$

where we have for the mean motion $n = 2\pi/P$; similarly the equation for the ellipse is

$$r = \frac{a(1 - e^2)}{1 - e \cos \theta} = a(1 + e \cos \theta)$$

where a is of course the semimajor axis. After combining these equations, integrating, and inverting, we find

$$\theta = nt - 2e \sin nt.$$

With a little trigonometric manipulation and by discarding terms in e^2 and higher, we have

$$x = r \cos \theta = a(\cos nt + \tfrac{3}{2}e - \tfrac{1}{2}e \cos 2nt)$$

$$y = r \sin \theta = a(\sin nt - \tfrac{1}{2}e \sin 2nt).$$

These equations correspond to motion with period P in circle of radius a displaced by $\tfrac{3}{2} ae$ from the sun, together with a circular epicyclet of radius $\frac{1}{2}$ *ae* and period $\frac{1}{2}$ *P*. This is essentially the model for planetary longitude adopted by Copernicus in his *De revolutionibus*, although he did not strictly maintain the 3:1 ratio between the displacement of the main circle and the radius of the epicyclet. (For example, the Copernican ratio for the Mars model is 2.92:1).

I should like to acknowledge the late Sir Harold Jeffreys, who showed me this demonstration without realizing how precisely it described the Copernican model.

Notes and References

[1] Professor Edward Rosen's Copernican biography in his *Three Copernican Treatises* (New York, 1971) has provided an authoritative source for details of Copernicus's life, and I have borrowed from him several felicitous turns-of-phrase as well as English translations of some of the Latin texts (which I have generally abridged). Other writings that have been particularly stimulating include J. Ravetz, *Astronomy and Cosmology in the Achievement of Nicolaus Copernicus* (Wroclaw, 1965), P. Duhem, *To Save the Phenomena* (Chicago, 1969), and L. A. Birkenmajer, *Mikolaj Kopernik* (Cracow, 1900) (English translation under joint preparation by J. Dobrzycki and myself). Also useful are A. Koyré, *The Astronomical Revolution* (Paris, 1973) and *The Great Books of the Western World* (Chicago, 1952) volume 16, which contains an English translation of *On the Revolutions* as well as Ptolemy's *Almagest*.

[2] Reprinted as selection 13 of this anthology; see esp. p. 246. The reader is reminded that we now know that the annotations in the Vatican and Prague copies of *De revolutionibus* are not by Tycho Brahe, but by Paul Wittich—see the Preface to this anthology.

Did Copernicus Owe a Debt to Aristarchus?

D uring Copernicus's lifetime, in the early years of the sixteenth century, very little was known in western Europe of Aristarchus. Our best source for Aristarchus's heliocentric ideas, the *Sand-reckoner* of Archimedes, was not published until 1544, the year after Copernicus died.[1] In our modern era of library catalogs, reference indices, and data retrieval systems, it is easy but anachronistic to imagine that Copernicus could have consulted a manuscript in Italy when he was beginning to learn Greek. In fact, Copernicus relied almost entirely on printed works for his information, and there is only a single case where we know for sure that he used a manuscript source.[2] Even the encyclopedist Giorgio Valla, who published various translations of Archimedes, did not mention the *Sand-reckoner* in his *De expetendis et fugiendis rebus opus* (Venice, 1501).[3]

Of course, Copernicus profited indirectly from Aristarchus's work on the sizes and distances of the Sun and Moon: the basic idea of using the size of the Earth's shadow as measured during a lunar eclipse was transmitted through Hipparchus and is found, although in a drastically revised form, in Ptolemy's *Almagest*, and the diagram for this method appears in Book IV, Chapter 19 of Copernicus's *De revolutionibus*. However, the question of greatest interest is how much Copernicus might have known directly of Aristarchus, particularly concerning his heliocentric ideas.

Selection 10 reprinted from *Journal for the History of Astronomy,* vol. 16 (1985), pp. 36–42.

In the autograph manuscript of Copernicus's *De revolutionibus*, still preserved at the Jagiellonian Library in Cracow, the name of Aristarchus of Samos appears six times.[4] I shall first mention five relatively uninteresting entries in the part of his manuscript submitted for publication. In three places, Copernicus by mistake attributes to Aristarchus a value of the obliquity of the ecliptic that should have been credited to Erathosthenes. The mistake was due to his misinterpretation of the word "Archusianus" appearing in the 1515 *Almagest* (I,12, f. 9v), which was Gerhard of Cremona's attempt to render the Arabic transcription of Eratosthenes's name.[5] In a fourth citation, Copernicus originally mentioned Aristarchus in connection with the motion of precession, but before publication he correctly changed the entry to Aristyllus. The fifth citation includes Aristarchus in a list of those who believed that the year was exactly $365\frac{1}{4}$ days long, a not very helpful and probably erroneous statement.[6] Thus, the name of Aristarchus appears only four times in *De revolutionibus* as it was finally printed in 1543, and at least three of these are erroneous.

There is, however, a sixth reference to Aristarchus in the manuscript, far and away the most interesting one because it refers to the Greek astronomer's cosmology, but it was crossed out before publication. What Copernicus had written in the passage is as follows:[7]

> And if we should admit that the motion of the Sun and Moon could be demonstrated even if the Earth is fixed, then with respect to the other wandering bodies there is less agreement. It is credible that for these and similar causes (and not because of the reasons that Aristotle mentions and rejects), Philolaus believed in the mobility of the Earth and some even say that Aristarchus of Samos was of that opinion. But since such things could not be comprehended except by a keen intellect and continuing diligence, Plato does not conceal the fact that there were very few philosophers in that time who mastered the study of celestial motions.

The foregoing passage says very little about the Aristarchan cosmology. Had Copernicus known more, he surely would have been happy to mention it, since he needed all the support that he could muster for his own unorthodox views, and since he quotes with enthusiasm other possible geokineticists from Antiquity with less reputable credentials. In a curious way, Copernicus's intellectual heritage is closely rooted to the island of Samos, but to Pythagoras rather than Aristarchus; he repeatedly cites the Pythagoreans, and he knew full well their geokineticism because of Aristotle's protests in *De caelo*. In order to examine the question of what debt, if any, Copernicus owed to Aristarchus, we must inquire about the reasons for this deletion and we must examine in detail what knowledge Copernicus had about his illustrious Greek predecessor.[8]

Throughout most of his life, Copernicus worked in comparative isolation on his astronomical system. By the time he was in his sixties, he had little hope, or intention, of publishing his voluminous manuscript. However, in the spring of 1539 a young astronomer from Wittenberg arrived, eager to learn more detail of Copernicus's radical cosmology. Inflamed with enthusiasm for the new system, Georg Joachim Rheticus urged the aging Copernicus to put the finishing touches on his manuscript so that it could be taken to Germany for publication. In this process, a number of editorial changes were made in the manuscript, including the one that had eliminated the earliest reference to Aristarchus.

Originally, Copernicus had laid out his treatise in seven major sections or books. The first dealt entirely with cosmology, ending with the fleeting reference to Aristarchus and with Copernicus's own longer Latin translation of the so-called "Letter from Lysis," a brief work dealing with Pythagorean philosophy. The translation may simply have been one of those youthful exercises used by Copernicus to improve his skill in Greek. In any event, upon second thought, Copernicus decided that the "Letter from Lysis" was not really germane to his purpose, so the entire section was withdrawn and the original Book II on mathematical methods was joined directly onto Book I on cosmology.

Various suggestions have been advanced as to why Copernicus eliminated the "Letter from Lysis" (and therefore also the brief reference to Aristarchus) from his text, but it seems to me that no elaborate explanation is required. The "Letter from Lysis," emphasizing the desire of the Pythagoreans to keep their philosophy secret, may have been appropriate for a manuscript destined to gather dust on the shelves of the cathedral library, but in a printed treatise it offered an awkward and largely irrelevant digression. From an editorial viewpoint, it was eminently sensible to remove the material and to paraphrase some of the ideas in the introductory preface to the Pope. This is precisely what Copernicus did.[9]

When Copernicus reworked the omitted material into the dedicatory letter to Pope Paul III, he mentioned the "Letter from Lysis" in the first paragraph. However, at a later point where Copernicus mentions his precursors, he by then felt it would be more authoritative to quote his source in the original Greek. Thus Copernicus set down several Greek sentences from the *Opinions of the Philosophers*, a work attributed to Plutarch. In translation they read:[10]

> Some think that the Earth remains at rest. But Philolaus the Pythagorean believes that, like the Sun and Moon, it revolves around the Fire in an oblique circle. Heraclides of Pontus and Ecphantus the Pythagorean make the Earth move, not in a progressive motion, but like a wheel in a rotation from west to east about its own center.

That this passage is not really from Plutarch but from a pseudononymous work by Aetius of Antioch is of no consequence to us here—what matters is that Copernicus chose to quote this particular paragraph, probably because in the Latin edition it was conspicuously labeled "concerning the motion of the Earth."[11] Because the name of Aristarchus simply did not appear here, the direct reference to Copernicus's intellectual ancestor was quite inadvertently omitted. In fact, Aetius (Pseudo-Plutarch) had actually mentioned Aristarchus a few pages earlier in his *Opinions of the Philosophers*, but in a place where he discussed eclipses. The text says: "Aristarchus counts the Sun among the fixed stars; he has the Earth moving around the ecliptic and therefore by its inclinations he wants the Sun to be shadowed."[12] This vague and confusing passage may well have been the only hint Copernicus had about Aristarchus as an architect of a heliocentric system. Had Copernicus been able to search the literature more carefully, he might have found two other references to Aristarchus in genuine works of Plutarch, one of which clearly witnessed to the heliocentric idea in Antiquity. In the *Platonic Questions*, Query VIII, Plutarch writes:[13]

> Ought the Earth . . . . be understood to have been devised not as confined and at rest, but as turning and whirling about in the way set forth later by Aristarchus and Seleucus, by the former only as an hypothesis, but by Seleucus beyond that as a statement of fact?

In *On the Face in the Orb of the Moon*, Plutarch mentions Aristarchus of Samos three times, twice in connection with his treatise *On Sizes and Distances*, but in the third as follows:[14]

> Thereupon Lucius laughed and said: "Oh, sir, just don't bring suit against us for impiety as Cleanthes thought that the Greeks ought to charge Aristarchus the Samian with impiety on the ground that he was disturbing the hearth of the Universe because he sought to save the phenomena by assuming that the Heaven is at rest while the Earth is revolving along the ecliptic and at the same time is rotating about its own axis.

Had Copernicus known of these references, particularly the latter one, he likely would have quoted it. But there is not a shred of evidence that Copernicus knew anything about Aristarchus as a heliocentrist except for the single rather cryptic passage on eclipses in the *Opinions of the Philosophers*.

How was it that Copernicus even saw the *Opinions of the Philosophers*? It was printed in Greek in Venice in 1509 in the *Plutarchi opuscula*, a work that also contained both of the genuine Plutarchian references to Aristarchus's cosmology. Professor Edward Rosen has assumed that Copernicus consulted this dense and bulky edition, noting only the passage from the *Opinions* and overlooking the other references to Aristarchus.[15] There is, however, no firm evidence that Copernicus either owned or had access to the 1509 Greek text, nor do we know how fluent he was in reading Greek. It seems to me more likely that he would have first found the information in a Latin source, and, indeed, a 1516 Latin edition of the *Opinions of the Philosophers* with numerous convenient subheadings was available in the library of the cathedral where Copernicus worked. Copernicus himself wrote: "I first found in Cicero that Hicetas supposed the Earth to move. Later I also discovered in Plutarch that certain others were of this opinion."[16]

Perhaps by "later" Copernicus meant "after 1518," when the nine items including the Latin Pseudo-Plutarch were bound together in the volume owned by the cathedral library.[17] Because he had the idea for the heliocentric system by 1514 if not earlier, Copernicus would then have read Aetius's *Opinions of the Philosophers* only after he had firmly and independently grasped the heliocentric hypothesis. On the other hand, he could possibly have seen it even earlier in Giorgio Valla's *De expetendis et fugiendis rebus opus* of 1501. Valla used his translation of the *Opinions of the Philosophers*, including the enigmatic passage about Aristarchus, for Books XX and XXI of his own *Opus*. We know that Copernicus consulted this section of the copy of Valla's *Opus* in the Cathedral Chapter's library when he wrote about the shape of the Earth at the end of *De revolutionibus* I,3. It is rather hard to imagine, however, that those vague lines would have been the catalyst for Copernicus's revolutionary thinking.

Much later, when Copernicus was making the final revisions of his *De revolutionibus*, he could have sought the original Greek text for the passage he knew in Latin from the Basel 1531 Greek edition of the *Opinions*.[18] If this scenario is correct, then Copernicus never saw the genuine Plutarchian references to Aristarchus.

What other evidence do we have concerning how and where Copernicus arrived at his heliocentric cosmology? We simply do not know if Copernicus accepted heliocentrism as an undergraduate in Cracow in the early 1490s, as a graduate student in Italy between 1496 and 1503, or after his subsequent return to Poland. Nor is there any clear-cut path that led Copernicus to his heliocentric cosmology. Like Aristarchus's conception nearly two millennia earlier, Copernicus's cosmology was a great adventure of the mind, a mental construction not forced by any observations and in fact contrary to the immediate senses.

Space does not permit an examination of the beautiful argument, based on one of the few extant early manuscript sources from Copernicus, that the Polish astronomer arrived at his heliocentric system through a Tychonic geo-heliocentric system.[18] If, indeed, Copernicus groped his way to the heliocentric synthesis via an intermediate geocentric scheme, then it is all the more unlikely that knowledge of Aristarchus had any practical influence on his work. Copernicus, like most modern scientists, presents his finished plan and says almost nothing about the pathway to his discovery. Yet there is no question but that harmonious bonds of commensurability, linking the phenomena together "as if by a golden chain," were ultimately persuasive in Copernicus's mind: he liked the idea of placing the Sun, clearly a unique body among the planets, in a unique central place, and he was impressed by the rhythmic regularity possible with the heliocentric arrangement—with Mercury, the fastest planet, revolving nearest the Sun, with Saturn, the slowest, placed at the farthest position, and with the Earth falling in the natural sequence between Venus and Mars. The elegance of this arrangement he could not have discovered directly in any possible citations from Aristarchus. There is no question but that Aristarchus had the priority of the heliocentric idea. Yet there is no evidence that Copernicus owed him anything.[19] As far as we can tell, both the idea and its justification were found independently by Copernicus.

It is not really the task of the historian of science to assess the comparative originality of these two scientific giants. The heliocentric cosmology was convincing neither to the contemporaries of Aristarchus nor to those of Copernicus, but Copernicus had the good luck to be born not only at a time when science was beginning to reach, so to say, a critical mass, but also at a time when scientific works were beginning to be printed; therefore his arguments survived and convinced a later generation of astronomers. For better or for worse, scientific credit goes generally not so much for the originality of the concept as for the persuasiveness of the arguments. Thus, Aristarchus will undoubtedly continue to be remembered as "The Copernicus of Antiquity," rather than Copernicus as "The Aristarchus of the Renaissance."

Notes and References

[1] The famous passage in the *Sand-reckoner* reads: "But Aristarchus brought out a book consisting of certain hypotheses, wherein it appears, as a consequence of the assumptions made, that the universe is many times greater than the universe just mentioned. His hypotheses are that the fixed stars and the Sun remain unmoved, that the Earth revolves about the Sun in the circumference of a circle, the Sun lying in the middle of the orbit, and that the sphere of the fixed stars, situated about the same centre as the Sun, is so great that the circle in which he supposes the Earth to revolve bears such a proportion to the distance of the fixed stars as the centre of the sphere bears to its surface." The translation is from Sir Thomas Heath, *Aristarchus of Samos* (Oxford, 1913), p. 302.

[2] These are the three observations of Mercury by Bernard Walther, not published until 1544.

[3] Even though Valla failed to mention the *Sand-reckoner*, he did in fact own the oldest and most complete manuscript of Archimedes, namely Greek manuscript A, which is the source for our text of the *Sand-reckoner*. Furthermore, Valla had seen a translation made by Jacobus Cremonensis around 1450; Jacobus's own copy of the translation went to the Marciana in Venice in 1468, where it remains today (Marciana f.a.327). A copy of this translation was sent to Nicholas of Cusa, and Regiomontanus made another copy around 1462 which was the basis for the printed Basel edition of 1544. I am indebted to Marshall Clagett for these details, found in his *Archimedes in the Middle Ages*, vol. 3, part III, *The Medieval Archimedes in the Renaissance, 1450–1565* (Philadelphia, 1978), chap. 2.

[4] This is established with a complete word-in-context index prepared by Heribert Nobis at the Copernicus Forschungsstelle in Munich, and with the index and notes in Edward Rosen's translation, Nicholas Copernicus, *On the Revolutions* (Warsaw-Cracow, 1978). The references are found in Book III, chaps. 2, 6, and 13.

[5] See N. M. Swerdlow and O. Neugebauer, *Mathematical Astronomy in Copernicus' De Revolutionibus* (New York, 1984), note 11 on p. 133.

[6] Probably taken from Censorinus, *De die natali liber*, chap. 19; this small work appeared in at least eight collections printed during Copernicus's lifetime.

[7] Based on the translation on p. 513 in Owen Gingerich, "From Copernicus to Kepler. Heliocentrism as Model and Reality," *Proceedings of the American Philosophical Society*, vol. 117 (1973), pp. 513–22 [reprinted as selection 16 in this anthology].

[8] This task is rendered much easier by the article and notes of Rosen cited in refs. 4 and 15, and by the review by Byron Emerson Wall, "Anatomy of a Precurser, The Historiography of Aristarchos of Samos," *Studies in the History and Philosophy of Science*, vol. 6 (1975), pp. 201–28.

[9] I feel that it is unfortunate that the two most recent English translators of Copernicus's book, A. M. Duncan (1976) and Edward Rosen (1978), have disregarded his decision and have inserted the canceled "Letter from Lysis" directly into the text rather than into an appendix.

[10] Aetius Amidenus [Pseudo-Plutarch], *De placitis philosophorum* III.13; Hermann Diels, *Doxographi Graeci* (Stuttgart, 1929); translation from Rosen, *On the Revolutions*, p. 5.

[11] *De motione terrae*, fol. 20v, *De philosophorum placitis* (Strassburg, 1516); I am indebted to Jerzy Dobrzycki for access to a microfilm of the Uppsala copy of this rare edition—see below, ref. 17.

[12] *Ibid.* fol. 14, *Aristarchus solem haerentibus stellis annumerat terram vero circulum solis orbem versat itaque eius inclinationibus obumbrari solem voluit.* The Greek text (from Diels, *Doxographi Graeci*, 355) is: Ἀρίσταρχος τὸν ἥλιον ἵστησι μετὰ τῶν ἀπλανῶν, τὴν δὲ γῆν κινεῖ περὶ τὸν ἡλιακὸν κύκλον καὶ κατὰ τὰς ταύτης ἐγκλίσεις σκιάζεσθαι τὸν δίσκον.

[13] Harold Cherniss (trans.), *Plutarch's Moralia*, vol. 13 (Cambridge, Mass., 1976), pp. 77, 79.

[14] Based on Harold Cherniss and William C. Helmbold (trans.), *Plutarch's Moralia*, vol. 12 (Loeb Library, Cambridge, Mass., 1957), p. 55.

[15] Edward Rosen, "Aristarchus of Samos and Copernicus," *Bulletin of the American Society of Papyrologists*, vol. 15 (1978), pp. 85–93, esp. pp. 89–90.

[16] Rosen translation, *On the Revolutions*, p. 4.

[17] This collection of short works is still found in the Uppsala University Library—see item 42, p. 380, in Pawel Czartoryski, "The Library of Copernicus," *Science and History: Studies in Honor of Edward Rosen Studia Copernicana*, vol. 16) (Wroclaw, 1978), pp. 355–96.

[18] See Noel M. Swerdlow, "The Derivation and First Draft of Copernicus's Planetary Theory," *Proceedings of the American Philosophical Society*, vol. 117 (1973), pp. 423–512.

[19] After I presented this paper in Samos in 1980, Professor L. Biermann pointed out Eugen Brachvogel's "Nikolaus Koppernikus und Aristarch von Samos," *Zeitschrift für die Geschichte und Altertumskunde Ermlands*, vol. 35 (1935), pp. 703–67. Brachvogel's conclusions, more philosophically based, are similar: "Kein Faden, nichts, führt da von Aristarch zu Koppernikus. Was dieser an der Weite des Himmels erschaute, hat er für sich allein erspäht."

"Crisis" versus Aesthetic in the Copernican Revolution

I n a chapter in *The Structure of Scientific Revolutions* entitled "Crisis and the Emergence of Scientific Theories," Thomas Kuhn states: "If awareness of anomaly plays a role in the emergence of phenomena, it should surprise no one that a similar but more profound awareness is prerequisite to all acceptable changes of theory. On this point historical evidence is, I think, unequivocal. The state of Ptolemaic astronomy was a scandal before Copernicus's announcement."[1] A paragraph later he elaborates:

> For some time astronomers had every reason to suppose that these attempts would be as successful as those that had led to Ptolemy's system. Given a particular discrepancy, astronomers were invariably able to eliminate it by making some particular adjustment in Ptolemy's system of compounded circles. But as time went on, a man looking at the net result of the normal research effort of many astronomers could observe that astronomy's complexity was increasing far more rapidly than its accuracy and that a discrepancy corrected in one place was likely to show up in another.

The existence of an astronomical crisis facing Copernicus in the early

Selection 11 reprinted from *Vistas in Astronomy*, ed. by, A. Beer and K. Strand, vol. 17 (1975), pp. 85–95.

1500s is presupposed by one author after another; perhaps it is most vividly expressed by de Vaucouleurs, who writes:[2]

> The [Ptolemaic] system was finally overthrown as a result of the complexity which arose when an ever-increasing number of superimposed circles had to be postulated in order to represent the ever-multiplying inequalities in the planetary motions revealed by observational progress.

Nevertheless, my own researches have convinced me that this supposed crisis in astronomy is very elusive and hard to find, at least in the places where we are normally told to look. As a simple but powerful example of what I have in mind, let me cite the work of two leading ephemeris makers of the sixteenth century, Johannes Stoeffler and Johannes Stadius.

Stoeffler was born in 1452. When late in life he became professor of mathematics at Tübingen, he already enjoyed a virtual monopoly with the ephemerides prepared by himself and Jacob Pflaum; these had continued through 1531 those of Regiomontanus. At Tübingen he extended his calculations to 1551, and these were published there posthumously in 1531.[3]

At about the same time that Stoeffler died (1530), Johannes Stadius was born (1527). In the 1560s Stadius taught mathematics at Louvain, and later he worked in Paris. Stadius was the first computer to adopt the Copernican parameters for a major ephemeris.[4] His own tables were, in effect, the successors to Stoeffler's, and their users included Tycho Brahe.

In this modern age of refined planetary theory and of electronic computers, it has become possible to calculate with fair precision where the planets really were in the sixteenth century, and hence I have been able to graph the errors in the planetary positions predicted by Stoeffler and by Stadius. These error patterns are as distinctive as fingerprints and reflect the characteristics of the underlying tables. That is, the error patterns for Stoeffler are different from those of Stadius, but the error patterns of Stadius closely resemble those of Maestlin, Magini, Origanus, and others who followed the Copernican parameters (see Figures 1 and 2).[5]

The first result of this comparison is the fact that the errors reach approximately the same magnitude before and after Copernicus. In the Regiomontanus and Stoeffler ephemerides, the error in longitude for

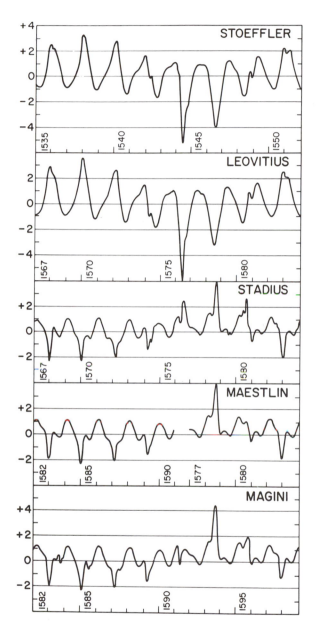

FIGURE 1. *The errors in the predicted longitude for Mars in the Alfonsine-based ephemerides of Stoeffler and Leovitius and three Copernican-based ephemerides. Some of, but not all, the typographical or obvious computation errors of Stadius have been corrected. Note the close agreement in error patterns after intervals of 15 and 32 years. (Drawn by Barbara L. Welther.)*

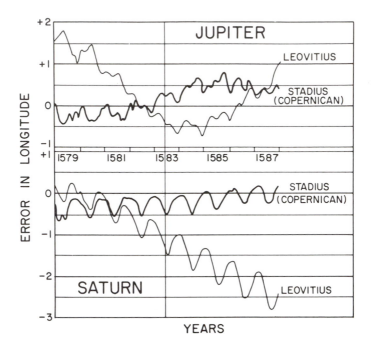

FIGURE 2. *Errors in predicted longitudes of Jupiter and Saturn near the time of their great conjunction in May 1583, a conjunction closely observed by Tycho Brahe. There is evidently much computational noise in Stadius's positions for Jupiter. (Drawn by Barbara L. Welther.)*

Mars is sometimes as large as 5°. However, in 1625, the Copernican errors for Mars reached nearly 5°, as Kepler complained in the Preface to his *Rudolphine Tables.*[6] And in Tycho's observation books, we can see occasional examples where the older scheme based on the *Alfonsine Tables* yielded better predictions than could be obtained from the Copernican *Prutenic Tables.* Now if the scandalous crisis of Ptolemaic astronomy was its failure to predict planetary positions accurately, Urania was left with nearly as much of a crisis on her hands after Copernicus.

Many simple historical accounts of the Copernican revolution emphasize not the accuracy but the simplicity of the new system, generally in contrast to the horrendous complex scheme of epicycles-upon-epicycles supposedly perpetrated by pre-Copernican astronomers. This tale reached its most bizarre heights in the 1969 *Encyclopaedia Britannica,*[7] where the article on astronomy states that by the time of Alfonso in the thirteenth century, 40 to 60 epicycles were required for each planet! More typically, we find what Robert Palter has called the "80-34 syndrome"—the claim that the simpler Copernican system required only 34 circles in contrast to the 80 supposedly needed by Ptolemy.[8] The Copernican count derives from the closing statement of his *Commentariolus*: "Altogether, therefore, 34 circles suffice to explain the entire structure of the universe and the entire ballet of the planets."[9] By the time Copernicus had refined his theory for his more mature *De revolutionibus*, he had rearranged the longitude mechanism, thereby using six fewer circles, but he had added an elaborate precession-trepidation device as well as a more complicated latitude scheme for the inner planets. Even Copernicus would have had difficulty in establishing an unambiguous final count.[10] A comparison between the Copernican and the classical Ptolemaic system is more precise if we limit the count of circles to the longitude mechanisms for the Sun (or Earth), Moon, and planets: Copernicus requires 18, Ptolemy 15.[11] Thus, the Copernican system is slightly more complicated than the original Ptolemaic system.

The 80-34 myth claims that the original simplicity of the Ptolemaic system was lost over the course of the ensuing centuries. "Theory patching was the order of the day," writes one recent author. The 80 circles presumably resulted from the piling of one epicycle on another, reminiscent of the lines

Great fleas have little fleas
    upon their backs to bite 'em,
And little fleas have lesser fleas,
    and so *ad infinitum*.

Astronomers have been fond of this view, because of the parallel between epicycles-on-epicycles and an analysis by Fourier series.[12] Nevertheless, this contrast between the simplicity of the Copernican system and the complexity of the detailed Ptolemaic mechanisms proves to be entirely fictitious.

Consider Stoeffler once more, the successor of Regiomontanus and the most successful ephemeris maker of his day. If improvements were available in a patched-up scheme of epicycles-on-epicycles, surely Stoeffler would have used them. Two extensive sets of calculations allowed me to investigate this possibility.

First, I recomputed the thirteenth-century *Alfonsine Tables*, showing that they are based on a pure Ptolemaic theory—that is, with an eccentric, equant, and single epicycle for the superior planets. The parameters were almost all identical to those originally adopted by Ptolemy, but the precessional motion had been augmented by trepidation, an improvement irrelevant to the discussion of epicycles-on-epicycles. Second, I used the *Alfonsine Tables* to generate a daily ephemeris for three centuries;[13] these positions agreed so closely with those published by Stoeffler that I am forced to conclude he used the unembellished Ptolemaic system, as transmitted through the *Alfonsine Tables*.[14]

Thus, this second result of investigating the ephemerides indicates that only a simple, classical Ptolemaic scheme was used for the prediction of planetary positions in 1500. I am convinced that the complex, highly embroidered Ptolemaic system with all the added circles is a latter-day myth. To support my view, there are at least two more good arguments, although I can mention them only in passing. First, the most sophisticated understanding of the Ptolemaic system in the fifteenth century is reflected in the tract against Cremonensis,[15] in which Regiomontanus picks faults with an anonymous medieval work, the *Theorica planetarum*. One receives the impression here that, in 1464, astronomers were once again just able to comprehend Ptolemy but scarcely able to improve on his work. Second, the astonishing, almost complete absence of recorded observations before 1450 again suggests that pre-Copernican astronomers had little basis for adding those mythical epicycles-on-epicycles. I simply cannot believe Kuhn's statement that "as time went on, a man looking at the net result of the normal research effort of many astronomers could observe that astronomy's complexity was increasing far more rapidly than its accuracy and that a discrepancy corrected in one place was likely to show up in another."

I am willing to grant that Copernicus's cosmology represents, in a certain profound sense, a simplification, but I refuse to concede that the Ptolemaic theory had by the beginning of the sixteenth century reached a complex, patched-up state nearing collapse. In terms of the detailed mechanism for any particular planet, it would have been very difficult for Copernicus's contemporaries to distinguish between the two schemes on the basis of complexity.

Where, then, is the astronomical crisis that Copernicus faced? Kuhn goes on to say:[16]

> By the early sixteenth century an increasing number of Europe's best astronomers were recognizing that the astronomical paradigm was failing in application to its own tra-

ditional problems. That recognition was prerequisite to Copernicus' rejection of the Ptolemaic paradigm and his search
for a new one. His famous preface still provides one of the
classic descriptions of a crisis state.

This preface is the last extant piece of Copernican prose, written just
before the publication of his book. A polemical passage, it attempts to
justify his radical departure from traditional cosmology and to protect
his work from future detractors. If one believes astronomy was at the
point of crisis, then it is perhaps possible to read it as a classic description of a crisis state.

On the other hand, I believe that an alternative reading is preferable.
After criticizing the alternative system of homocentric spheres and,
indirectly, Ptolemy's equant, Copernicus says:[17]

> Nor have they been able thereby to discern or deduce the
> principal thing—namely the design of the universe and the
> fixed symmetry of its parts. With them it is as though one
> were to gather various hands, feet, head and other members,
> each part excellently drawn, but not related to a single body,
> and since they in no way match each other, the result would
> be monster rather than man.

This "fixed symmetry of its parts" refers to the fact that, unlike in the
Ptolemaic scheme, the relative sizes of the planetary orbits in the
Copernican system are fixed with respect to each other and can no
longer be independently scaled in size. This is certainly one of the most
striking unifications brought about by the Copernican system—what I
would call a profound simplification. Clearly, this interlinking makes
the unified man, and in contrast the individual pieces of Ptolemy's
arrangement become a monster.

What has struck Copernicus is a new cosmological vision, a grand
aesthetic view of the structure of the Universe. If this is a response to
a crisis, the crisis had existed since A.D. 150. Kuhn has written that the
astronomical tradition Copernicus inherited "had finally created a
monster," but the cosmological monster had been created by Ptolemy
himself.

In this view, there is no particular astronomical reason why the
heliocentric cosmology could not have been defended centuries earlier,
and it is in fact shocking that Copernicus, with the accumulated experience of fourteen more centuries, did not come up with a substantial
advance in predictive technique over the well-honed mechanisms of

Ptolemy. The debased positivism that has so thoroughly penetrated our philosophical framework urges us to look to data as the foundation of a scientific theory, but Copernicus's radical cosmology came forth not from new observations but from insight. It was, like Einstein's revolution four centuries later, motivated by the passionate search for symmetries and an aesthetic structure of the universe. Only afterward the facts, and even the crisis, are marshalled in support of the new world view.[18]

But why, if all this is true, did a Copernicus come in the sixteenth century, and not in the fourteenth or even the tenth century? Were the astronomical questions in Cracow in 1492 particularly conducive to challenging the old order? I have no doubt but that the growing problems of precession, trepidation, and the motion of the eighth sphere acted as a spur to Copernicus's thinking about astronomy. His attack on this problem demonstrates his unusual level of technical ability, which had certainly been rare in the Middle Ages. Copernicus's examination of precession may have led him to consider a moving Earth.[19] Nevertheless, the heliocentric system is scarcely a necessary consequence of the observation of precession.

No, I believe that it was something outside astronomy in the European intellectual climate in the sixteenth century that set the stage for the introduction of a new paradigm; as Professor Benjamin Nelson put it in an earlier paper in this symposium—it had something to do with "societies, communities, and communications." In his words, the flowering of new world views must be considered within the context of complex sociocultural structures. The sixteenth century was manifestly an age of change. While Copernicus was a student at Cracow, Columbus set sail across an unknown ocean. The new explorations made Ptolemy's time-honored geography obsolete. Discoveries of classical authors brought in a new humanism with fresh Neoplatonic ideals. Even the traditional authority of the Church was to crumble before the challenge of Luther and the reformers.

A powerful catalyst for these changes was the explosive proliferation of printing.[20] As a student in Cracow, Copernicus could secure and annotate his own printed set of *Alfonsine Tables* as well as Regiomontanus's *Ephemerides*. Later, probably in Italy, he obtained Regiomontanus's *Epitome of Ptolemy's Almagest*; the close paraphrases of many of its passages in the *De revolutionibus* show the formative role this book played in his researches. Still later, the first full printed *Almagest* of 1515 provided another useful source of data.[21] Ultimately, it was the printed edition of his *De revolutionibus* that prevented his ideas from falling into oblivion.

In many ways, the world was ready for an innovative view of the cosmos. Copernicus, with both the intellect and the leisure to fashion a new cosmology, arrived on the scene at the very moment when the increased flow of information could both bring him the raw materials for his theory and rapidly disseminate his own ideas. An imaginative thinker striving to uncover fresh harmonies in the universe, he also achieved the technical proficiency to command respect for his mathematics and his planetary tables. One can easily argue that Copernicus was not the equal of Ptolemy or of Kepler in mathematics, although for his day he stood well above his contemporaries. Yet as a sensitive visionary who precipitated a scientific revolution, Copernicus stands as a cosmological genius with few equals. In celebrating his birth, we celebrate the man who, perhaps unwittingly, is the founder of modern science.

Notes and References

[1] Thomas S. Kuhn, *The Structure of Scientific Revolutions* (Chicago, 1962), pp. 67–68.

[2] Gérard de Vaucouleurs, *Discovery of the Universe* (London, 1957), pp. 32–33.

[3] In 1474 in Nuremberg, Regiomontanus printed his own ephemerides for 1475 through 1506, and these were reissued by various printers, including Ratdolt in Venice. Stoeffler and Pflaum issued their ephemerides in Ulm in 1499 for the years 1499 to 1531, with the title *Almanach nova plurimis annis venturis inservientia*, and these were repeatedly reissued by Liechtenstein in Venice. I have not yet ascertained if they recalculated the overlapping period from 1499 to 1506. Stoeffler's 1531 edition in Tübingen, with the title *Ephemeridum opus*, was edited by the successor to his professorial chair, Phillip Imsser; these tables were also promptly reprinted by Liechtenstein in Venice.

Edward Sherburne gives a charming account of the death of Stoeffler in the biographical appendix to his *The Sphere of Marcus Manilius* (London, 1675), p. 46:

> "His death, or the occasion thereof at least, was very remarkable (if the Story be True). Having found by calculation, that upon a certain Day his life was like to be endangered by some ruinous accident, and the day being come, to divert his thoughts from the apprehension of the danger threatening him, he invites some Friends of his into his Study, where, after discourse, enticing into some dispute, he, to decide the controversie reaches for a Book, but the Shelf on which it stood being loose came down with all the Books upon him, and with its fall so bruised him, that he died soon after of the hurt, Voss. in Addend. ad Scient. Mathemat. But the whole Story of his Death, of which some make Calvisius the Author, is false by the Testimony of Jo. Rudolphus Camerarius Gen-

itur. 69. Centur. 2. who had it from Andraas Ruttellius his Audi-
tour; for he died of the Plague at Blabira Feb. 16. 1531 in the 78th
year of his Age, happening (according to Calculation if you will
believe it) from the Direction of $\odot$ to σ."

[4] Copernicus's own almanac was never printed and is now lost (see Edward Rosen,
"Nicholas Copernicus, a Biography," in his *Three Copernican Treatises*, 3rd ed.,
New York, 1971, pp. 374–75). Rheticus published an ephemeris for a single year,
1551, based on the tables in *De revolutionibus*. E. Reinhold published an ephemeris
for 1550 and 1551, using his Copernican-based *Prutenicae tabulae* (Tübingen,
1551); subsequent workers generally adopted Reinhold's tables as their avenue to
the Copernican parameters. Stadius's *Ephemerides novae* (Cologne, 1556) included
predictions for 1554–70, and later editions carried the tables through 1600. A post-
humous edition went to 1606, but the additional years were probably appended by
the publisher from the Alfonsine-based ephemerides of Leovitius. Stadius published
his own planetary tables, *Tabulae Bergenses aequabilis et apparentis motus orbium
coelestium* (Cologne, 1560), but these were essentially a plagiarism of the *Prutenicae tabulae*. Lynn Thorndike (*A History of Magic and Experimental Science*, vol.
5, New York, 1941, pp. 303–4) quotes Tycho Brahe's estimate of Stadius as having
been "more facile than accurate," an opinion apparently shared by Maestlin and
Magini, who eventually produced major alternative ephemerides of their own.

[5] The ephemerides used for the figures are Johannes Stoeffler, *op. cit.*; Cyprian Leowitz, *Ephemeridum novum atque insigne opus ab anno 1556 usque in 1606 accuratissime supputatum* (Augsburg, 1557); Johannes Stadius, *Ephemerides novae et epac-
tae ab anno 1554 ad annum 1600* (Cologne, 1570); Michael Maestlin, *Ephemerides
novae ex tabulis Prutenices anno 1577 ad annum 1590 supputatae* (Tübingen, 1580);
G. A. Magini, *Ephemerides coelestium motuum secundum Copernici observationes
supputatae* (Venice, 1582). The comparisons were made against the computed longitudes in Bryant Tuckerman, *Planetary, Lunar, and Solar Positions A.D. 2 to A.D.
1649, Memoirs of the American Philosophical Society*, vol. 59 (Philadelphia, 1964).
The figure was prepared by Barbara L. Welther.

Additional error graphs from sixteenth- and seventeenth-century ephemerides can
be found in Owen Gingerich, "The Mercury Theory from Antiquity to Kepler,"
Actes du XIIᵉ Congrès International d'Histoire des Sciences, vol. 3 A (1971) pp.
57–64 [reprinted as selection 23 in this anthology], and "Kepler's Place in Astronomy," *Vistas in Astronomy*, ed. by A. and P. Beer, vol. 18, 1974, pp. 261–78 [re-
printed as selection 19 in this anthology].

[6] "Johannes Kepler: Preface to the Rudolphine Tables," translated by Owen Gingerich and William Walderman, *Quarterly Journal of the Royal Astronomical Society*,
vol. 13 (1972), pp. 360–73, see especially p. 367.

Tycho frequently compared his own observations to the predictions from the
Alfonsine and Copernican tables, usually to the advantage of Copernicus. A particularly favorable comparison occurred at the time of the great conjunction of Jupiter
and Saturn in 1583 (see Figure 12), although by 20 August 1584, Tycho's comparison for Jupiter showed the two schemes equally in error, and by 21 December 1586,
the Alfonsine calculation was decidedly better, especially in latitude. Frequently, the
Copernican latitudes proved inferior, even when the longitude excelled—for exam-

ple, for Saturn on 24 January 1595. Tycho compared lunar positions in December 1594, and toward the end of the month the Alfonsine-based Leovitius ephemeris was superior. The most conspicuous Copernican errors found by Tycho occurred during the August opposition of Mars in 1593, exceeding 5°; this configuration repeated in 1625 when Kepler noted the large errors during the particularly close approach of Mars. Tycho's investigations are published in J. L. E. Dreyer (ed.), *Tychonis Brahe Dani Opera Omnia*, vols. 10–13 (Copenhagen, 1923–6.

[7] "Astronomy. I. History of astronomy. B. Mediaeval astronomy," *Encyclopaedia Britannica*, vol. 2 (Chicago, 1969); p. 645:

> "King Alfonso X of Castile kept a number of scholars occupied for ten years constructing tables (the Alphonsine tables, c. 1270) for predicting positions of the planetary bodies. By this time each planet had been provided with from 40 to 60 epicycles to represent after a fashion its complex movement among the stars. Amazed at the difficulty of the project, Alfonso is credited with the remark that had he been present at the Creation he might have given excellent advice. After surviving for more than a millennium, the Ptolemaic system had failed; its geometrical clockwork had become unbelievably cumbersome and without satisfactory improvements in its effectiveness."

[8] Robert Palter, "An Approach to the History of Early Astronomy," *History and Philosophy of Science*, vol. 1 (1970), pp. 93–133. Palter traces the 80-34 myth back as far as Arthur Berry's *A Short History of Astronomy* (London, 1898).

[9] Edward Rosen, "Nicholas Copernicus, a Biography," in his *Three Copernican Treatises*, 3rd ed. (New York, 1971), p. 90.

[10] According to Ernst Zinner, *Entstehung und Ausbreitung der Coppernicanischen Lehre* (Erlangen, 1943), pp. 186–87, Copernicus should have included precession, the regression of the lunar nodes, and the change of solar distance in his count in the *Commentariolus*, thus getting a total of 38 circles. Arthur Koestler, in *The Sleepwalkers* (London, 1959), pp. 572–73, attempted to count the circles in *De revolutionibus*, but he overlooked the fact that Copernicus had by then replaced the so-called Ṭūsī couple in the longitude mechanisms by an eccentric, thereby listing at least six unnecessary circles; on the other hand, he could have claimed that the motion of the apsidal lines for Mercury and the superior planets each required a circle.

[11] Copernicus replaced the Ptolemaic mechanism for varying the size of Mercury's orbit with a Ṭūsī couple, and he also accounted for the apsidal motion of the Earth's orbit with two circles. If the apsidal motions for Mercury and the superior planets are counted, then Copernicus required 22 circles for the motions in longitude.

[12] A letter to *Physics Today*, vol. 24, no. 12, p. 11 (December, 1971) remarked that 400 years ago *The Physical Review* might have been full of such papers as "A Ten Epicycle Fit to the Orbit of Mars," and a review article on radio galaxies in *The Astronomical Journal*, vol. 77 (1972), p. 541, summarized with "The question is, 'Are we drawing too many epicycles?' "

[13] E. Poulle and O. Gingerich, "Les positions des planètes au moyen âge: application

du calcul électronique aux tables Alphonsines," *Académie des inscriptions et belles lettres comptes rendu des séances* (1968), pp. 531–48.

[14] Recently, I found in the Badische Landesbibliothek in Karlsruhe what I believe to be Stoeffler's personal manuscript copy of these tables, which he may have used in calculating his ephemerides. It is Codex Ettenheim-Münster 33, 93r–198r. I wish to thank the director, Dr. Kurt Hannemann, for showing me this manuscript. See Karl Preisendanz, *Die Handschriften des Klosters Ettenheim-Münster*, vol. 9 in *Die Handschriften der Badischen Landesbibliothek in Karlsruhe* (Wiesbaden, 1932).

[15] Johannes Regiomontanus, *Disputationes contra Cremonensia deliramenta* (Nuremberg, 1474 or 1475). According to Ernst Zinner, *Leben und Wirken des Joh. Müller von Königsberg*, 2nd ed. (Osnabrück, 1968), p. 335, Regiomontanus wrote the tract in August 1464.

[16] Kuhn, *op. cit.*, p. 69.

[17] N. Copernicus, *De revolutionibus orbium coelestium*, fol. iii(v) (Nuremberg, 1543). Edward Rosen suggests for "certain symmetriam" the term "true symmetry." I believed that "fixed" conveys a slightly better nuance in this context, but now I would tranlate it "sure commensurability."

[18] See Gerald Holton, "Einstein, Michelson, and the 'Crucial Experiment'," *Isis*, vol. 62 (1969) pp. 133–97.

[19] J. R. Ravetz, in *Astronomy and Cosmology in the Achievement of Nicolaus Copernicus* (Wroclaw, 1965), argues that studies of precession may have led to the Copernican cosmology. L. Birkenmajer, in *Mikolaj Kopernik* (Cracow, 1900), suggested that the deficiencies in the Ptolemaic lunar model may have started Copernicus on the road to the heliocentric system. Important as these may have been in the development of Copernicus's technical proficiency, there is no convincing argument that these studies would have led to a Sun-centered cosmology.

[20] See Owen Gingerich, "Copernicus and the Impact of Printing," *Vistas in Astronomy*, vol. 17 (1975), pp. 201–20. See also E. L. Eisenstein, "The Advent of Printing and the Problem of the Renaissance," *Past and Present*, no. 45 (1972), pp. 19–89.

[21] A detailed discussion of Copernicus's use of these books is found in L. Birkenmajer, *Mikolaj Kopernik* (Cracow, 1900). A useful list of books owned by, or available to, Copernicus is found in L. Jarzębowski's *Biblioteka Mikolaja Kopernika* (Toruń, 1971). An earlier list of the Copernican books now found in Sweden is E. Barwiński, L. Birkenmajer, and J. Łos, *Sprawozdanie z Poszukiwań w Szwecyi* (Cracow, 1914), pp. 94–119. (Since this article was written, an even better reference by P. Czartoryski is available—see ref. 17 of the preceding selection.)

Early Copernican Ephemerides

The earliest Copernican ephemerides are particularly interesting because they predate the publication of Erasmus Reinhold's *Prutenicae tabulae coelestium motuum* in 1551. Reinhold's tables, based on those in *De revolutionibus*, rapidly became the accepted foundation for planetary predictions, and after 1551 no printed ephemeris went back directly to Copernicus's tables.[1] Reinhold's ephemerides for 1550 and 1551 were calculated from his *Prutenicae tabulae*, but in contrast, Georg Joachim Rheticus computed his ephemeris for 1551 directly from the clumsier tables provided in *De revolutionibus*. Thus the comparison of the Reinhold and Rheticus predictions for 1551 gives a unique opportunity to see how the original Copernican predictions differ from those of the *Prutenicae tabulae*.

The *Prutenicae tabulae* and Reinhold's *Ephemerides*

Reinhold's *Prutenicae tabulae* offers a straightforward recipe for the computation of solar, lunar, and planetary positions. All of the necessary starting numbers (radices), tables of mean motion, and tables of corrections are readily at hand in this compendium. For an electronic recomputation, however, Reinhold's tables were bypassed and each position was calculated directly from Reinhold's parameters and the underlying theory. Because of the ambiguities of interpolation, positions generated by hand from the tables will sometimes round off differently by 1′ compared to the direct and more accurate machine

Selection 12 reprinted from *Science and History: Studies in Honor of Edward Rosen, Studia Copernicana*, vol. 16 (1978), pp. 403–17.

procedure. (On rare occasions when an inferior planet is moving very rapidly, discrepancies of 2' or more can arise.)

Reinhold's *Ephemerides duorum annorum 50. et 51. supputatae ex novis tabulis astronomicis* (Tübingen, 1550) confirms the foregoing statements. My recomputation offers completely convincing proof that Reinhold calculated his ephemeris from his new and as yet unpublished *Prutenicae tabulae*. The vast majority of the positions agree within 1', even the rapidly moving Moon. This substantiates both the computational procedure and the difference in longitude of 47 minutes of time between Königsberg (the meridian of the *Prutenicae tabulae*) and Wittenberg (the meridian of the ephemerides), a value adopted from the *Prutenicae tabulae*. It should be noted that these ephemerides are computed for noon at Wittenberg, whereas the *Prutenicae tabulae* are for midnight at Königsberg.

The Copernican Tables and Rheticus's Ephemerides

In the same year Reinhold offered his ephemerides, Rheticus published his own *Ephemerides novae seu expositio positus diurni siderum et συσχηματισμῶν praecipuorum ad annum redemtoris nostri Jesu Christi Filii Dei MDLI* (Leipzig, 1550). Were Reinhold and Rheticus so out of touch by this time that they were unaware of this duplication of work? Or was there a deliberate rivalry between them? In any event, if there are substantial differences in the positions between the *Prutenicae tabulae* and *De revolutionibus*, we might expect their ephemerides to differ, and specifically we would expect the ephemeris of Rheticus to follow more closely the original Copernican tables.

An astronomer wishing to compute planetary positions from the tables in *De revolutionibus* would have found the difficulty of operation in marked contrast to the ease of the *Prutenicae tabulae*. A number of essential quantities are not readily found; the radices, for example, are well hidden in the text. At least some of the early owners of *De revolutionibus* who wished to calculate planetary positions left telltale tracks behind in the marginal annotations of the tables: they added the relevant radices at the head of each of the mean motion tables. *De revolutionibus* offered a second hurdle by giving only raw data on the motions of the apsidal lines, omitting any tables or standard computational procedure, a deficiency fully corrected by Reinhold in the *Prutenicae tabulae*. This feature in particular makes the Copernican calculation of positions somewhat ambiguous. Beyond this, Copernicus's printed tables sometimes differ from the ones in the manuscript, a point discussed in the appendix to this paper.

The computer immediately reveals clearly discernible differences between the predictions of the *Prutenicae tabulae* and those from *De revolutionibus*; these are most marked for Mercury, where the differences can exceed a degree. For the Sun, the agreement is always better than 1'; for the Moon and superior planets the differences seldom exceed 5', but for Venus they occasionally reach 20'. It is perhaps not surprising that the inferior planets have the largest disagreement, for the autograph manuscript of *De revolutionibus* shows that at precisely this point Copernicus had made frequent revisions; if he felt unsatisfied about any part of his work, it must surely have been here (and no doubt also in the ensuing latitude theory).

The ephemeris of Rheticus does not agree precisely with the Copernican computation, although his numbers are generally fairly close. The degree to which Rheticus followed Copernicus is well shown on two dates when Mercury is particularly discrepant with the two systems:

1551	*March 21*	*July 19*
Copernicus (computer)	11° 24'	130° 7'
Rheticus, *Ephemerides*	11 25	130 11
Prutenicae tabulae (computer)	10 23	129 2
Reinhold, *Ephemerides*	10 28	129 6

The Actual Errors of the "Ephemerides"

Whether the differences between the Copernican and Prutenic calculations are significant depends on the actual accuracy of the predictions. It is nowadays relatively easy to calculate the positions of the Sun and planets in the sixteenth century to an accuracy of 0.01. To find the position of the Sun, Moon, and all five bright planets at a given moment requires the evaluation of approximately 500 sines and cosines, and can be done on a CDC 6400 computer in about a quarter of a second. In this research I have used the subroutines kindly provided by Professor Peter Huber. Huber generally followed the computational procedure adopted by Bryant Tuckerman for his volumes on planetary, lunar, and solar longitudes,[2] but in several cases he made changes to improve the accuracy. The direct computer procedure eliminates the numerical errors that occasionally result from the tedious interpolation and differencing with the Tuckerman Tables.

Table 1 gives the results of the comparison with these modern val-

TABLE 1. "Observed" (Modern Computed) versus Computed Positions

1551	Table	Sun	Δ	Moon	Δ	Saturn	Δ	Jupiter	Δ	Mars	Δ	Venus	Δ	Mercury	Δ
Jan. 10	Modern	299° 43'		339° 24'		311° 1'		81° 48'		91° 16'		285° 48'		283° 14'	
	Copernicus	299 27	16'	338 49	35'	311 18	−17'	81 43	5'	92 56	−1° 40'	285 32	16'	284 17	−1° 3'
	Rheticus	299 25	18	338 54	30	311 18	−17	81 43	5	92 55	−1 39	285 27	21	284 10	−0 56
	Prutenicae	299 26	17	338 47	37	311 21	−20	81 51	−3	92 53	−1 37	285 36	12	284 38	−1 24
	Reinhold	299 27	16	338 46	38	311 21	−20	81 51	−3	92 53	−1 37	285 35	13	284 37	−1 23
Mar. 31	Modern	19 39		317 50		319 41		85 7		106 34		25 32		1 58	
	Copernicus	19 25	14	318 20	−30	320 9	−28	84 55	12	106 43	−9	25 10	22	2 14	−16
	Rheticus	19 23	16	318 30ª	−40	320 8	−27	84 53	14	106 44	−10	25 5	27	2 13	−15
	Prutenicae	19 25	14	318 18	−28	320 13	−32	85 3	4	106 44	−10	25 14	18	1 28	30
	Reinhold	19 26	13	318 17	−27	320 14	−33	85 2	5	106 43	−9	25 13	19	1 28	30
June 29	Modern	106 6		54 22		320 48		103 17		155 18		135 28		132 43	
	Copernicus	106 4	2	55 10	−48	321 20	−32	103 12	5	155 29	−11	134 25	1° 3	132 25	18
	Rheticus	106 0	6	55 12	−50	321 21	−33	103 9	8	155 29	−11	134 21	1 7	132 15	28
	Prutenicae	106 4	2	55 9	−47	321 25	−37	103 15	2	155 30	−12	134 28	1 0	132 24	19
	Reinhold	106 4	2	55 8	−46	321 26	−38	103 15	2	155 29	−11	134 28	1 0	132 23	20
Sep. 27	Modern	193 16		158 57		315 20		120 34		213 33		239 37		206 22	
	Copernicus	193 12	4	159 17	−20	315 29	−9	120 31	3	214 1	−28	239 9	28	206 17	5
	Rheticus	193 8	8	159 25	−28	315 31	−11	120 32	2	214 0	−27	239 3	34	206 12	10
	Prutenicae	193 12	4	159 13	−16	315 34	−14	120 32	2	214 1	−28	239 10	27	206 38	−16
	Reinhold	193 11	5	159 12	−15	315 34	−14	120 31	3	213 59	−26	239 10	27	206 39	−17
Dec. 26	Modern	284 11		273 48		319 45		120 24		279 30		277 14		270 44	
	Copernicus	283 56	15	274 6	−18	319 57	−12	120 25	−1	279 13	17	281 15	−4 1	271 45	1
	Rheticus	283 52	19	274 15	−27	319 57	−12	120 27	−3	279 12	18	281 2	−3 48	271 36	52
	Prutenicae	283 56	15	274 3	−15	320 1	−16	120 28	−4	279 11	19	280 50	−3 36	272 5	21
	Reinhold	283 56	15	274 2	−14	320 1	−16	120 28	−4	279 10	20	280 53	−3 39	272 4	20

ªRheticus's ephemerides give 318°50' for the lunar position on March 31, clearly a typographical error.

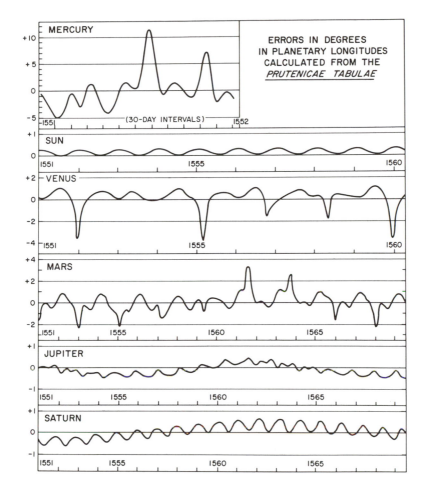

FIGURE 1. *Errors in degrees in planetary longitudes calculated from the Prutenicae tabulae.*

ues for selected dates in 1551. Figure 1 shows the errors in the *Prutenicae tabulae* over an extended period.

The solar positions contain both an annual periodic error of 16′, arising from an error of 30° in the position of the apsidal line and a small error in the eccentricity and a linear displacement error that increases with time; the latter stems from a mean motion that is slightly too small. The differences between Copernicus's and Reinhold's parameters are completely negligible compared to the actual errors for the solar predictions.

For the Moon we would expect to find errors of ±40' because Copernicus (as well as Ptolemy) did not account for the so-called "variation" in his lunar model. The maximum errors found during 1551 occasionally exceed 50' and average just under 30' (based on 36 positions at 10-day intervals). The Copernican and Prutenic calculations differ insignificantly.

As might be expected, among the superior planets both the discrepancies and the absolute errors were largest for Mars and least for Saturn. Positions computed at 100-day intervals for 30 years showed that the results from the *Prutenicae tabulae* and *De revolutionibus* tended to agree within 10'. Nevertheless, there were spectacular exceptions; for example, at the Martian opposition in the autumn of 1563, the discrepancy between the two procedures exceeded 20' while the actual error exceeded 2°.

Reinhold's and Copernicus's calculations for Venus throughout 1551 showed differences of only a few minutes until Venus approached inferior conjunction at the end of the year. In December the differences exceeded 20' and the absolute errors approached 4°, as may be seen in Table 1. Calculations made at 50-day intervals for subsequent years suggest that December of 1551 was the worst possible case. Of all the planets, Mercury had both the largest discrepancies and the largest actual errors, reaching nearly 10° in July of 1551. These large errors generally fall when Mercury is invisible, as I have demonstrated elsewhere.[3]

Agreement with Copernicus's Own Observations

Erasmus Reinhold, in the dedication to his *Prutenicae tabulae*, praised Copernicus as "the most learned man whom we may call a second Atlas or a second Ptolemy," but nevertheless criticized the astronomical tables in *De revolutionibus*, saying "the computation is not even in agreement with his observations on which the foundation of the work rests." Reinhold's statement suggests that it would be useful to compare the Copernican observations not only with his retroactive predictions, but also with the actual positions. The results for the 16 "modern" planetary positions reported by Copernicus are shown in Table 2.

Copernicus used three oppositions for each superior planet and three elongations of Mercury in order to derive the positions of the apsidal lines, the eccentricities, and the size of the epicyclet. In each case one of these observations was used together with an ancient ob-

TABLE 2. Copernicus's Planetary Observations in *De revolutionibus*

Date, Frauenburg Time		Precessed Position (Copernicus observed)	Computed (Copernicus)	Δ o−c	Computed (Reinhold)	Δ o−c	Actual Position (computed)	Δ o−c	Reference (De revolutionibus)
Saturn									
1514 May 5	10:48 P.M.	232° 38'	232° 36'	+2'	232° 41'	−3'	232° 18'	+20'	V, 6
1520 July 13	12:00 noon	300 43	300 37	+6	300 42	+1	300 20	+23	V, 6
1527 Oct. 10	6:24 A.M.	27 29	27 26	+3	27 29	0	27 33	−4	V, 6
1514 Feb. 24	5:00 A.M.	236 14	236 8	+6	236 13	+1	235 43	+31	V, 9
Jupiter									
1520 Apr. 30	11:00 A.M.	227 46	227 43	+3	227 46	0	226 55	+51	V, 11
1526 Nov. 28	3:00 A.M.	75 55	75 50	+5	75 57	−2	76 6	−11	V, 11
1529 Feb. 1	7:00 P.M.	141 7	141 4	+3	141 8	−1	140 42	+25	V, 11
1520 Feb. 18	6:00 A.M.	232 27	232 26	+1	232 26	+1	231 55	+32	V, 14
Mars									
1512 June 5	1:00 A.M.	262 46	262 55	−9	262 48	−2	262 26	+20	V, 16
1518 Dec. 12	8:00 P.M.	90 19	90 28	−9	90 22	−3	89 2	+1° 17	V, 16
1523 Feb. 22	5:00 A.M.	160 40	160 41	−1	160 44	−4	158 23	+2 17	V, 16
1512 Jan. 1	6:00 A.M.	218 41	218 37	+4	218 37	+4	218 38	+2	V, 19
Venus									
1529 Mar. 12	7:30 P.M.	36 35	36 38	−3	36 41	−6	36 59	−24	V, 23
Mercury									
1491 Sept. 9	5:00 A.M.	163 30 [a]	163 10	+20	163 30	0	163 33	−3	V, 30
1504 Jan. 9	6:30 A.M.	273 20	273 23	−3	273 25	−5	273 35	−15	V, 30
1504 Mar. 18	7:30 P.M.	26 55	26 58	−28	27 00	−30	26 31	−1	V, 30

[a]Bernard Walther's original observations of Mercury, 163°23', 273°15', and 26°30', were modified by Copernicus as shown.

servation to confirm the mean motion. For each superior planet he gave a fourth observation to establish the relative size of its orbit. For Venus, because in his day the rhythmic repetition of Venus's positions always put it in configurations that prevented the derivation of a modern apsidal line, he omitted a new determination of that parameter, but he used a single modern observation to check the eccentricity and ratio of orbital sizes, and to derive the mean motion.

For all the planets except Mercury, Copernicus achieved his goal of representing the observations within 10′, as may be seen in the first difference column, where the observations are compared with the positions computed from the tables of *De revolutionibus*. Even for Mercury the errors are not too bad, ±20′.

In the second column of differences we can inspect the modest improvement effected with Reinhold's *Prutenicae tabulae*. In fact, Reinhold's results are worse in nearly as many cases as he improves, and his largest error, 30′ for Mercury, considerably exceeds Copernicus's. Although it is clear in many cases that Reinhold was the more fastidious calculator, he was deluding himself if he thought he had made any significant improvement over Copernicus's prediction of planetary positions.

More surprising, however, is the comparison of Copernicus's data with the actual planetary positions. With the exception of Mercury, where he used Bernard Walther's observations, his data are always far poorer than his success in fitting it. For Mars two of the three reported oppositions are extremely bad, with errors of 1° and 2°, respectively. All three Martian oppositions are tersely reported in *De revolutionibus*, and the manuscript does not give any clues as to the sources of error. Precisely when and where an opposition occurs depends on the solar theory, for one must know the position of the sun as well as the position of the planet. For Copernicus, as well as for Ptolemy, it was necessary to reckon the position of the fictitious mean Sun and not the true Sun; this is related to the circumstance that Copernicus used the center of the Earth's orbit, rather than the physical Sun, as the center for his system. Hence, all these "observations" are at least one step removed from the actual sighting of the planet in the sky.

Although Copernicus matched the parameters of his planetary models very successfully to the chosen data, he was apparently unaware of the poor quality of his observational base. During Copernicus's lifetime Bernard Walther had made numerous planetary observations besides those of Mercury that Copernicus borrowed; an examination of Walther's many Mars observations shows that with respect to the starry background (that is, leaving aside the problems of precession and the coordinate framework) he achieved a consistency of 4' (root-mean-square error).[4] Evidently Copernicus (and Reinhold as well) failed to test the theory against other observations. As I have stated elsewhere, "This, in turn, suggests that the entire exercise was carried out primarily to show that the heliocentric cosmology was compatible with reasonable planetary predictions rather than to reform the accuracy of astronomical predictions. The evidence illuminates the mentality of a gifted theoretician for whom the observational foundations of his science held only a secondary interest."[5]

Hilary of Wiślica's "Ephemeris"

A possible Copernican ephemeris, even earlier than those of Reinhold and Rheticus, was never printed but is found as a manuscript in the Jagellonian Library (MS 608, pp. 125–151); it bears the following title page: ΕΦΗΜΕΡΙΞ *pro anno Domini. 1.5.4.9. ex tabulis Nicolai Copernici, pro finitore Cracoviensi per Hilarium a Wislicza Artium Ingenuarum Magistrum, Astronomiae Professorem Ordinarium, diligenter supputata.* (A later hand has altered "finitore" to "meridiano".) Ostensibly a Copernican ephemeris, it has attracted passing notice from L. A. Birkenmajer and E. Zinner.[6] A. Birkenmajer, in carefully examining Hilary's credentials, concluded that there was no other evidence that he had any real appreciation of, or sympathy for, Copernicus.[7]

Perhaps not surprisingly, Hilary's ephemeris is strictly Ptolemaic, based on the *Alfonsine Tables*, and his planetary positions have nothing to do with Copernicus. Because the *Alfonsine Tables* and *Prutenicae tabulae* show their own distinctive error patterns, positions derived from one scheme will differ rather noticeably from those of the other, and thus even an elementary comparison of the positions on a few dates will reveal the basis of the calculation. Table 3 shows positions from Stoeffler's Alfonsine-based ephemeris, from Hilary's, and from my own computer-generated Copernican positions.[8]

The positions are given for local noon; Stoeffler's meridian is Tübingen, approximately one hour farther west than Cracow or Königsberg. In that time interval the Moon advances approximately 30' and the Sun about $2\frac{1}{2}'$, so Hilary evidently assumed an incorrect longitude difference between Cracow and Toledo (the meridian of the *Alfonsine Tables*). With the assumption of a 24-minute longitude difference between Hilary's and Stoeffler's meridians, the numbers are in almost

TABLE 3. Planetary Longitudes

1549		Sun	Moon	Jupiter	Saturn	Mars	Venus	Mercury
Jan. 1	Stoeffler	21° 24′	16° 52′	19° 36′	12° 48′	20° 27′	4° 49′	28° 9′
	Hilary	21 24	16 40	19 35	12 48	20 27	4 49	28 9
	[Copernicus]	20 46	15 54	20 4	11 49	21 5	5 9	19 52
June 1	Stoeffler	19 56	6 50	27 46	14 25	4 31	5 29	29 24
	Hilary	19 55	6 39	27 48	14 25	4 31	5 29	29 33
	[Copernicus]	19 45	6 2	28 21	13 35	5 51	4 15	30 42
Dec. 31	Stoeffler	20 8	12 5	29 37	16 47	28 57	28 25	29 16
	Hilary	20 7	11 52	29 37	16 47	28 57	28 24	29 15
	[Copernicus]	19 30	11 5	29 57	15 45	27 59	27 41	27 48

(For brevity in tabulation the zodiacal sign has been ignored.)

perfect agreement, and the fact that Hilary of Wiślica's ephemerides were based on the *Alfonsine Tables* is established beyond any doubt.

Postscript

After preparing the foregoing article, I stumbled upon a manuscript that, for a fleeting moment, promised to be another early Copernican ephemeris. It is part of a notebook of quotations about Copernicus gleaned from various printed sources by Jan Brożek (1585–1652), professor of astronomy at Cracow; the manuscript is now found in the Schönbornsche Bibliothek, Schloss Weissenstein, Pommersfelden. After a passage copied from Magini, Brożek has written, "Has ephemerides ego habeo, post mortem clarissimi D. Doctoris Stanislai Jacobeii Curzeloviensis, praeceptoris mei clarissimi." There follow 26 pages of tables neatly headed "Canones mediorum seu Equalium motuum in annis aggregatis Julianis," immediately after which in a different ink (but possibly Brożek's hand) appears "per D. D. Sylvestrum Roguski ex N. Coperni. . . ." Unfortunately these tables are not an ephemeris. They are apparently a set of Copernican mean motions designed to avoid the use of the Egyptian year and the Copernican "commutation," and they have been made independently of Reinhold's similar ones, which form a subset of his *Prutenicae tabulae*. The present tables are insufficient by themselves for calculating an ephemeris and would presumably have been used in conjunction with the prosthaphaeresis tables in *De revolutionibus* itself.

Appendix: The Copernican Tables and Parameters

The Copernican system requires six or seven parameters for calculating the longitude of each planet, plus five parameters of precession and trepidation. These are the mean motion $\Delta\lambda$ and its radix λ_0; the mean apsidal motion $\Delta\omega$ and its radix ω_0; the orbital eccentricity e; the size of the auxiliary circle or epicyclet, r; and the ratio of Earth's orbit to the planetary orbit, R (or its reciprocal, so that this parameter is always less than unity). The precessional parameters are the rate of precession $\Delta\pi$, and its radix π_0; the rate of the anomaly of precession $\Delta\eta$ and its radix η_0; and the amplitude of trepidation ξ. The values used for the computer calculation are shown in Table 4.

As stated earlier, Copernicus does not explicitly give the apsidal motions; since these motions are small and do not affect the results strongly, they can be safely derived from Copernicus's data, together with the positions of the apsidal lines at the beginning of the Christian era.

In every case Copernicus explicitly states the eccentricities, and for the Moon and planets these can be confirmed by using them to recompute the tables of prosthaphaereses. For the sun an interesting discrepancy exists between the autograph manuscript and the Nuremberg edition of 1543; the editors of both the Toruń (1873) and Munich (1949) editions,[9] in attempting to go back to some original state, ended up with a bastard scheme. To elucidate this situation, I shall use Copernicus's own diagram and lettering from *De revolutionibus* III, 20 (Figure 2). In this model for the apparent solar motion, the Sun is at D; the center of the Earth's orbit, G, revolves clockwise on the small auxiliary circle EFG with the same period as the anomaly of precession(3434 years). Simultaneously, but very slowly, the center of this auxiliary circle, C, revolves counterclockwise with the motion of the apsidal line ACB. The effective eccentricity of the Earth's orbit is hence DG, and this quantity varies from a minimum DF to a maximum DE. (Let us designate the mean eccentricity DC by e and the radius of the small auxiliary circle by r.) Furthermore, the motion on the small circle changes the apsidal line by the angle ADK. The six columns of the *Tabula prosthaphaereseon Solis* are then:

C_1, C_2 the independent variable;

C_3 the change in the direction of the solar anomaly generated by the motion of the auxiliary circle, reaching a maximum value such that

$$\text{Max}(C_3) = \sin^{-1}(r/e);$$

TABLE 4. Planetary parameters from Copernicus and Reinhold[a]

	Mean Motion	Δλ	η₀	Δω	ω₀	e	r	R
Sun	59' 8" 11‴ 22⁗	0.985 608 20"/day	272°52	0.985 589 64"/day	211°32	0.036 9	0.004 8	—
	59 8 11 22 16 11	0.985 608 20	272.4976	0.985 588 75	211.6506	0.036 95	0.004 75	—
Moon	12° 11' 26" 41‴ 31⁗	12.190 747 76	209.97	13.064 983 80	207.12	0.023 7	0.109 7	—
	12 11 26 41 29 57 50	12.190 747 68	209.9730	13.064 983 33	207.2244	0.023 66	0.109 75	—
	Commutation			Commutation				
Saturn	57 7 44 5	0.033 459 64	66.70	27.485 009×10⁻⁶/day	225.00	0.057 0	0.028 5	0.109 0
	57 7 44 4 22 22	0.033 459 71	66.6975	27.914 700	225.0014	0.056 95	0.028 5	0.109 11
Jupiter	54 9 3 49	0.083 090 51	174.25	8.855 137	154.06	0.045 8	0.022 9	0.191 6
	54 9 3 47 30 57	0.083 090 64	174.234 2	8.237 464	153.9319	0.045 75	0.022 9	0.199 62
Mars	27 41 40 22	0.524 032 40	34.15	21.434 68	107.75	0.097 3	0.050 0	0.658 0
	27 41 40 23 18 34	0.524 032 33	34.1297	21.874 914	107.7589	0.096 87	0.050 0	0.657 72
Venus	36 59 27 57	0.616 518 31	126.75	—	48.33	0.0164	0.0104	0.719 3
	36 59 28 0 7 18	0.616 518 53	126.775 6	—	48.35	0.0164	0.0104	0.719 3
Mercury	3° 6 24 13 39	3.106 729 92	46.40	43.634 03	187.54	0.073 6	0.021 2	0.357 3
	3 6 24 14 5 35 48	3.106 731 91	46.883 6	44.021 300	186.580 3	0.073 51	0.021 2	0.357 34
			Precessional Parameters					
	Δπ 3.8205×10⁻⁵/day	π₀ 5° 32'	Δη 2.872,16×10⁻⁴/day		η₀ 6° 45'		ξ 1° 10'	
	3.8205	5° 32' 24"	2.872,16		6° 40' 27"		1° 11' 22" 30"	

[a]Copernican parameters are on the first of each pair of lines. The Copernican mean motion is derived from the "commutation" (specified in *De revolutionibus* and shown in the first column) by subtracting the latter from the solar motion. Reinhold's mean motion is taken directly from the sexagesimal form in the *Prutenicae tabulae*, but his commutation is shown here for comparison.

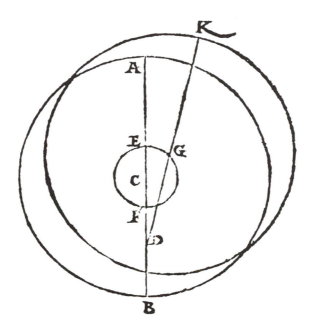

FIGURE 2. *Copernicus's diagram and lettering for the apparent solar motion from* De revolutionibus *III, 20.*

C_4 the proportional eccentricity normalized to base 60 between the maximum and minimum values, that is, $60(DG - DF)/(DE - DF)$;

C_5 the longitudinal effect of the orbital eccentricity at minimum value, comparable to the equation of the center for the planets, reaching a maximum value such that
$$\text{Max}(C_5) = \sin^{-1}(r/e);$$

C_6 the change in C_5 when the orbital eccentricity is at a maximum.

From the foregoing it is clear that e and r can be deduced from the maximum values of columns 3 and 5 in the table of prosthaphaereses. In Copernicus's manuscript in III, 21 the values of $368\frac{1}{2}$ and $47\frac{1}{2}$ were originally given for e and r, respectively (as well as 1289 for the normalized ratio r/e), and these numbers correspond to the maximum values of 7°24′ and 1°50′ in the manuscript table on fol. 103v. However, in the text (fol. 101v) Copernicus canceled the fractions ($\frac{1}{2}$) and indicated with a unit digit under the original numbers that his new values were $(47 + 1 =)$ 48 and $(368 + 1 =)$ 369, which agree with a new marginal value of the normalized ratio, 1300. Copernicus did not re-place the table of prosthaphaereses, but he did indicate between the columns that the new maximum of C_3 would be 7°28'. (Note that C_5 would remain unchanged because the difference $e - r$ was still the same.) The printed text in the 1543 Nuremberg edition follows the changes, giving the numbers 48, 369, and 1300; and, furthermore, C_3 has been recalculated (conceivably by Rheticus) to agree with these values, thus reaching a maximum of 7°28'. (A few values near the maximum of C_5 have also been raised by one minute to agree with what was clearly the first way in which Copernicus wrote out the manuscript table on fol. 103; this change produced a slight inconsis-tency in the printed version.)

The editors of the 1949 Munich edition established an inconsistent version by adopting the new numbers from the manuscript text, plus the original table that Copernicus clearly intended to revise. In this they followed the inconsistencies of the 1873 Toruń edition.

The editing of C_5 is so complex that a brief table is necessary:

C_1	Manuscript Original form	Manuscript Altered	1543 Nuremberg	1873 Toruń	1949 Munich
84	1° 50'	1° 49'	1° 50'	1° 49'	1° 49'
87	1 50	1 50	1 50	1 50	1 50
90	1 51	1 50	1 51	1 51	1 50

It is perhaps interesting to note in passing that Copernicus's original form of C_5 corresponds to what he supposed to be the actual eccen-tricity in his time, 0.003 23, rather than the minimum value, 0.003 21, adopted in his alteration. I should also point out that, because C_3 enters the calculation only in establishing the direction of the apsidal line and not in the longitude directly, the final positions are not much affected by adopting the printed values or (as Reinhold in effect did) the original manuscript values.

One of the most sensitive parameters is the mean motion, especially when the longitudes are calculated with respect to a radix many cen-turies earlier at the beginning of the Christian era. Inconsistent values both in the manuscript and in the printed text, plus the existence of tables in the printed text clearly revised according to Copernicus's intentions, once again demonstrate that editors must understand as-tronomical as well as paleographical principles.

In the case of Venus, for example, the text (fol. 144v in the manuscript, fol. 135 in the 1543 edition) gives the daily mean motion as

$$36'\ 59''\ 28'''\ 35'''',$$

in agreement with the daily mean motion table (f. 147v in the manuscript, fol. 139 in the printed edition). However, at the bottom of the annual mean motion table in the manuscript (fol. 147), Copernicus corrected the entries for 30 and 60 years, with numbers corresponding to

$$36'\ 59''\ 27'''\ 57''''.$$

Since the new numbers give better results, Copernicus must have intended the correction to be carried out for the whole table; someone (Rheticus?) did this for the 1543 Nuremberg edition, and Rheticus clearly followed the revised table in his own ephemeris. In this case the change produced an inconsistency, inasmuch as neither the text value nor the daily mean motion tables were changed. (Copernicus probably realized that the daily mean motion had only a small effect, and so he didn't worry about changing it.) For Venus the Toruń as well as the Munich editors rather reasonably included both versions of the table, although the printed values must obviously be preferred in making calculations.

In the manuscript text, the daily mean motions of Mars and Mercury are given respectively as

$$27'\ 41''\ 40'''\ 8''''\ \text{and}\ 3°\ 6'\ 24''\ 7'''\ 40''''.$$

These are given in the manuscript tables as

$$27'\ 41''\ 40'''\ 22''''\ \text{and}\ 3°\ 6'\ 24''\ 13'''\ 40''''.$$

The 1543 Nuremberg edition prints the latter values consistently for both the text and tables, but, not unexpectedly, both the Toruń and Munich editors revert to the inconsistent form given in the manuscript. Since early users of the Copernican scheme would in any event have used the tabular form, these have been adopted for the modern machine calculations.

Notes and References

[1] Additional background information on Reinhold is found in a companion paper to this one, O. Gingerich, "The Role of Erasmus Reinhold and the *Prutenic Tables* in the dissemination of Copernican Theory," *Studia Copernicana*, vol. 6 (Wroclaw, 1973), pp. 43–62, 123–125 [reprinted as selection 13 in this anthology].

[2] See especially B. Tuckerman, *Planetary, Lunar, and Solar Positions A.D. 2 to A.D. 1649*, vol. 59 of *Memoirs of the American Philosophical Society* (Philadelphia, 1964).

[3] O. Gingerich, "The Mercury Theory from Antiquity to Kepler," *Actes du XIIe Congrès International d'Histoire des Sciences, 1968*, vol. 3A (Paris, 1971), pp. 57–64 [reprinted as selection 23 in this anthology].

[4] Reported by Dr. John Free, a student in my seminar. The actual root-mean-square error is 15' for 30 Mars observations of Walther recorded in J. Schöner, *Scripta clarissimi mathematici M. Joannis Regiomontani* (Nuremberg, 1544).

[5] O. Gingerich, "Remarks on Copernicus's Observations," *The Copernican Achievement*, ed. by R. Westman (Berkeley and Los Angeles, 1975).

[6] L. A. Birkenmajer, *Stromata Copernicana* (Cracow, 1924), p. 81; E. Zinner, *Entstehung und Ausbreitung der coppernicanischen Lehre* (Erlangen, 1943), pp. 270, 286. Hilary's ephemeris is no. 14696 in J. C. Houzeau and A. Lancaster, *Bibliographie Générale de l'Astronomie* (Brussels, 1887-89; reprint London, 1964).

[7] A. Birkenmajer, "Hilary de Wiślica, etait-il zelateur du système heliocentrique de Copernic a Cracovie," *Studia Copernicana*, vol. 4 (Wroclaw, 1972), pp. 721–60.

[8] A. Birkenmajer shows that Hilary had the ephemerides of both Stoeffler and Pitati in his estate when he died. I have taken the positions from J. Stoeffler, *Ephemeridum opus* (Venice, 1532), but Stoeffler's numbers are reprinted in P. Pitati, *Almanach novum* (Venice, 1542).

[9] *Nicolai Copernici Thorunensis de revolutionibus orbium caelestium libri VI*, ed. by M. Curtze (Torun, 1873); *Nikolaus Kopernikus Gesamtausgabe*, vol. 2: *De revolutionibus orbium caelestium, Textkritische Ausgabe*, ed. by F. Zeller and C. Zeller (Munich, 1949).

Erasmus Reinhold and the Dissemination of Copernican Theory

Copernicus's *De revolutionibus* proclaims on the title page of the editio princeps, "You also have very convenient tables from which you can most easily calculate [the positions of the planets] for any time. Therefore buy, read, use."

The tables of *De revolutionibus* compare not unfavorably with those in the *Almagest*, but just as Ptolemy soon had the *Handy Tables*, so too the Copernican tables were soon expanded into a handier form in Erasmus Reinhold's *Prutenicae tabulae* of 1551. Thus, the 1566 Basel printing of *De revolutionibus*, almost a page-by-page reprint of the 1543 Nuremberg edition, omits the advertisement on the title page but adds a testimonial from Reinhold himself: "All posterity will gratefully remember the name of Copernicus, by whose labor and study the doctrine of celestial motions was again restored from its near collapse. Under the light kindled in him by a beneficent God, he found and explained much which from antiquity till now was either unknown or veiled in darkness."

Reinhold's name, through his handy tables, was thus closely linked with Copernicus and Copernicanism. Reinhold was the senior mathe-

Selection 13 reprinted from *Studia Copernicana*, vol. 6 (1973) pp. 43–62, 123–25.

matics professor at the University of Wittenberg, charged with teaching astronomy. His younger colleague, the professor of lower mathematics, Georg Joachim Rheticus, taught the geometry and arithmetic. Reinhold must have learned the details of the new astronomy soon after Rheticus returned from his visit to Copernicus in 1540. In his 1542 commentary to Peurbach's *Theoricae novae planetarum*, Reinhold wrote: "I know of a modern scientist who is exceptionally skillful. He has raised a lively expectancy in everybody. One hopes that he will restore astronomy." And later, "I hope that this astronomer, whose genius all posterity will rightly admire, will at long last come to us from Prussia. . . ."[1]

With these credentials, Reinhold has long been considered not only one of the first Copernicans but, through his *Prutenicae tabulae*, one of the leading influences in the Copernican revolution. Historians were therefore taken by surprise when, in 1960, the late Aleksander Birkenmajer published a short account giving evidence for Reinhold's geocentric cosmology.[2] In this article, I shall try to give a new assessment of both the role and the efficacy of the *Prutenic Tables* in the reform of astronomy, as well as an evaluation of Reinhold's cosmology and its pervasive influence.

The *Prutenic Tables*

BIBLIOGRAPHY. The *Prutenicae tabulae coelestium motuum* was first published by Ulrich Morhard in Tübingen in 1551. It was reissued by Morhard's widow in 1562 with a new title page, but with only about half the signatures reset, the remainder of the book being made up of the old stock.[3] Another edition appeared in Tübingen in 1571, published by Oswald and Georg Gruppenbach, with corrections by Michael Maestlin. The book in the first three forms is now quite rare, and only about a dozen copies altogether are found in the United States. The final, and more common, form was edited by Caspar Strubius and printed in Wittenberg in 1585 by Matthew Welack. All editions include three folding tables, one for sexagesimal multiplication and two giving examples for the computation of planetary positions.

Besides these editions, another work appeared in 1560 so closely based on the *Prutenic Tables* that by modern standards it would be considered outright plagiarism: Johannes Stadius's *Tabulae Bergenses aequabilis et adparentis motus orbium coelestium* (published by the heirs of Arnold Birckmann, Cologne). Stadius, in dedicating his book to Bishop Robert de Berg, mentions that Copernicus's new work was dedicated to Pope Paul III; he praises Reinhold in a line but fails to note even the existence of the *Prutenic Tables*. Stadius owned a copy of *De revolutionibus*—it is still preserved in the library of the West Point Military Academy in New York—but the evidence shows decisively that he did not independently work from Copernicus's treatise. This fact must have been recognized by the anonymous author of a manuscript[4] written in French sometime after 1582, who clearly labeled his account "selon de la Doctrine de Nicolas Copernic et d'Erasme Reinholde," yet who modeled his canons and the form of his tables closely after those of Stadius. That he gives no credit to Stadius is poetic justice.

THE NATURE OF THE TABLES.   In the introduction to his own *Rudolphine Tables*, Johannes Kepler gives a succinct description of Reinhold's tables:[5]

> For the tables ought to be handy canons, easy to use; the authors of the Alfonsine and other tables have aided this handy use even by the form of their books, the numerical tables being bound together and very short instructions being placed at the beginning. The book of Copernicus, on the other hand, has the tables dispersed throughout the text among the demonstrations in the manner of the Ptolemy's Syntaxis [Almagest]. Thus it happens that the mind of anyone desiring to use the tables is distracted by the text, and the work deprives itself of its own chief usefulness. . . . With this in mind Reinhold undertook the work, and he shows that he wore himself out in this huge and disagreeable task. If you wish to know his purpose, it is very laudable indeed: the definite knowledge of the motions, the length and starting point of the year, the equinoxes, solstices, eclipses, and the great conjunctions, so that from the most sublime collection of these things, the wisdom and goodness of the Creator might shine forth.

Thus, the *Prutenic Tables* offer what was claimed for *De revolutionibus*: "Very convenient tables from which you can most easily calculate (the positions of the planets) for any time." Anyone who tries quickly to calculate a planetary position with Copernicus's treatise as a sole guide finds himself in a frustrating position. Whereas the tables of mean motions and corrections are neatly organized, the radices (that is, the starting positions at a known time from which later positions can be found) are buried in the text. Furthermore, Copernicus found that the fundamental line of aphelion for each planet slowly changed over the centuries, but he offered no simple method to find this line of apsides for a specific time.

Reinhold overcame these difficulties with a systematic organization of tables for the conversion of dates, tables of radices, and expanded mean motion and correction tables. Furthermore, he added a redundancy of tables so that almanac computers familiar with the older *Alfonsine Tables* could perform the steps in an analogous manner. Prefacing the whole work are 68 leaves of Latin instructions on the detailed use of the tables. Examination of a computational example shows most clearly the nature of the tables and illuminates several fine points regarding the differences between Reinhold, Copernicus, and Alfonso's table makers.

A SAMPLE CALCULATION.   A twentieth-century astronomer, coming to this historical problem *de novo*, would probably assume that Copernicus treated the Earth like the other planets and that after he had found the heliocentric longitudes he would proceed to combine them. Such a proper heliocentric approach, however, was not realized until the work of Kepler. In *De revolutionibus* and in the *Prutenic Tables*, the mean Sun receives special treatment, and the result enters directly into the calculation of the geocentric longitude for each planet. Computationally, it is exactly as if each planet has a large epicycle, but Copernicus simply renames this the "parallax of the orbit," that is, the effect of viewing the planet from a moving station.

Throughout his explanatory canons, Reinhold uses as his paradigm the position of Saturn at the birth of the Duke of Prussia, 17 May 1490. Following the customary procedure of the *Alfonsine Tables*, the year, month, day, and hour are converted into sexagesimal days by use of convenient tables found in the short second section of the volume: 2, 31, 6, 33; 27, 20. (This is the sexagesimal equivalent of 543,993, the days since 1 January A.D. 1.)

The advantage of the sexagesimal form becomes apparent in the next step. Copernicus calculated the daily motion of Saturn as $57'7''44'''$. . . . From this, the motion in 2 days is known, as well as 3, 4, 5, . . . , 60 days. The motion in 60 days is of course $57°7'44''$. . . , one sexagesimal place larger than for a single day. With a 60-entry table of mean motion in longitude, the mean motion since 1 January A.D. 1, can be readily found by a sum of numbers with the appropriate sexagesimal point shift:

Sexagesimal	Mean motion of Saturn				Apogee of Saturn				Commutation			
2	0^s	54°	35'	48"	0^s	12°	3'	33"	15^s	28°	8'	45"
31	2	14	6	15		3	6	55	30	59	46	16
6	0	12	2	44			0	36	5	42	46	24
33 days		1	6	15				3		31	25	15
27				54							25	42
20				1								19
Position 1 Jan. A.D. 1	1	6	41	51	3	45	0	5	3	25	48	0
Mean longitude	4	28	33	48					2^s	8°	20'	41"
Longitude of apogee	4	0	11	12←	4^s	0°	11'	12"				
Eccentric anomaly		28°	22'	36"								

(Leading places are dropped off because they represent complete circuits of the planet.)

The procedure for the mean longitude duplicates the scheme of the *Alfonsine Tables*, but the motion of the apogee is a Copernican innovation (with a curiously geocentric vestige in its nomenclature). Reinhold has provided the tables for the apogee, which are lacking in *De revolutionibus*. Similar sexagesimal procedures must be followed for the precession, trepidation, Sun, and solar anomaly. The first two of these mimic the Alfonsine routine exactly, but with the size of the oscillatory trepidation greatly diminished. The solar anomaly, however, is new and results from Copernicus's treatment of the changing direction of apogee. Because the general technique is identical with the handling of trepidation, it would have posed no serious difficulty for the computer (See the Appendix, p. 245).

Given the mean longitude of Saturn and of the Sun, corrections must be made for Saturn's intrinsic variations of angular speed in its orbit as well as for the Earth's orbital position. These are accomplished with the "Canones Prosthaphaereseon Saturn." Following O. Neugebauer, we can designate the columns in order as follows:

C_1 independent variable;

C_3 effect of orbital eccentricity;

C_4 proportional parts, the proportional effect on the maximum orbital parallax as a function of position in planetary orbit;

C_5 effect of orbital parallax when the planet is farthest from the Sun;

C_6 the change in C_5 when the planet is nearest the Sun.

We can now formulate the remaining procedure for the computation of the true longitude λ of the planet. For the given moment t, we already have the following elements:

λ_0 longitude of apogee,
α eccentric anomaly,
γ longitude of mean Sun $-$ longitude of mean planet.
(Copernicus, with a certain aesthetic economy, omits the tabulation of the planets' mean longitudes, because they can always be obtained from this last quantity, which he calls the commutation.) Then the true longitude $\lambda = \lambda(t)$ is given by

$$\lambda = \lambda_0 + \alpha + C_3(\alpha) + C_5(\gamma') + C_4(\alpha)C_6(\gamma'),$$

where

$$\gamma' = \gamma - C_3(\alpha).$$

The Copernican procedure represents one important simplification over the method of Ptolemy and Alfonso. Copernicus tabulated C_4 (the effect of orbital parallax) for apogee, whereas Ptolemy used a mean position. Consequently Ptolemy had to make a choice in correcting for either a nearer or a farther position; thus, his scheme required an additional column with its attendant extra rules of addition or subtraction.

The numbers for Saturn, as Reinhold tabulates them on his foldout sheet, go as follows[6]:

1	λ_0	4^s	$0°$	$11'$	$12''$
2	α	0	28	22	35
3	$C_3(\alpha)$		2	57	3
4	$\alpha + C_3$		25	25	32
5	$C_4(\alpha)$			2	34
6	$C_5(\gamma')$		4	45	29
7	$C_6(\gamma')$		0	37	5
8	C_4C_6		0	1	35
9	$C_5 + C_4C_6$		4	47	4
10	$\gamma_0 + \alpha + C_3$	4	25	36	44
11	Sum(9) + (10)	4	30	23	48
12	Precession		26	59	29
13	λ	4^s	$57°$	$23'$	$17'' = 297°\ 23'\ 17''$

The final step is needed because Copernicus's tables refer the longitude to the actual first star of Aries rather than to the traditional equinoctial position.

REINHOLD VERSUS COPERNICUS.   In the preface to the *Prutenic Tables*, Reinhold writes:[7]

Copernicus, the most learned man whom we are able to name other than Atlas or Ptolemy, even though he taught in a most learned manner the demonstrations and causes of motion based on observation, nevertheless fled from the job of constructing tables, so that if anyone computes from his tables, the computation is not even in agreement with his observations on which the foundation of the work rests. Therefore first I have compared the observations of Copernicus with those of Ptolemy and others as to which are the most accurate, but besides the bare observations, I have taken from Copernicus nothing other than traces of demonstrations. As for the tables of mean motion, and of prosthaphaereses and all the rest, I have constructed these anew, following absolutely no other reasoning than that which I have judged to be of maximum harmony.[7]

Even a preliminary glance at the *Prutenic Tables* confirms that Reinhold undertook a massive recalculation. In the tables of prosthaphaereses, *De revolutionibus* tabulates the corrections for every third degree and to the nearest minute of arc. The *Prutenicae tabulae* gives the same quantities for every degree and to seconds of arc. Furthermore, as mentioned, Reinhold added mean motion tables for the apogees, a tabulation completely lacking in Copernicus's treatise.

A more thorough comparison of the tables—in fact, a complete recalculation with a high-speed electronic computer—shows that the accuracy of Reinhold's calculations systematically exceeds that of Copernicus. The most glaringly inept calculation in the tables of *De revolutionibus* is the proportional parts column, C_4, for Jupiter, where Copernicus seems to have even less numerical control than usual; because this quantity depends on the difference between two similar quantities, one must carry more significant figures than usual. Reinhold's recalculation is excellent:

Proportional Parts for Jupiter

	Copernicus	*Reinhold*	*CDC 6400 Computer*
30°	2′ 50″	3′ 6″	3′ 5″
60°	13 10	12 5	12 4
90°	26 57	25 55	25 53
120°	41 50	41 46	41 44
150°	55 15	54 54	54 51

These results raise the question as to whether Reinhold used any-

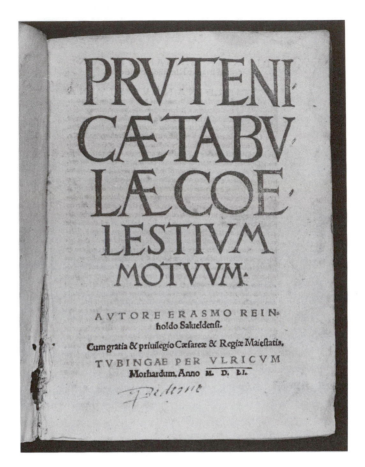

FIGURE 1. *The title page of the* Prutenic Tables. *Collection of R. B. and M. S. Honeyman.*

thing from Copernicus besides the observations and the discovery that planets slowly change their lines of apsides. After taking recourse to the electronic computer once again, we can answer this in the affirmative in a most remarkable way.

This detective work depends on a subtle difference between Ptolemy and Copernicus in the calculation of the tables of prosthaphaereses. In the Ptolemaic system, a device called an equant helps achieve the nonuniform motion of a planet in its orbit; the equant is an axis of uniform angular motion off-center in the orbital circle (see Figure 2). Copernicus strongly objected to this device, which he felt violated the

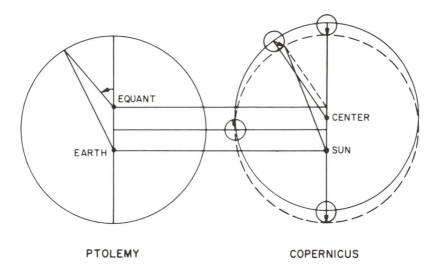

PTOLEMY   COPERNICUS

FIGURE 2. *Copernicus's replacement of the Ptolemaic equant with an epicyclet. If the distance between the earth and equant according to Ptolemy is 2e, then the epicyclet radius is e/2.*

principle that planetary movements should be composed of uniform circular motions. Thus, he replaced the equant with a small epicyclet. The Copernican vector linkage is shown on the right of Figure 2; the epicyclet revolves to produce a regular trapezoid. The upper dashed line therefore moves in angles exactly like the Ptolemaic equant. However, the Copernican vectors do not trace out an exact copy of the Ptolemaic deferent circle; in fact, the Copernican model bulges out at the sides.

The differences in the correction for orbital eccentricity, C_3, are very small—for Mars about 2′ of arc, which just allows us to be sure that Copernicus did actually compute his tables this way, since his are given in *De revolutionibus* to minutes of arc. Because the *Prutenic Tables* include seconds of arc, it is much easier to be definitive about Reinhold's method of computation. Reinhold changes the Copernican eccentricity, but the computations show that he retains the epicyclet mechanism exactly.

Notice that the Copernican eccentricity is $\frac{3}{2}$ times the Ptolemaic eccentricity, and the equivalence of parameters is maintained if the epicyclet eccentricity is half the Ptolemaic eccentricity. However, Copernicus deviates from the 3:1 ratio for the eccentricity:epicyclet, as shown in the following table (for easier comparison the Ptolemaic eccentricities have been scaled by $\frac{3}{2}$):[8]

	Ptolemy	Copernicus		Reinhold	
	Eccentricity	Eccentricity	Epicyclet	Eccentricity	Epicyclet
Saturn	0.0854	0.0854	0.0285	0.08542	0.0285
Jupiter	0.0687	0.0687	0.0229	0.06862	0.02287
Mars	0.15	0.1460	0.05	0.145300	0.05
Venus	0.0312	0.0246	0.0104	0.0246	0.0104
Mercury	0.075	0.0736	0.0212	0.07350	0.02119

It is interesting to see how closely Reinhold follows Copernicus, not only with respect to the mechanism, but also in his keeping the planetary epicyclets the same or almost the same even when he adjusts the central eccentricity. Thus, we see that Reinhold owed a great debt to Copernicus beyond his new observations.

The relative size of the epicycle to deferent, or Earth's orbit to planet's orbit, is tabulated here:

	Ptolemy	Copernicus	Reinhold	Modern
Saturn	0.10833	0.1090	0.10911	0.1048
Jupiter	0.1917	0.1916	0.19062	0.192
Mars	0.658	0.658	0.65772	0.656
Venus	0.72	0.7192	0.71930	0.723
Mercury	0.375	0.3573	0.35733	0.387

ADOPTION OF THE *PRUTENIC TABLES*.   Reinhold himself was the first to use the new tables for an ephemeris, which he computed for 1550 and 1551. As was customary in such a situation in those days, Reinhold offered glowing praise to Copernicus; but he did not take the occasion to mention the Sun-centered world view.

In 1556, John Feild in England, at the request of John Dee, published his *Ephemeris anni 1557 currentis iuxta Copernici et Reinholdi canones*, which warned the reader of the *Alfonsine Tables* whose mistakes became daily more apparent. He added: "Wherefore, I have published this Ephemeris for the year 1557, following in it Copernicus and Erasmus Reinhold, whose writings are established and founded on true, sure and plain demonstrations."[9] Feild followed this with his *Ephemerides trium annorum. 58. 59. et. 60. ex Erasmi Reinholdi tabulis.*

The availability of Johannes Stadius's *Ephemerides novae et exactae ab anno 1554 ad 1570*, published in Cologne in 1556, apparently discouraged further independent calculations in England. Stadius praised Copernicus's work and Reinhold's tables without mentioning the heliocentric cosmology. He did, however, print a letter from Gemma Frisius, who not only severely criticized the *Alfonsine Tables* and lauded the *Prutenic Tables* but also mentioned that the Copernican heliocentric hypothesis gave a better understanding of planetary distances and certain features of retrograde motion.

Gemma added, however, that those who objected to the ephemerides because of the underlying hypothesis understood neither causes nor the use of hypotheses. "For these are not posited by the authors as if this must exist this way and no other." He further remarks: "Nay, even if someone wished to refer to the sky those motions that Copernicus assigns to the earth, he could do so and according to the very canons of calculation. However, it did not please that most learned and prudent man, on account of his invincible intellect, to invert the entire order of his hypotheses, and so he was content to have posited those that sufficed for the true discovery of the 'phenomena.' " Gemma's position, as we shall see in the second part of this paper, agreed closely with Reinhold's.

In 1559 (and 1560), Stadius republished the book, retaining the original prefatory material and adding ephemerides for six more years. By 1570, however, when he published an extended version for 1554 to 1600, he mentioned Copernicus and Reinhold only in passing and omitted any reference either to the *Prutenic Tables* or to his own *Tabulae Bergenses*, leaving the reader uninformed as to the basis for his calculations. He did, however, retain the letter from Gemma.

The major post-Copernican ephemeris still based on the *Alfonsine Tables* was the *Ephemeridum novum atque insigne opus ab anno 1556 usque in 1606 accuratissime supputatum* of Cyprian Leowitz, a gigantic folio published in 1557 in Augsburg. His silence with respect to Copernicus is baffling, since in a companion work published the previous year, *Eclipsium omnium ab anno domini 1554 usque in annum domini 1606 accurata descriptio*, he mentions Copernicus with the apellation "mathematici clarissimi" at least five times.[10] Tycho Brahe recorded a conversation in 1569 in which Leowitz declared that the Copernican results agreed better with the observations for the superior planets, and solar eclipses, but not for lunar eclipses or the inferior planets.[11]

In his written account, Brahe proceeded to contradict this from his own observations and to give a harsh judgment on Leowitz's competence. He also stated that Leowitz had calculated an extensive ephemeris based on the *Prutenic Tables*, but had never published them. In any event, the reasons for Leowitz's preference for the *Alfonsine Tables* in the 1550s are not clear; the best account has been given by A. Birkenmajer, who finds evidence for an anti-Copernican bias in Leowitz.[12]

Another extensive Alfonsine-based ephemeris was published in Venice in 1555 by J. B. Carello, *Effemeridi per anni 17 . . . (dall'anno 1554 all'anno 1580)*; soon after, three successive Latin editions were issued, and one of these was cited by Tycho in his observing records. In 1563 G. Moleti issued *l'Efemeridi per anni XVIII (dall'anno 1563 all'anno 1580)*. The publisher, Vincenzo Valgrisio, is the same for both Moleti and Carello (in the 1558 and 1563 editions); so are the woodblocks and the planetary positions, although the horoscopes at the beginning of each year differ slightly. It is inconceivable that the calculations were independent.[13]

All other major ephemerides from 1555 to 1610 were based on the *Prutenic Tables*. These included M. Maestlin's *Ephemerides novae ab anno 1577 ad annum 1590 supputatae ex Tabulis Prutenicis* (Tübingen, 1580), G. A. Magini's *Ephemerides coelestium motuum secundum Copernici hypotheses Prutenicasque Reinhold tabulas accuratissime supputatae* (Venice, 1582), and D. Origanus's *Novae motuum coelestium ephemerides Brandenburgicae calculo duplici luminarium, Tychonico & Copernicaeo* (Frankfurt-an der Oder, 1609). Origanus's earlier volume of ephemerides (Frankfurt an der Oder, 1599) contains within its title. . . *incipientes ab anno 1595, quo I Stadii maxime aberrare incipiunt . . .*, referring to the rather bizarre and unexplained circumstance that in 1595 the Stadius ephemerides switched from the *Prutenic Tables* to the *Alfonsine Tables*.

It is a curious fact that until the systematic observations of Tycho Brahe, there was relatively little way to distinguish between the accuracy of the *Alfonsine Tables* and *Prutenic Tables* (see Figure 3). Why the *Prutenic Tables* were so widely adopted is a topic as yet unexplored. Quite possibly, the great conjunction of Jupiter and Saturn in 1563, which happened to be far more accurately predicted by the *Prutenic Tables*, had a particularly important influence.[14] Eclipses, which could be observed without instruments, might have played a similar role, although to my knowledge no specific evidence on this point has been discussed.

Although none of the Prutenic-based ephemerides more than hinted at the heliocentric cosmology, they displayed Copernicus's name more

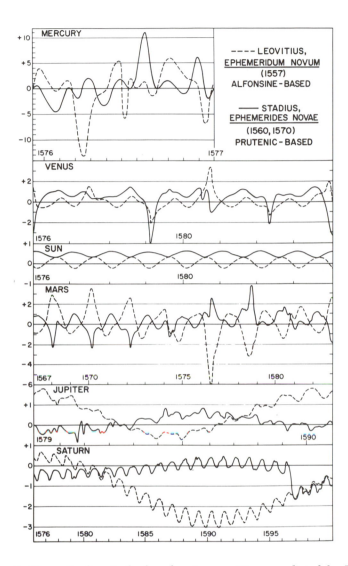

FIGURE 3. *Errors in longitude for planetary positions predicted by Leovitius and by Stadius. The displacement of the Sun in the Stadius scheme is real and not an accident of the geographical longitude chosen for the ephemeris. Note that in 1596 Stadius reverts to the* Alfonsine Tables, *as shown here for Saturn (see text for additional remarks). These graphs were produced by Barbara L. Welther using Bryant Tuckerman's* Planetary, Lunar and Solar Positions A.D. 2–A.D. 1649 *(Philadelphia, 1964).*

or less prominently. Thus, the *Prutenic Tables* and their derivative ephemerides were an important avenue for spreading Copernicus's reputation as a mathematical astronomer. Since the tables did not disseminate the word about the Copernican cosmology, however, the heliocentric system could not have been transmitted by the ephemerides alone.

The key source for the propagation of the heliocentric idea was undoubtedly *De revolutionibus* itself. The first edition of at least 400 copies was followed in 1566 by a second of about the same size. Modern claims that *De revolutionibus* was an unread book are false, as my systematic search for sixteenth-century annotations in the copies is showing. The book may not have been read very intelligently, especially the long technical parts that gave *De revolutionibus* its reputation as the most important new astronomical treatise of its age. Yet the cosmology was eminently readable, and found near the beginning. Thus, we may well imagine a symbiotic relation between the practical tables that brought Copernicus fame as a mathematical astronomer, and the great *De revolutionibus* that introduced the Copernican system itself.

Reinhold's Attitude toward the Heliocentric Cosmology

Kepler, in describing the *Prutenic Tables*, says "Copernicus insisted on absurd hypotheses, which Reinhold believed would have offended and frightened off the readers. He therefore decided that he should leave out any mention of the strange suppositions as well as the copious and tedious demonstrations, and publish the tables themselves separately in the form of a handbook, after correcting and calculating them more diligently, so that they might represent more exactly the fundamental observations on which Copernicus built his structure."

Even though the *Prutenic Tables* are silent about their cosmological presuppositions, Reinhold's favorable remarks contributed significantly to the growing reputation of Copernicus. Reinhold was in a key position in the decade following the death of Copernicus; not only was he the leading mathematical astronomer of his generation, but he became Dean of the University of Wittenberg in the summer of 1549 and Rector in the winter of 1549–50. Thus, his personal opinion of the heliocentoric cosmology was potentially of some consequence, and even his lack of a published endorsement is instructive. Clues about his attitude can be gleaned from his writings.

A careful reading of Reinhold's printed works shows that the Wittenberg astronomer must have been in close agreement with the anonymous introduction added to *De revolutionibus* by Andreas Osiander, which stated that the "hypothesis need not be true nor even probable; if they provide a calculus consistent with the observations, that alone is sufficient."[15] Reinhold, like Osiander, saw the task of the astronomer as model-making. Pierre Duhem, in a short but perceptive analysis made already in 1908, demonstrated from Reinhold's published books that this was the viewpoint he held. Among the various passages cited by Duhem, the following from the *Prutenic Tables* seems to treat the Copernican hypotheses simply as geometric devices for the construction of astronomical tables, devices similar in nature to the Ptolemaic ones: "It should be known that the diurnal movement of a planet is the sum of two parts: The first is the true movement of the epicycle, which Copernicus sometimes calls the Earth's movement and sometimes apparent motion; the other is the true movement of the planet by which it is itself animated, for instance, according to the customary Ptolemaic hypotheses, the movement of an epicycle along the circumference."[16]

It is beyond the scope of this article to cite additional texts, but suffice it say that a similar conclusion was reached by L. A. Birkenmajer, by E. Zinner, by A. Birkenmajer, and by others.[17]

Besides the printed works, we can gain additional insight into Reinhold's attitudes from a small manuscript legacy. By a stroke of good fortune, his own annotated copy of *De revolutionibus* has been preserved; I take this opportunity to describe it for the first time, and I find it irresistible to describe the circumstances of its discovery.

REINHOLD'S *DE REVOLUTIONIBUS.*   In November 1970, en route to Edinburgh, I visited Dr. Jerome Ravetz of Leeds. Among the Copernican matters we discussed was the surmise that there are probably more people alive today who have studied *De revolutionibus* carefully than at any time in the generation immediately following Copernicus. In fact, the sixteenth-century scholars who might really have understood the book could be counted on the fingers of two hands: Rheticus, Reinhold, Schöner, Stadius, Maestlin, Magini, Clavius, Kepler, and possibly Galileo.[18]

Two days later, in the Crawford Collection of the Royal Observatory in Edinburgh, I had an opportunity to examine a beautifully preserved copy of the 1543 *De revolutionibus.*[19] To my astonishment, the book was handsomely annotated throughout in several colors by someone familiar with the astronomical literature available in the 1540s. The annotations included citations to Rheticus's *Narratio prima*, Regiomontanus's *Epitome*, Ptolemy, Werner, Valla, and Do-

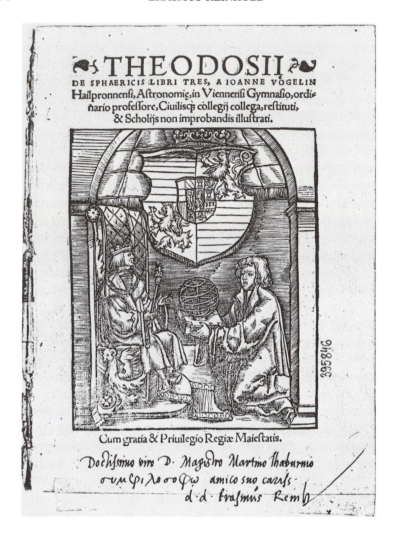

FIGURE 4. *A specimen of Reinhold's writing: the presentation inscription on the Theodosius* De sphaericis *in the Wroclaw University Library.*

menico Maria. Furthermore, the owner was evidently much interested in the practical problems of planetary positions; however, references to the *Prutenicae tabulae* were conspicuously absent, suggesting that the comments were inscribed before 1551.

My first inclination was to doubt our hypothesis that so few sixteenth-century scholars might have comprehended Copernicus's work, but further examination opened the possibility that the Crawford *De revolutionibus* might have been the personal copy of one of the leading astronomers of the 1540s.

Besides the general nature of the annotations, two specific features suggested the name of Erasmus Reinhold. In the back of the volume, following several leaves of chronological tables, there is a manuscript sexagesimal multiplication table resembling a printed foldout page in the *Prutenic Tables*. Furthermore, many of the manuscript notes in the star catalog match changes introduced in the *Prutenicae tabulae*. For example, the longitudes of stars 20, 21, 30, and 31 in Draco are corrected by 90°. In the margin beside the Lucida of Lyra is written FIDUCULA, and same appears as a printed marginal note in the *Prutenic Tables*. The situation is similar with respect to PRIMA for the first star of Aries. The most striking agreement is the marginal addition of the words "Trium in dorso," "Media," and "Reliqua" to three successive stars in Sagittarius, a change followed in the *Prutenic Tables*.

A more subtle clue appears near the bottom of fol. 64, where a marginal note reads, "Albategnius. Inter Ptolemaij e. Albategnij observationem anni hinc colliguntus 749. At ipse ponit anno 742." Although Copernicus used the name "Albategnius" in his autograph manuscript, the word was systematically changed to "Machometus Aratensis" in the printed text. As L. A. Birkenmajer pointed out, Reinhold uses "Albategnius" elsewhere, possibly indicating his familiarity with Copernicus's original text (cf. ref. 17).

Further confirmation is provided by the original binding, blind-stamped pigskin over boards, bearing the date 1543 and the initials E R S. Reinhold almost invariably referred to himself as Erasmus Reinholdus Salveldensis, thus in agreement with the initials. The roll-stamp used on the binding showing figures of Christ, St. Paul, Moses, and St. John is by an unidentified binder with initials *I H*, who presumably worked in Saxony and whose material has been described by Konrad Haebler.[20]

The most convincing test, however, has been the comparison of the distinctive half-cursive, half-printed style of handwriting. Figures 4, 5, and 6 allow the reader to see this for himself for a small fraction of the material. The most extensive sample of Reinhold's writing is the Berlin manuscript discussed in the next section. Like the annotated *De revolutionibus*, it lacks a confirming signature, and therefore the present investigation also enhances its authenticity.[21]

After the exciting sleuthing required to find the annotated *De revolutionibus* and to attribute the ownership to Reinhold, it is

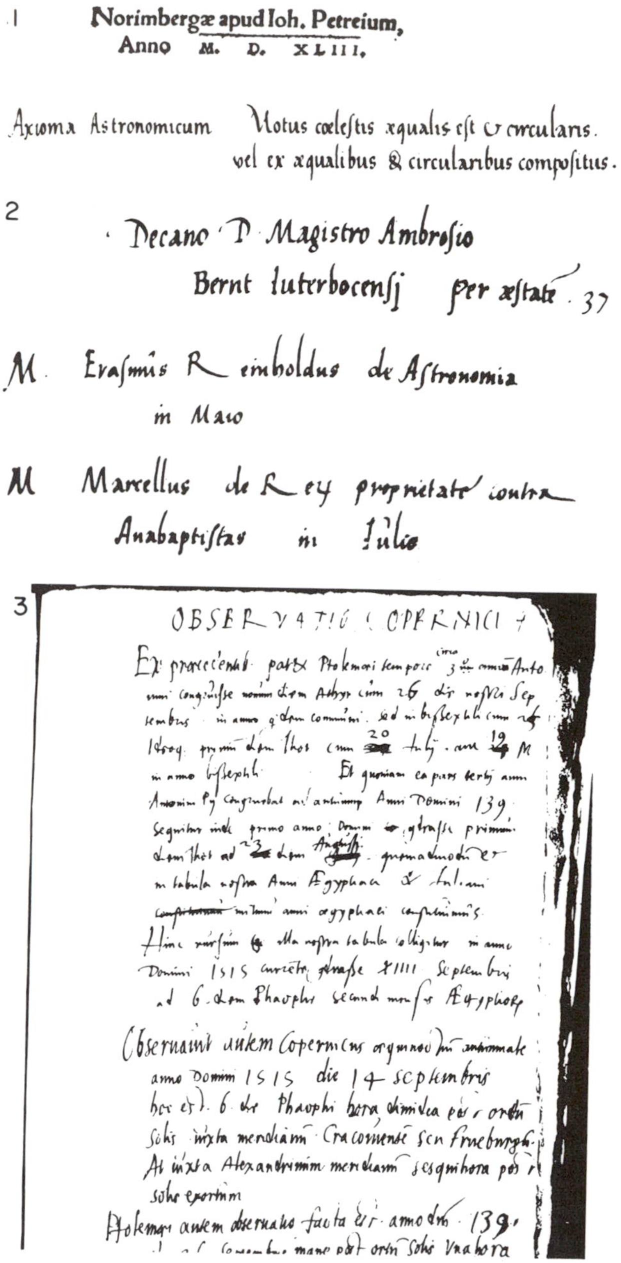
1
Norimbergæ apud Ioh. Petreium,
Anno M. D. XLIII.
Axioma Astronomicum Motus cœlestis æqualis est & circularis.
vel ex æqualibus & circularibus compositus.
2
Decano D. Magistro Ambrosio
Bernt Iuterbocensi per æstate 37
M. Erasmus Reinholdus de Astronomia
in Maio
M Marcellus de Rey proprietate contra
Anabaptistas in Iulio
3
OBSERVATIO COPERNICI

FIGURE 5. *Three specimens of Reinhold's writing: (1) from the title page of Reinhold's* De revolutionibus *in Edinburgh; (2) from the Wittenberg University Dean's Book, folio 148v; (3) a much less formal style from Berlin Latin 2° 391, folio 4.*

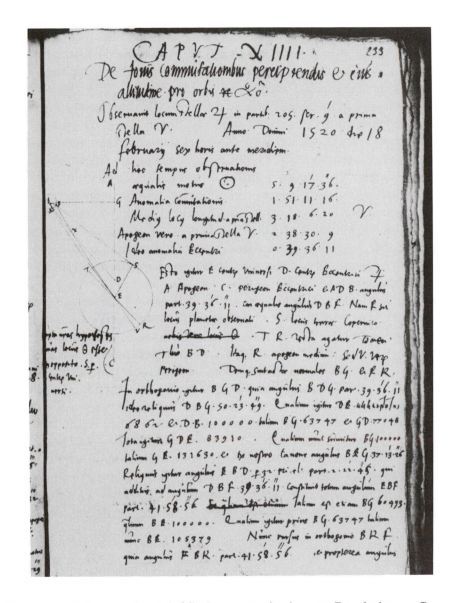

FIGURE 6. *Folio 233 of Reinhold's* Commentarius in opus Revolutionum Copernici *(Berlin manuscript Latin 2° 391), containing the suggestive marginal annotation as well as the orbital parameters for Jupiter.*

disappointing to find no clear opinion about Copernicus's cosmology expressed in the marginal notes. If, however, Reinhold viewed astronomers merely as makers of mathematical models, then the lack of such an opinion is completely expected. The nature of the annotations provides a further subtle confirmation of this idea.

On the title page of his *De revolutionibus*, Reinhold neatly inscribed a Latin phrase fully consonant with this impression: "Axioma Astronomicum: Motus coelestis aequalis est et circularis vel ex aequalibus et circularibus compositus." The lines paraphrase the title of Copernicus's Chapter 4—"The movement of celestial bodies is uniform, circular and eternal, or compounded of circles." As the recomputation of the *Prutenicae tabulae* revealed, Reinhold adopted the technical model-making aspects of Copernicus's work in full detail. The motto on the title page thus bears witness to what Reinhold considered important in *De revolutionibus*, namely, the ingenuity of its mathematical constructions.

The great cosmological Chapter 10 of Book I of Copernicus's treatise passes under Reinhold's hand almost unmarked. Where Copernicus notes roughly that Venus completes her circuit in nine months and Mercury in 80 days, Reinhold has penned corrections in the margin, 225 days and 88 days. And alongside the lines where Copernicus mentions his natural explanation of the lengths of retrograde arcs, Reinhold has scribed a decorative line for emphasis. These, together with two brief and innocuous marginal index notes earlier in the chapter, stand in marked contrast to the extensive notations in Books III, IV, and V. The pattern of annotations leaves no doubt as to where Reinhold's interests lay.

COMMENTARIUS IN OPUS REVOLUTIONUM COPERNICI. In the preface to the *Prutenic Tables* Reinhold remarks, "I have explained the true causes and the arrangement of the individual calculations in the commentary that I have written on the *De revolutionibus* of Copernicus [*opus revolutionum Copernici*]." Unfortunately, Reinhold died of the plague before he could publish this work or a dozen others mentioned in the imperial charter of 1549, "Diploma Caesareum Concessum Erasmo Rheinholt Salveldensi," printed at the beginning of the *Prutenicae tabulae.* Also included in this list was a treatise entitled "Hypotyposes orbium coelestium." Astronomers of the following centuries not only assumed that these works were lost, but also that, had they existed, the author would be revealed as an outstanding disciple of Copernican cosmology. Both assumptions are false. Around 1900

L. A. Birkenmajer and the Berlin librarian Valentin Rose independently discovered a manuscript in the Berlin Staatsbibliothek, Latin 2° 391, entitled "Commentarius in opus Revolutionum Copernici," which they attributed to Reinhold. It is, as they conjectured, an autograph manuscript, as the wealth of handwriting samples now known proves beyond doubt. The document, extending between fols. 1–63 and 187–259, has the character of well-organized mathematical notes, surely containing far more interations for planetary parameters than Reinhold could have put in print.

The completeness of the manuscript is a moot point. It is laid out according to Copernicus's chapters in *De revolutionibus*, and the absence of a chapter may simply mean Reinhold had no comments. For example, concerning Book III, Chapter 17, he says "Nihil habet difficultatis sed per sese planum et perspicuum est," and the rest of the page is blank.[22] On the other hand, the section on Saturn stops before the calculations agree with the parameters adopted in the printed tables, and the following page begins "♃ Initium vide in alto libro." In any event, no commentary is found here on Book I, where Copernicus introduces his cosmology, nor is there any direct discussion of world models, although Reinhold calculates for the Ptolemaic equant in each case before he goes on to the Copernican epicylet.

The manuscript is not devoid of cosmological interest, however, for as A. Birkenmajer discovered, there is a tantalizing marginal correction on folio 233 in the commentary on Book V, Chapter 14, concerning the relative size of orbits of the Earth and Jupiter[2] (see Figure 6). Reinhold has exactly copied the diagram from *De revolutionibus*, and writes "Nam F sit locus planetae observati, S locus terrae Copernicae, nobis vero locus ☉" ("For if F is the place of the planet observed, S is the place of the earth for Copernicus, for us the place of the true sun"); the last four words are struck out, and under the diagram a marginal note reads "Juxta nostras hypotheses novas, locus ☉ esset ex oppositio S per E, centrum universi" ("According to our new hypotheses, the place of the sum would be opposite S through E, the center of the universe"). On the basis of this note, Birkenmajer concluded (1) that Reinhold had arrived at a new planetary model neither Ptolemaic nor Copernican and (2) the system was in all likelihood what we might call "proto-Tychonian."

The clue remains tantalizing because it is no more than a hint—the diagram becomes wrong if a Tychonic system is intended, and later in the manuscript, under Mars, there is no corresponding comment at the similar place where point S again is identified with the Earth. Nevertheless, Birkenmajer's interpretation agrees with everything else we learn about Reinhold's approach to planetary models.

HYPOTYPOSES ORBIUM COELESTIUM.   In the catalog of the Cambridge University Library there appears an anonymous work attributed to Reinhold, *Hypotyposes orbium coelestium, quas appellant theoricas planetarium congruentes cum tabulis Alphonsinis et Copernici, seu etiam tabulis Prutenicis: in usum scholarum publicatae* (Strassburg, 1568).[23] That this might be a posthumously printed work of Reinhold, seems reasonable in the light of the title of a proposed book listed at the beginning of the *Prutenic Tables*: "Hypotyposes orbium coelestium, quas vulgo vocant Theoricas Planetarium, congruentes cum tabulis Astronomicis supra dictis." However, the book was reissued in Cologne in 1573, with sheets from the same printing, but with a slightly different title[24] and a new preface by C. Dasypodius (of Strassburg clock frame), which stated that the author was unknown, although possibly Reinhold.

Meanwhile, the identical text appeared in Wittenberg in 1571 under the title *Hypotyposes astronomicae, seu theoriae planetarum. Ex Ptolemaei et aliorum veterum doctrina ad observationes Nicolai Copernici & canones motuum ab eo conditos accomodatae.*[25] This time, however, Caspar Peucer claimed the authorship, and in an indignant preface declared that, unknown and unbidden by him, Dasypodius had printed the work. "I would prefer that Dasypodius would publish his own and not someone else's works," he wrote, "or at least not without the consent of the one to whom they belong."

Peucer's preface to this rare little text is an extraordinary commentary on the state of the Copernican hypotheses as taught in the great academic center, Wittenberg—the home of the *Prutenic Tables* and the spot from which Rheticus had gone to Poland to encourage the publication of *De revolutionibus*. Peucer complains of the rude state of the Alfonsine calculations, as well as the "offensive absurdity, so alien to the truth, of the Copernican theories." The proper solution, he contends, is the Ptolemaic model made consistent with recent observations. This is implied by Peucer's own title much more clearly than by the Dasypodius version.[26]

Peucer goes on to say:[27]

> Among these masters whom we remember, and whom Germany admires, there remains only Georg Joachim Rheticus whose genius is remarkable, obviously formed and created by nature for just these arts, and whose skill is matched

by no one. If a suitable Maecenas should offer help, whose patronage and support would loosen his straits and free him from other occupations so that he could devote himself exclusively to cultivating and completing his studies, then I have no doubt but that something great would result.

Of Erasmus Reinhold, my teacher, to whom I owe my eternal gratitude—a man well-versed, not only in mathematics but in universal philosophy, and very careful besides—brilliant testimonies to this care exist and therefore his studies were correct and deserving the highest praise. He conceived of the greatest things which he surely would have attacked and completed if a longer life had been granted him. Among others, he often promised us new hypotheses of motions, having grown weary of the Peurbachians'. Unfortunately the other works that he was contemplating were impeded by the elaboration of the *Prutenic Tables,* which do exist, by their confirmation, which was somewhat weak, and by his premature death, which tore from us the fruits of his work that would have been handed down to posterity from his careful and unflagging study.

When Reinhold died, Peucer relates, the University at Wittenberg asked him to transfer from medicine to mathematics in order to carry on Reinhold's studies. "I had understood from Erasmus that there was need for new hypotheses in doctrine of celestial motions," he continues. "I preferred the observations and canons of Copernicus to the *Alfonsine Tables* for many reasons, but I sensed the Copernican hypotheses were not to be introduced into the schools in this way." This, then, was the genesis of Peucer's textbook. Later, within the text proper, Peucer takes his position even more plainly: "We stand firm in the tracks of Ptolemy and the other ancients, omitting all the recent hypotheses of Copernicus, which he, having followed Aristarchus of Samos and certain others of the ancients, employed according to a particular plan of his own."[28]

How completely Peucer's own aims reflect those of Reinhold we can never know. At least one interesting trace remains, however. Reinhold's 1542 commentary on Peurbach, the text that contained the notable anticipations of Copernicus's opus, was issued posthumously in a revised edition in 1553 (and in at least nine subsequent editions). The revision would presumably have given Reinhold an opportunity to mention Copernicus's work in the commentary, and at the outset, in the solar theory, our expectations are rewarded by repeated references to Copernicus. But, beginning with the lunar theory, citations to Copernicus abruptly vanish.

A note from Caspar Peucer, inserted at the end of the solar theory, explains that the first section was revised and for the most part written anew with wonderful care—as Reinhold often said, "Second thoughts are wiser." Unfortunately, the premature death of Reinhold (at age 42) cut off the remaining revision, as well as the incomplete work on ephemerides, eclipse tables, and an "explicationem demonstrationum Copernici." From our point of view, these "second thoughts" are conservative indeed. None of the seven references to Copernicus in Reinhold's commentary on the solar theory would give the reader any clue that Copernicus believed in a fixed sun.

Peucer's *Hypotyposes* runs along similar lines to Reinhold's commentary, including even some identical phrases. Perhaps Peucer borrowed these from Reinhold when writing his own notes, but alternatively his work may have been closely based on a manuscript left behind by Reinhold. Like Reinhold, Peucer became an influential professor at Wittenberg. His own opinions undoubtedly followed the philosophical pattern established by Reinhold; we can notice that Peucer, like Reinhold, respected Copernicus as a mathematical astronomer even though he rejected or passed in silence over the Copernican cosmology.[29]

As a mathematical astronomer, Peucer's abilities did not rival those of his teacher, and there is no evidence that he did anything as creative in astronomy as envisioning a proto-Tychonian universe. Nevertheless, it is interesting to note that Tycho Brahe visited Wittenberg on at least four occasions (in 1566, 1568, 1575, and 1598), studying under Peucer on the first occasion and staying in a house owned by Peucer during the 1598 visit; also, Tycho visited Reinhold's son in Saalfeld in 1575 and there saw Reinhold's manuscripts.[30] We are tempted to imagine that Tycho's own cosmological views grew from seeds planted at Wittenberg by a tradition that honored Copernicus, but which followed Osiander's admonition that it is the duty of the astronomer "to conceive and devise hypotheses, since he cannot in any way attain the true causes."

In retrospect we see Erasmus Reinhold's role in the dissemination of the Copernican theory as both influential and ambiguous. Because of the quite apparent links between the *Prutenicae tabulae* and *De revolutionibus*, any success of the one enhanced the reputation of the other. Reinhold's favorable opinion even became part of the advertisement for the second edition of *De revolutionibus* in 1566. On the other hand, the astronomy lecturing at the great Wittenberg University reflected Reinhold's own silence on the heliocentric world view. Copernicus's cosmological insights were thus reduced to "absurd" hypotheses, but at least they were not yet forcibly suppressed. Like winter wheat, Copernicus's ideas lay dormant but virile, waiting to grow in the vernal climate of the seventeenth century.

Appendix: Sample Calculation of Precession, and Solar and Lunar Longitudes

It seems useful to show in detail the rather similar calculations for these three quantities, each of which involves first the tabulation of a mean quantity and an associated anomaly. In his foldout sheet, Reinhold uses the tables for years, months, days, etc., but here let us follow the sexagesimal days procedure for 17 May 1490:

Sexagesimal	Precession of the Equinox		Simple Anomaly
2	16° 30′ 16″		2^s 4° 5′
31	4 15 49		32 3
6	50		6
33 days	5		1
Position 1 Jan. A.D. 1	5 32 24		6 40
	26 19 24		2^s 42° 55′
Trepidation	40 5	Doubled	5^s 25° 50′
	26° 59′ 29″		

The doubled anomaly becomes the argument in the "prosthaphaereseon aequinoctialis" table and the trepidation (called merely "Praecessionis aequinocto") is read out and added as shown.

Now use the sexagesimal days for the mean motions of the solar longitude and anomaly:

Sexagesimal	Simplicis Solis	Anomaliae Solis
2	26^s 22° 44′ 32″	16^s 14° 20′
31	33 13 52 30	33 11 42
6	5 54 49 8	5 54 49
33 days	32 31 30	32 31
27	26 37	27
20	20	
Position 1 Jan. A.D. 1	4 32 29 52	3 31 39
	0^s 36° 54′ 29″	5^s 25° 29′

The procedure for the Sun involves a more elaborate table, prost-haphaereseon solis. If we follow the style of Neugebauer,[31] we have for the Copernican columns:

C_1 independent variable;

C_3 change in direction of the solar anomaly as a consequence of the small central circle that changes the orbital eccentricity, with the anomaly of precession as an argument;

C_4 proportional parts in the change of orbital eccentricity;

C_5 longitudinal effect of orbital eccentricity at minimum eccentricity;

C_6 the change in C_5 when the orbital eccentricity is at a maximum.

The true longitude λ of the Sun can now be obtained. For the given moment t we have the following elements:

$$\psi \text{ anomaly of precession,}$$

$$\lambda_M \text{ mean longitude of Sun,}$$

$$\eta \text{ solar anomaly.}$$

Then the true longitude $\lambda = \lambda(t)$ is given by

$$\lambda = \lambda_M + C_5(\eta') + C_4(\psi) \cdot C_6(\eta')$$

where

$$\eta' = \eta + C_3(\psi).$$

In the example on Reinhold's foldout sheet we have:

η	5^s	$25°$	$28'$	$56''$
$C_3(\psi)$		2	28	5
η'	5	27	57	1
$C_4(\psi)$			1	31
λ_M	0	36	54	29
$C_5(\eta')$			57	10
$C_6(\eta')$			16	18
$C_4 \cdot C_6$				25
λ	0	37	52	4
Precession		26	59	29
λ	1^s	$4°$	$51'$	$33''$

The first longitude of the Sun is measured from the first star of Aries according to Copernicus's custom, the last from the apparent equinox.

Similarly, find the positions of the difference in lunar and solar mean longitudes, and the lunar anomaly:

Sexagesimal	Lunae medius a sole	Anomaliae lunaris
2	53^s 22° 59′ 56″	47^s 52° 48′
31	54 47 26 29	0 52 8
6	13 8 40 9	18 23 24
33 days	6 42 17 41	7 11 9
27	4 5 29 9	5 53
20	4 4	4
Position 1 Jan. A.D. 1	3 29 58 23	3 27 13
	5^s 36° 55′ 51″	5^s 52° 39′
Doubled	5 13 51 42	

If we designate the elements for time t (again following Neugebauer[32])
as

η mean longitude of Moon — Sun

γ lunar anomaly

then the true longitude of the Moon is

$$\lambda = \eta + \lambda_M + C_5(\gamma') + C_4(2\eta) \cdot C_6(\gamma')$$

where $\gamma' = \gamma + C_3(2\eta)$ and the Copernican columns are:

C_1 independent variable;

C_3 motion in the smaller secondary epicycle (which changes the effective size of the large epicycle, thus in effect changing the orbital eccentricity, the so-called evection) as seen from the center of the large epicycle;

C_4 proportional parts in the change of epicycle size;

C_5 effect of the large epicycle at minimum size;

C_6 the excess over C_5, when the large epicycle is at maximum size.

In the example on Reinhold's foldout sheet we have:

γ	5^s	52°	39′	11″
$C_3(2\eta)$		− 10	21	19
γ'	5	42	17	52
$C_4(2\eta)$			11	8
η	5	36	55	51
2η	5	13	51	42
$C_5(\gamma')$		1	23	10
$C_6(\gamma')$			40	31
$C_4 \cdot C_6$			7	31
$C_5 + C_4 \cdot C_6$		1	30	41
$\eta + C_5 + C_4 \cdot C_6$	5	38	26	32	True Moon from mean Sun
Mean Sun		36	54	29
Precession		26	59	29
γ	0^s	42°	20′	30″	True Moon from apparent equinox.

Notes and References

[1] From the preface of *Theoricae novae planetarum Georgii Purbachii Germani ab Erasmo Reinholdo Salveldensi pluribus figuris auctae* (Wittenberg, 1543). Translation quoted from Pierre Duhem, *To Save the Phenomena* (Chicago, 1969), p. 72.

[2] A. Birkenmajer, "Le commentaire inédit d'Erasme Reinhold sur le 'De Revolutionibus' de Nicolas Copernic," in *La science au seizième siècle* (Paris, 1960), pp.171–77.

[3] J. C. Houzeau and A. Lancaster, *Bibliographie Générale de l'Astronomie* (reprint London, 1964), under the item 12727 lists another reissue, identical except for title page, at Wittenberg, in 1561. I have not been able to locate a copy of this issue and apparently E. Zinner only repeated the Houzeau–Lancaster reference in his *Geschichte und Literatur in Deutschland zur Zeit der Renaissance* (Stuttgart, 1964), item 2270.

[4] Bibliothèque Nationale MSFR12288; I am indebted to Professor I. B. Cohen for drawing my attention to this manuscript.

[5] Owen Gingerich and William Walderman (trans.), "Johannes Kepler: Preface to the Rudolphine Tables," *Quarterly Journal of the Royal Astronomical Society,* vol. 13 (1972), p. 366.

[6] In the explanatory introduction, Reinhold gives only Saturn as an example; the other planets as well are given on the foldout sheet, but the examples for both Venus and Mercury are marred by several computational errors.

[7] Erasmus Reinhold, *Prutenicae tabulae coelestium motuum* (Wittenberg, 1585). The first paragraph is quoted from the second page of the dedication to the Duke of Prussia, the second facing page *3 in the author's preface. I wish to thank Molly Sanderson Campbell for the translation. Incidentally, Reinhold says that the *Prutenic Tables* are named for both the Duke of Prussia and for Copernicus.

[8] Reinhold's numbers in this and the following table were deduced during the analysis with the CDC 6400 computer, and, with the exception of Saturn's parameters, were subsequently verified from the Reinhold manuscript discussed in the second part of this paper. Apparently some of Reinhold's work was recorded elsewhere, including the final stages of the Saturn calculations and the initial stages for Jupiter.

[9] See Rev. Joseph Hunter, "Some particulars of the Life of John Feild, The Proto-Copernican of England," Gentleman's Magazine, May (1834). Francis Johnson, in his generally excellent book *Astronomical Thought in Renaissance England* (Baltimore, 1937), errs on p. 134 in supposing that Feild's work was "a revision of Reinhold's Prutenic Tables, reduced to the position of London for the convenience of English astronomers." The error is repeated by Christine Schofield on p. 292 of "The Geoheliocentric Mathematical Hypothesis," *British Journal for the History of Science,* vol. 2 (1965), pp. 291–96.

[10] According to Edward Rosen, "Was Leovitius an Opponent of Copernicus?" In his *Three Copernican Treatises,* 3rd ed. (New York, 1971), pp. 301–2. Most of Rosen's opinions are contradicted by Birkenmajer's article (see ref. 12).

[11] J. L. E. Dreyer, *Tychonis Brahe Opera Omnia,* vol. 3 (Copenhagen, 1916), pp. 221–22.

[12] A. Birkenmajer, "Was Leovitius an Opponent of Copernicus" (in Polish), *Kwartalnik Historii Nauki & Techniki* of the Polish Academy of Sciences, vol. 4, (1959), pp. 32–34; I am indebted to P. Czartoryski and J. Dobrzycki for an advance copy of the French translation in *Studia Copernicana*, vol. 4 (1972), pp. 767–78.

[13] Among Leowitz's extant unpublished manuscripts is "Ephemerides compendiosae quadringentorum annorum incipientes a. 1349 ex extendentes se usque ad a. 1750," Vienna Nationalbibliothek manuscript 10786. Another item of possible interest, cataloged next to Tycho's papers, is "Tabulae colligendorum ex ephemeridibus motuum planetarum ad quodius tempus propositum," manuscript 10719. I have not yet examined either manuscript. See *Tabulae Codicum Manu Scriptorum in Bibliotheca Palatina Vindobonensi Asservatorum VI* (Vienna, 1873).

[14] J. C. Houzeau and A. Lancaster, *op. cit.* ref. 3, state under item 14779, Moleti, that the calculations are made with the *Prutenic Tables*, and this is repeated in J. C. Houzeau, *Vade-Mecum de l'Astronomie* (Bruxelles 1882) p. 925. The confusion may have arisen from the fact that Moleti used the *Prutenic Tables* for the times of new and full Moon and for eclipses, according to E. Zinner, *Entstehung und Ausbreitung der Coppernicanischen Lehre* (Erlangen, 1943) p. 279. The following positions for 26 September 1563 show that Carello and Moleti agree with the Alfonsine-based Leovitius:

	Sun	Moon	Saturn	Jupiter	Mercury
Leovitius	12° 18'	27° 49'	4° 8'	4° 7'	4° 41'
Carelli	12 18	27 22	4 8	4 3	4 40
Moleti	12 18	27 22	4 8	4 3	4 40
Stadius	12 11	28 00	1 57	4 9	8 15

[15] See J. L. E. Dreyer, *Tycho Brahe* (reprint New York, 1963), pp. 18–19.

[16] Edward Rosen, *op. cit.* (ref. 10), p. 25.

[17] Pierre Duhem, *op. cit.* (ref. 1), p. 73.

[18] L. A. Birkenmajer, *Mikolaj Kopernik* (Craców, 1900), especially pp. 622–633. E. Zinner, *op. cit.* (ref. 3); A. Birkenmajer, *op. cit.* (ref. 2).

[19] After examining over 100 copies of the 1543 *De revolutionibus*, I realize that the list is too small and would have to include Casper Peucer (whose annotated copy is preserved at the Paris Observatory), Henry Savile (the donor of a copy in the Bodleian Library, Oxford), John Dee, Thomas Digges, and quite a few others.

[20] I should like to thank Professor H. A. Brück, Astronomer Royal for Scotland, and Mrs. Vivienne Lawson, Librarian of the Royal Observatory Edinburgh, for placing a microfilm of the Crawford *De revolutionibus* at my disposal for further study.

[21] Konrad Haebler, *Rollen und Plattenstempel des XVI. Jahrhunderts*, vol. 1 (Leipzig, 1928), pp. 188–89. I am greatly indebted to Howard Nixon of the British Museum for this identification.

[22] Other specimens of Reinhold's writing, including his signature, are (1) in vol. 1 of the Dean's Book of the University of Wittenberg, now in the Halle University Archives, including fols. 138v, 139, 143, 148v–151v; (2) a presentation inscription on Theodosius's *De sphaericis libri tres* (Vienna, 1529), no. 395846 in the Wroclaw University Library; (3) a presentation inscription in a formal hand on his *Prutenicae tabulae* (1551), a copy later owned by Tycho Brahe (now no. 14 J 176 in the Prague

Univ. Library); (4) several lines and two signatures in a copy of his edition of Ptolemy's *Mathematicae constructions liber primus* (Wittenberg, 1549), owned by R. B. and M. S. Honeyman, San Juan Capistrano, California; (5) extensive marginal annotations in the *Epytoma in almagestum Ptolemei* (Venice, 1496), no. 1196 in the library of the Paris Observatory; and (6) a 204-leaf manuscript commentary on Euclid, no. Latin 4° 32 in the Deutsche Staatsbibliothek, Berlin (see V. Rose, *Verzeichniss der lateinischen Handschriften der Königlichen Bibliothek zu Berlin*, vol. 2, part 3 (1905), p. 1366, no. 81.

[23] I am indebted to Dr. J. Dobrzycki for a microfilm of this manuscript, as well as for a translation of L. A. Birkenmajer's chapter on this material.

[24] See H. M. Adams, *Catalogue of Books Printed on the Continent of Europe, 1501– 1600, in Cambridge Libraries* (Cambridge, 1967), item R327. Houzeau and Lancaster, *op. cit.*, list this in item 2639 and in a note attribute it to Reinhold. Their variant title and date apparently derive from Johann Ephraim Scheibel, *Astronomische Bibliographie* (Breslau, 1786), under the year 1565. That this was an error from J. Clessius, *Elenchus consummatissimus librorum* (1602) Scheibel already understood, as indicated under the year 1568.

[25] *Absolutissimae orbium coelestium hypotyposes, quas Planetarum Theoricas vocant: congruentes cum Tabulis Alphonsinis & Copernici, seu etiam tabulis Prutenicis. Opus non minus elegans, quam utile, quo ea quae ad Astronomiam percipiendam necessaria sunt evidentissime explicantur, planetarum etiam motus qui qualesue sint, quibusue afficiantur passionibus erudite traditur: omnia in gratiam studiosae iuventutis in lucem aedita.* Houzeau and Lancaster, *op. cit.*, list this correctly, as item 12736, but the book is octavo, not quarto.

[26] *Ibid.* item 12737, under Peucer; the Strassburg imprint of 1571 is quite possibly a ghost and is not listed in M. U. Chrisman, *Bibliography of Strasbourg Imprints, 1480–1599* (New Haven, 1982).

[27] The Peucer text is only one of many to take this stance, the best known being G. A. Magini's *Novae coelestium orbium theoricae congruentes cum observationibus N. Copernici* (Venice, 1589). A particularly early one is J. Garcaeus, *Tractatus brevis et utilis, de erigendis figuris coeli* (Wittenberg, 1555 etc). Only slightly different is Albertus Leoninus, *Theoria motuum coelestium referens doctrinam Copernici ad mobilitatem solis eamque sequentes hypotheses cum nova de motu ipsius terrae sententia & hypothesi* (Cologne, 1583); this latter book is so extremely rare that several writers have incorrectly supposed that it is a ghost.

[28] Peucer, *op. cit.* (fols. 2v–3). I wish to thank Ann Wegner Brinkley for a most useful preliminary translation of this material.

[29] C. Peucer, *op. cit.* in text (see note 25), pp. 37–38; in the Dasypodius edition (note 24), p. 34.

[30] Peucer, in his *Elementa doctrinae de circulis coelestibus et primo motu* (Wittenberg, 1551), gives special prominence to Copernicus in his chronological list of astronomers at the beginning of the book. See comments by Lynn Thorndike, *History of Magic and Experimental Science*, vol. 6 (New York, 1941), p. 11, and by P. Duhem, *op. cit.* (ref. 1), pp. 75–76. On the other hand, I have found at Caius College a set of manuscript astronomy notes taken at courses in Wittenberg in 1566–70 while Peucer was at the peak of his astronomical career, and although Copernicus is occasionally mentioned, students were apparently carefully shielded from the helio-

centric cosmology. See my article in *Proceedings of the American Philosophical Society*, vol. 117 (1973) pp. 513–22, 1973 [reprinted as selection 16 in this anthology].

[31] J. L. E. Dreyer, *op. cit.* (ref. 14), pp. 21–23, 29, 83, 271.

[32] O. Neugebauer, "On the Planetary Theory of Copernicus," in *Vistas in Astronomy*, ed. by A. Beer, vol. 10 (Oxford, 1968), pp. 89–103.

[33] O. Neugebauer, *The Exact Sciences in Antiquity* (Providence, 1957), pp. 196–98.

De revolutionibus:
An Example of Renaissance Scientific Printing

Nicholas Copernicus's *De revolutionibus orbium coelestium* is the most important astronomical book of the sixteenth century. Typographically dull and formidably technical, it nevertheless remains one of few truly great landmarks in scientific thought. *De revolutionibus* was published in Nuremberg in the spring of 1543, the year in which Copernicus died, and a reprint was issued in Basel 23 years later, in 1566.

Copernicus wrote his book at the cathedral in Frombork (Frauenburg), far from any academic center and far from any major printer. He would never have seen his work printed except for the intervention of a young professor of astronomy from Lutheran Wittenberg, Georg Joachim Rheticus. Rheticus traveled to this northernmost Catholic diocese in Poland in 1539, bringing as gifts to Copernicus three bound volumes containing five scientific works.[1] Three of the titles had been published by Johannes Petreius in Nuremberg, who had recently established himself as the leading scientific printer north of the Alps.

The 67-year-old Copernicus, who had never had any students, warmed to the young disciple, and by 1540 gave him permission to publish in nearby Gdansk the now exceedingly rare *Narratio prima* or

Selection 14 reprinted from *Print and Culture in the Renaissance,* ed. by Gerald P. Tyson and Sylvia S. Wagonheim (Newark, Delaware, 1986), pp. 55–73.

"First Report." The favorable reception accorded this brief tract persuaded Copernicus to let Rheticus take a copy of the treatise itself back to Germany for printing.

Probably from the very beginning Rheticus had it in mind to have Copernicus's book printed by Petreius in Nuremberg rather than in the academic center of Wittenberg. At that time in Wittenberg the principal printers were Joseph Klug, Hans Lufft, Georg Rhaw, and Nicholas Schirlentz,[2] who were generally so busy with Lutheran materials and occasional octavo university texts that they apparently could not cope with a large and complex scientific work. (The situation repeated itself a few years later when the senior professor of astronomy of Wittenberg, Erasmus Reinhold, published the *Prutenicae tabulae* based on Copernicus's work, not in Wittenberg but in Tübingen; if ever one would like to read the proofs near at hand, it would be for pages of numerical tables!) In contrast to the Wittenberg houses, Petreius had issued not only such large scientific works as Regiomontanus's *De triangulis* (1533), Apianus's *Instrumentum primi mobilis* (1534), and Witelo's *Optica* (1535) (the texts given to Copernicus), but also folio editions of Apianus's *Instrumentus sinuum* (1541), and Peurbach's *Tractatus super propositiones Ptolemaei* (1541).[3] Thus it was that in 1542 Rheticus procured an academic leave to oversee the printing of Copernicus's manuscript at Petreius's shop in Nuremberg.

Probably the first step was to commission the cutting of the 142 woodblock illustrations. These involved geometrical diagrams with letters on the blocks. Only six were used twice; two others (on fols. 153 and 158) were accidentally interchanged.

Second, it was necessary to acquire at least one hundred reams of paper. The sheets were 399 by 281 mm, a typical size known as pot paper, and the ratio of the sides was $\sqrt{2}:1$ so that repeated folding in half would have preserved the same proportion.[4] The sheets were printed in folio and gathered in pairs to give each lettered signature four recto folio numbers (i.e., eight pages). The paper has the common watermark P (only coincidentally Petreius's own initial), and Petreius used this particular stock for perhaps the first time. The thickness of the sheets varies considerably, occasionally giving the false impression of a thick-paper copy. Eventually, by signature P, Petreius temporarily ran out of this paper, and hence that signature had a much thinner stock with a goblet watermark. Part of the press run of signature Q was also on this stock; thereafter the original P-watermark paper was used.

As for type, Petreius employed primarily the Roman font that he

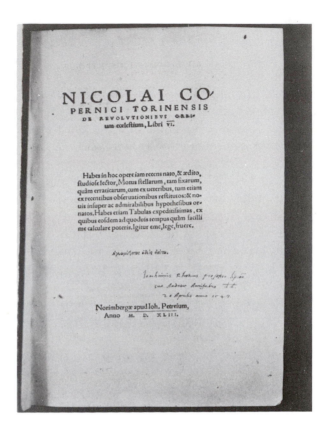

FIGURE 1. De revolutionibus *(Nuremberg, 1543). Title page with a presentation inscription from Copernicus's only disciple, Georg Joachim Rheticus, dated 20 April 1547.*

had used for many years, approximately 5 mm high, giving 37 lines to his page plus the running head and catchword. Most of the small woodblock initials (four lines high), as well as the larger initials (eight lines high) that open the chapters, had been regularly used by Petreius since the 1520s. They are considerably more elegant than those of most of his contemporaries, and have been ascribed to Hans Sebald Beham, a Nuremberg artist strongly influenced by Dürer.[5]

How long did the printing take? Rheticus went to Nuremberg early in May of 1542, and he autographed a presentation copy of the completed work on 20 April 1542.[6] Copernicus himself, some seven hundred kilometers to the northeast, did not receive the final sheets until the day of his death, 24 May 1543.[7] The printing must have started

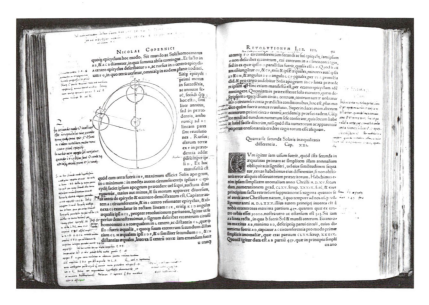

FIGURE 2. De revolutionibus *(Nuremberg, 1543). A well-annotated page from the copy owned by Erasmus Reinhold, teacher at Wittenberg in the 1540s. Permission of Astronomer Royal of Scotland, Royal Observatory, Edinburgh.*

immediately upon Rheticus's arrival, because he had corrected the first two signatures before the end of May in 1542.[8] This section needed only one simple woodblock diagram, and probably some weeks were required for additional blocks to be cut, so Rheticus took off most of June to visit his birthplace in Feldkirch. Therefore, the printing did not resume before the first part of July. By mid-October, when Rheticus left Nuremberg to take up his new position as a professor in Leipzig, the work was by no means finished and the proofreading was turned over to Andreas Osiander, the learned pastor of the Sankt Lorenz Kirche in Nuremberg.

At least at first, Petreius used only one press at a time for the work, as may be readily established through his repeated use of a very limited number of initials. The general rule is that one press could print both sides of a ream of paper (480–500 sheets) in a day, so with an edition of four to five hundred copies, completely efficient proofreading, and no delays for woodblocks, the printing of the 101 different sheets would have required a minimum of four months. As we have seen, the actual printing took at least twice as long as the minimum.

The book must have been printed sequentially from folio 1 to 196, with the six folios of unnumbered front matter printed last. At about the same time that the title page was struck off, Petreius issued an errata leaf for the first 146 folios of the book; we know that the errata leaf was issued almost simultaneously with the printing of the title page because some copies of the book (about one in twenty) have an errata leaf with a title page on the other side, with precisely the same setting as the title page of the edition itself. Now some of the corrections on the errata leaf are the sort that only an author himself would make, so we can conjecture that the printed signatures may have been sent in batches to Copernicus for his inspection.[9] Thus we could specify that if the mail took two weeks each direction, and was delayed two weeks in the sending and in the turnaround (which would tally with the month's delay in Copernicus's receiving the final copy), then the process would have taken at least two months, which is consistent with the 30-sheet gap between the end of the errata leaf and the finished book and the less-than-maximal rate of printing one sheet every two days. Incidentally, in a few copies of the book a standard set of errors is corrected by hand not just to folio 146, but to the very end.[10]

While the foregoing analysis has offered a plausible argument for a first edition of four to five hundred copies, we could possibly adjust the numbers to double this. Unfortunately there is no direct evidence surviving in Nuremberg concerning any of Petreius's press runs. Apparently the only comparable sixteenth-century data comes from the Plantin-Moretus press in Antwerp; their editions varied considerably in size, but for medical books and herbals, eight hundred constituted a typical run.[11] These probably could be expected to sell much better than a specialized treatise such as De revolutionibus. For perspective, the press run of the first edition of Newton's Principia is estimated at three hundred to four hundred copies.[12]

As an alternative approach for seeking the size of an edition, as well as a way to learn more about the audience of book buyers and the cost of the book, I now turn to an almost unique project to produce a complete census of Copernicus's De revolutionibus.

In the fall of 1970, when historians of astronomy were already anticipating the quinquecentennial of Copernicus's birth in 1973, I happened to discuss the readership of De revolutionibus with a colleague in England.[13] Little was known about this topic, although Arthur Koestler had branded it "the book nobody read" in his The Sleepwalkers. In our discussion we concluded that this dense treatise probably had very few readers who got all the way through.

Imagine my surprise when, a few days later at the Royal Observatory in Edinburgh, I had an opportunity to examine a first edition of Copernicus, and I found that it was thoroughly and perceptively annotated from beginning to end! It took some detective work to establish that the copy belonged to Erasmus Reinhold, the senior professor of astronomy at Wittenberg in the 1540s, and the leading astronomical pedagogue in the generation immediately following Copernicus.[14] His own annotations gave such a revealing insight into the way the book was seen and studied at the center of the Lutheran academia that I promptly resolved to examine as many other copies of De revolutionibus as possible in order to see what sort of manuscript annotations they might include.

At first, I merely recorded whether or not the books contained marginalia, but gradually the census gained a life of its own, and I began to record full details of binding and provenances as well as notes on the nature of the annotations, if any. Early on I decided to include the second edition (Basel, 1566) in addition to the first (Nuremberg, 1543). The third edition (Amsterdam, 1617) came after the Copernican system was well on its way to being accepted, and hence was of less interest even though important scientists such as Huygens and probably Descartes owned this seventeenth-century edition.

Eventually I aspired to completeness, a hopelessly unattainable goal that has rarely been attempted among major books. Only the Guttenberg Bible, the first folios of Shakespeare, and the elephant folios of Audubon's birds have had complete censuses attempted. There are surveys of locations of such books as Newton's *Principia*, Harvey's *De motu cordis*, and Vesalius's *De humani corporis fabrica*, and I have myself compiled over one hundred locations of Apianus's *Astronomicum Caesarem* (Ingolstadt, 1540), but these lists have not in general tried to be exhaustive.[15]

It is comparatively easy to locate one or several copies of any great book through a variety of aids, including union catalogs in several countries; it is quite another matter to locate all the copies in a given country. For example, the National Union Catalog lists only approximately one-third of the copies of *De revolutionibus* in the United States, not to mention that it includes two ghost locations of the first edition as well. The Adams catalog of sixteenth-century books in Cambridge libraries[16] is comparatively complete, but it does not include the third copy of the first edition held by Trinity College. (Trinity once had visions of auctioning their triplicate copy, but as the wealthiest of the Cambridge colleges, they could scarcely afford the publicity in what might be perceived as selling off their patrimony!)

In any event, the existence of such union catalogs provided an excellent starting point, but was only a beginning for building a complete census. The second step was a letter-writing campaign with *The World of Learning* as an indispensable guide for selecting large and/or old libraries that might own such a volume. Nevertheless, this key reference work has its own blind spots; it did not list the Eton library (which happens to have two copies of the first edition) on the grounds that Eton was merely a preparatory school. Similarly, important libraries in out-of-the-way places (such as Catania, Italy, or Görlitz, East Germany) were formerly overlooked. And even when a library was included, it could be passed by as unpromising. At one point I found some evidence that a library in Liverpool had a second-edition Copernicus, so I sent out some inquiries, and the university librarian replied, "I suppose it is the second edition you are particularly interested in for we do have two copies of the first!"

A third important means of tracing Copernicus's *De revolutionibus* has been through the marketplace. Many book dealers have willingly shared their knowledge of locations, both in institutions and in private collections. Recently, Dr. H. A. Feisenberger of Sotheby's enabled me to examine the record book of Ernst Weil, a rare book dealer who was active in London from 1934 till 1965, and who kept track of numerous science titles passing through the market during those decades.[17] This enabled me to discover that certain copies of *De revolutionibus* mentioned in book catalogs were in fact identical with ones I had already located, and it also showed me how Weil had combined two defective copies into a single complete book.

The published auction records are another important tool for locating books. In the process of examining these records, I was at one point surprised to note a copy in a Grolier binding, since I had not come across a Copernicus with that description. Further investigation showed that the book was auctioned as a "modern facsimile," a euphemistic way of designating a fake binding. Although the Grolier Club in New York has recorded the genuine Grolier bindings, they have not bothered about the fakes. However, I soon discovered that Howard Nixon, former Keeper of Printed Books at the British Museum, was an expert on this subject, and he was happy to be able to inform me that the volume in question had been bought by the Victoria and Albert Museum in London as an example of decorative art. Who

NICOLAI
COPERNICI TO-
RINENSIS DE REVOLVTIONI-
bus orbium cœleftium,
Libri v i.

IN QVIBVS STELLARVM ET FI-
XARVM ET ERRATICARVM MOTVS, EX VETE-
ribus atɋ recentibus obferuationibus, reftituit hic autor.
Præterea tabulas expeditas luculentasɋ addidit , ex qui-
bus eofdem motus ad quoduís tempus Mathe-
matum ftudiofus facillime calcu-
lare poterit.

ITEM, DE LIBRIS REVOLVTIONVM NICOLAI
Copernici Narratio prima, per M. Georgium Ioachi-
mum Rheticum ad D. Ioan. Schone-
rum fcripta.

Cum Gratia & Priuilegio CæfMaieft.
BASILEAE, EX OFFICINA
HENRICPETRINA

FIGURE 3. De revolutionibus *(Basel, 1566). Title page of the second edition in the National Library in Prague, long thought to have been annotated by Tycho Brahe, but the marginalia are now known to belong to Paul Wittich.*

would have guessed that the Victoria and Albert houses a first edition *De revolutionibus!*

As a result of this ongoing search, I have located just over 250 copies of the first edition, and about 290 of the second. It is difficult to make any estimate of the completeness of the census, but the auction records give an initial idea. Of approximately 40 first editions whose auctions are recorded since 1890, I have located virtually all the copies sold since 1910. During the past two years five apparently unrecorded copies of the first edition have turned up, primarily in the marketplace, but one proved on the basis of the census information to be a stolen copy. In the same period eight more copies of the second edition have come to light, primarily through a renewed search in Italy, where the previous effort had evidently not gone deep enough. Extrapolation of the present rate of discovery indefinitely into the future would lead to an absurd abundance of copies, but to estimate the number of undis-

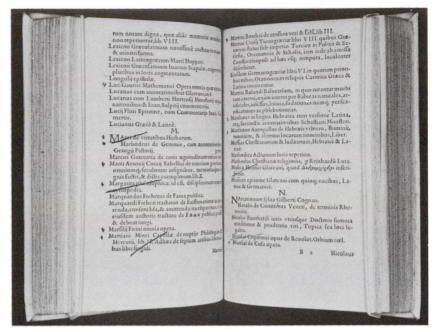

FIGURE 4. *1595 book catalog from the printer Heinrich Petri showing the second edition Copernicus still in print 29 years after its publication. Permission of Houghton Library, Harvard University.*

covered copies as fewer than 25 of each edition might be unwarranted optimism.

Not only have I established the locations, but for the most part I have actually gone to see the books, repeatedly crisscrossing Europe and North America. The process of examining the volumes myself has made possible a more consistent and reliable description than would otherwise have been the case. More than once the statement received in advance from the local librarian, that the book was unannotated, proved false on actual inspection. Furthermore, the opportunity to see many of the copies in the context of their collections has enabled me to give richer details concerning their history. The systematic examination of the copies has allowed at least partial provenances to be established for each volume, and this in turn leads to some inferences about the size of the edition.

How many copies were originally printed? One way to estimate this number is to use information from the extant copies. From those with known provenance, we can estimate the percentage that have survived. For instance, we can make a list of the sixteenth-century astronomers who probably would have owned a copy, and we can ask how many of these ownerships are now recorded in the existing copies. If we name 80 astronomers and find traces of 40 in the books, then we could conclude that perhaps half of all of the original copies are now located. This presumes that all our named astronomers really did own the book, that they left their ownership marks in their copies, and that these marks have not been defaced. It seems reasonable to include some sort of correction for these factors, but how much should it be?

In the actual analysis, with a sample of about 80 names, roughly 40 percent of the astronomers have copies located. Given the roster of about 250 known copies of each of the first two editions, and with some correction for the circumstances listed above, we can deduce printings of four hundred to five hundred copies, with the second being perhaps slightly larger than the first.

One factor that might influence whether a sixteenth-century astronomer actually owned Copernicus's book would be price and availability. The correspondence of astronomers in the sixteenth century is filled with attempts to acquire specific books, which was often difficult. The first edition of *De revolutionibus* probably stayed in print nearly twenty years. Houghton Library at Harvard has a 1595 book catalog from Heinrich Petri, the publisher of the second edition.[18] Copernicus's book is listed, but crossed out by hand, suggesting that the second edition went out of print just around this time, nearly thirty years after its 1566 printing.

The census has cast some light on the price of the volume. One bibliophile who always annotated his books with respect to price was Achilles Permin Gasser; his copy exists in the Vatican Library, but unfortunately for us, his book was presented to him free by Petreius! A first edition bought in Wittenberg in 1545 carries the price 1 florin;[19] it was part of a group, priced as follows:

Regiomontanus, *Epitome* (Basel, 1543)	14 g
Copernicus, *De revolutionibus* (Nuremburg, 1543)	1 f
Werner, *Geographiae Ptolemaei* (Nuremberg, 1514)	5 g
Peurbach, *Tractatus super propositiones Ptolemaei*	
Nuremberg, 1541)	3 g
	2 f 10 g

(The Peurbach was a much thinner folio printed by Petreius, and its price per page was twice as much as for the Copernicus.) Kepler's *De revolutionibus* contains an early clipping from another copy with a price of 28 groschen 6 pfennigs, but this price of over 2 florins could well have included a binding and possibly other works.[20] In 1570 Michael Maestlin of Tübingen paid $1\frac{1}{2}$ florins for a copy bound with the Regiomontanus *De triangulis*.[21] In the sixteenth century a florin was a sizable sum; when Rheticus was lured to Leipzig, he was offered the exceptionally high salary of 140 florins.[22] If we use that as a baseline, then a copy of the book cost well over $100 by present standards. There exists in Bamberg a small booklet of notes intended to be placed in specific positions in the margins of *De revolutionibus*; apparently the astronomer-copyist was too poor or simply unable to find the book for himself, and he wrote out the notes from his teacher's book for the time when he might have his own copy.[23]

The census does show that the distribution had definite geographic restrictions. The first edition was sold primarily throughout northern Europe. Virtually all of the copies reaching England, however, came with scholars returning from the Continent, and there is no evidence that the book was distributed through English booksellers. On the other hand, the Basel edition poured into Italy, France, and England. Thus in Oxford the Bodleian Library had three or four copies of the 1566 edition by 1622, but apparently did not get the *editio princeps* until it was secured as an antiquarian item in the nineteenth century. The second edition remains to this day comparatively rare in Germany.

Having noted the availability of the book in the sixteenth century, I now turn to the opposite side of the coin, the destruction of copies over the intervening years. Is accidental fire a great enemy of books? The evidence is surprisingly sparse on this means of attrition. A copy of Copernicus's book was definitely lost in the great fire in Copenhagen in 1728. Probably copies were burned in the great fire of London in 1666, but no specific losses have been documented. It is difficult to find reports of great libraries being burned, and despite the wooden floors in Oxford and Cambridge no major library fires seem to have occurred. It is sometimes reported that peasants looted and burned the mathematical library of John Dee (who listed two copies of *De revolutionibus*), but the evidence now seems in favor of selective and knowledgeable looting.[24] In America the Harvard Library burned in 1765, but the collection did not contain a Copernicus; the University of Virginia Library partly perished in 1895, and it is not known if the 1566 Copernicus now in the collection is the one originally ordered by Thomas Jefferson, or a replacement.

In wartime the situation is different. In the Thirty Years' War the fragmentary evidence suggests that books were moved—some were "liberated" and brought to Sweden, for example—rather than being destroyed outright. In modern times the peril of books seems more severe. A copy was probably lost in Strasbourg in the Franco-Prussian War; copies were lost at Arras in France and possibly an especially important one at the Louvain in World War I; others were destroyed in Douai, Dresden, Frankfurt, and Munich in World War II.

The *De revolutionibus* was sufficiently formidable that it is unlikely that more than a handful of copies were worn out from use. At a very early time (surely by the end of the sixteenth century) it was recognized as a classic, and therefore a valuable as well as expensive book, nor did it go out of date. Hence, Copernicus's treatise was spared the fate of numerous other works from this time that were simply thrown away. Such circumstances, for example, have now made Kepler's little annual calendars extremely rare or even nonexistent.[25]

Nevertheless, volumes that are rarely if ever used are also subject to slow destruction. A number of copies of *De revolutionibus* have been nibbled by mice, some are badly wormed, and copies kept in a humid Mediterranean environment, particularly the second edition, can be very badly browned. Water is perhaps a more dangerous enemy of books than fire; although the one documented case of pirates throwing a shipload of books overboard (because they were so outraged to find no better cargo[26]) is an extreme example, dampness and mildew can ruin books so badly that they become obvious candidates for the trash collector. I would not be surprised if more copies of *De revolutionibus* have been lost through neglect and dampness than from fire and bombs. However, the destruction rate could well have been so low that a majority of those printed still survive.

Perhaps the most significant discovery made in the course of the census is the fact that the book was far more widely read and studied than had previously been imagined. In this connection it was also interesting to find that the marginal annotations often go in families. Reinhold's notes were faithfully copied in part at least twice, and from these copies flowed additional generations of the same remarks. Thus, his opening title page motto, that "celestial motion is circular and uniform, or composed of circular and uniform parts," is now found in several copies, and other technical comments have propagated into a group of perhaps 15 books. A second family of annotations, of more mysterious origin and disseminated around 1580, is found in six or seven copies.

These and other groups of marginalia are generally concentrated in *De revolutionibus* Book III, where Copernicus grapples with the motion of the Earth or Sun. This part is relatively independent of the heliocentric cosmological framework. It is fascinating to discover how rarely the earlier, cosmological portions of the treatise are annotated with any serious degree of perception.

Another revealing group of manuscript changes concerns the title page and front matter of the book. After Rheticus had overseen the first part of the printing, he left to accept a new professorship in Leipzig, as mentioned above. When he finally received copies of the new work in April of 1543, he was dismayed to discover that an anonymous introduction had been added in Nuremberg. This advice to the reader stated that the goal of the work was simply to present a scheme for calculation and that the hypotheses need not be considered true nor even probable. Whether Rheticus was more offended by the philosophy of this statement or by the omission of introductory material that he had intended to include, we can never know. In any event, two existing presentation copies from Rheticus have the added introduction crossed off with a red crayon, and a further exemplar testifies to the former existence of yet a third copy treated in precisely the same way.[27] In turn, several other copies have the introduction crossed out, apparently following the example from Rheticus.

Whether Rheticus knew for sure who wrote the anonymous introduction is not quite clear. There is an interesting three-stage annotation in one of the *De revolutionibus* copies owned by Michael Maestlin, the astronomy professor at Tübingen, who is best remembered as Kepler's teacher.[28] He first wrote, "This preface was added by someone, whoever its author may be (for indeed its weakness of expression and choice of words reveal that it is not Copernicus's style). . . ." On the second page he later wrote, "I found the following words written somewhere among the books of Philip Apian (which I bought from his widow): 'On account of this letter Rheticus . . . became embroiled in a very bitter wrangle with the printer, who asserted that it had been turned over to him with the rest of the work. Rheticus, however, suspected that Osiander had prefaced it to the work. If Rheticus knew this for certain, he declared, he would rough up the fellow so violently that in the future he would mind his own business. . . . Nevertheless, Apian told me that Osiander had openly admitted to him that he had added this as his own idea.' "

Maestlin's third note says simply, "NB. I know for sure that the author of this letter was Andreas Osiander." How could Maestlin be so sure? Because his student Kepler had acquired a copy of *De revolutionibus* that had been presented by Petreius to a member of the

Nuremberg circle of scholars, and this book contained Osiander's name above the introduction.[29] Kepler later publicized this information by printing it in a paragraph on the verso of the title page for his *Astronomia nova* (Heidelberg, 1609). Meanwhile, however, other astronomers in contact with the Nuremberg or Wittenberg groups also found out, as early annotations attest. For example, Valentine Engelhart, who recorded that he bought his copy in Wittenberg in 1545, noted above the introduction that "Andreas Osiander wrote this preface, which Petreius printed without Rheticus's knowledge."

One other curious change goes along with the copies in which Rheticus crossed out the introduction, for he also drew his crayon through the last two words in the title *De revolutionibus orbium coelestium*. The same is true in the copy that Kepler had acquired; Rheticus's colleague Erasmus Reinhold and Valentine Engelhart treated their copies likewise. Precisely what is offensive about the words "celestial spheres" is particularly puzzling since Copernicus used a very similar expression at the beginning of his printed dedication to Pope Paul III; there he referred to his book almost synonymously as *De revolutionibus sphaerarum mundi*. Nevertheless, throughout the rest of the book (which had been printed earlier) the running title is always solely *De revolutionibus*, and the final three books each end with a phrase such as "Finis libri sexti Revolutionum." In one of his copies the sixteenth-century astronomer Johannes Praetorius wrote, "Rheticus affirmed that this preface has been added by Osiander. Also the title was changed contrary to the will of the author by the same person. For it should be 'de Revolutionibus.' Osiander added 'orbium coelestium'."[30] Perhaps Petreius, as printers are wont to do even to this day, decided together with Osiander that the title needed some improvement. The cognoscenti, however, recognized that Copernicus felt that the addition of "Heavenly Spheres" to the original "Concerning the Revolutions" deflected the implicit emphasis on the motions of the earth itself.

Yet another kind of manuscript markings has proved to be quite fascinating: the censorship according to the instructions issued by the Inquisition in 1620. Books were often placed on the *Index librorum prohibitorum* "until corrected," but only in this unique case did the Holy Congregation eventually spell out in full detail the dozen passages to be changed. Thus it is almost immediately possible to recognize which copies have been treated by the censor. I say "almost immediately" because in some instances the offending text was pasted over with a slip of paper on which the new text was inscribed, and if these have been removed by washing, it is a more subtle matter to detect the former censorship.

The census has revealed that about one copy in twelve was caught by the censors, and that these were primarily in Italy where perhaps 60 percent of all the copies were "corrected."[31] In France or Austria numerous copies in clerical institutions completely escaped the censorship. Surprisingly, no copies in Spain or Portugal were censored, and an investigation has shown that in fact the *De revolutionibus* was explicitly permitted in Iberia! Unexpectedly the census has confirmed what scholars had suspected—namely, that the Galileo affair was viewed elsewhere primarily as a local Italian imbroglio.

The detailed study of Copernicus's *De revolutionibus* has shown that this scientific classic is far more common than collectors and dealers had imagined, although this fact has not eroded the 1986 price of over $50,000 for a good copy of the first edition. Converging evidence suggests that a first edition of four hundred or even five hundred copies was printed in 1543, with a comparable number in the Basel edition of 1566. The book was widely distributed, so that it got into the hands of the majority of astronomy professors in the sixteenth century, as well as into major libraries that intended broad representation in their holdings. Copies were owned by the Venetian music theoretician Giuseppe Zarlino; by Escorial architect Juan de Herrera; by the Pléiade Pontus de Tyard; by humanists Johannes Sambucus and Pietro Francesco Giambullari; by the antiquaries John Aubrey and William Camden; by financier Johann Jakob Fugger, who bankrupted himself by book collecting; by Henry II, Philip II, George II, Sigismund II Augustus, Count Egmont, Elector Otto Heinrich, and by Duke August, whose library at Wolfenbüttel was the finest in Europe in the early eighteenth century; by Saint Aloysius Gonzaga and by Giordano Bruno; by Thomas Digges, Tycho Brahe, Galileo, and Kepler; and by a host of lesser-known medical doctors, astrologers, and dilletantes.[32] The royalty did not annotate their books, but many others did, leaving behind a precious record of the way in which the book was perceived and used during the scientific Renaissance.

Notes and References

[1] The volumes are now in the Uppsala University Library; see Pawel Czartoryski, "The Library of Copernicus," *Studia Copernicana*, vol. 16 (1978), pp. 355-401.

[2] See Josef Benzing, *Die Buchdrucker des 16. und 17. Jahrhunderts im deutschen Sprachgebiet*, vol. 12 of *Beiträge zum Buch- und Bibliothekwesen* (Wiesbaden, 1963), pp. 466–68.

[3] See Joseph C. Shipman, "Johannes Petreius, Nuremberg Publisher of Scientific Works, 1524–50, with a Short-Title List of His Imprints," in *Homage to a Bookman; Essays on Manuscripts, Books and Printing written for Hans P. Kraus on his 60th Birthday*, ed. by Hellmut Lehmann-Haupt (Berlin, 1967), pp. 147–62.

[4] Philip Gaskell, *A New Introduction to Bibliography* (Oxford, 1972), esp. pp. 57–77.

[5] Albert F. Butsch, *Die Bücherornamentik de Hoch- und Spätrenaissance* (1881; reprint Munich, 1921), 2:22 and Tafeln 97A and 97B; catalogers of Beham's work seem confused on this point, including Gustav Pauli, *Hans Sebald Beham* (Strasbourg, 1901) and F. W. H. Holstein, *German Engravings, Etchings and Woodcuts*, ca. 1400–1700, vol. 3, *Hans Sebald Beham* (Amsterdam, ca. 1955).

[6] This copy is illustrated in the Sotheby catalog *The Celebrated Library of Harrison D. Horblit Esq.*, part 1, item 240, 11 June 1974, and was owned by Haven O'More, Cambridge, Massachusetts when this article was written. It is now in a private collection in Italy.

[7] See item 503, Bishop Giese to Rheticus, 26 July 1543, in Marian Biskup, *Calendar of Copernicus' Papers, Studia Copernicana*, vol. 8 (1973).

[8] Johann Forster to Johann Schradi, 29 June 1542, letter edited by K. Ch. Forstemann in *Neue Mittheilungen aus dem Gebiet Historisch-antiquarischer Forschungen*, Vol. 2 (1836), p. 93, quoted in Karl Heinz Burmeister, *Georg Joachim Rheticus*, vol. 1 (Wiesbaden, 1967), p. 77.

[9] Noel M. Swedlow, "The Text of *De Revolutionibus*," *Journal of the History of Astronomy*, vol. 12 (1981), p. 44.

[10] Owen Gingerich, "An Early Tradition of an Extended Errata List for Copernicus' *De revolutionibus*," *Journal for the History of Astronomy*, vol. 12 (1981), pp. 47–52.

[11] Leon Voet, personal communication; see also his *The Golden Compasses*, vol. 2 (Amsterdam, 1972), pp. 169–73.

[12] I. Bernard Cohen, *Introduction to Newton's "Principia"* (Cambridge, Mass., 1971), p. 138.

[13] Owen Gingerich, "The Great Copernicus Chase," *American Scholar*, vol. 49 (1979), pp. 81–88; reprinted in *The Great Copernicus Chase and Other Adventures in Astronomical History* (Cambridge, 1992), pp. 69–81.

[14] Owen Gingerich, "The Role of Erasmus Reinhold and the Prutenic Tables in the Dissemination of the Copernican Theory," *Studia Copernicana*, vol. 6 (Wroclaw, 1973), pp. 43–62, 123–25 [reprinted as selection 13 in this anthology].

[15] Henry P. Macomber, "A Census of Copies of the 1687 First Edition and the 1726 Presentation Issue of Newton's *Principia*," *Papers of the Bibliographical Society of America*, vol. 47 (1953), pp. 269–300; Sir Geoffrey Keynes, *A Bibliography of the Writings of Dr. William Harvey, 1573–1657* (Cambridge, 1953); Michael Horowitz and Jack Collins, "A Census of Copies of the First Edition of Andreas Vesalius *De humani corporis fabrica* (1543)," *Journal of the History of Medicine and Allied Sciences*, vol. 39 (1984), pp. 198–221; my list of Apianus locations is planned for a future issue of the *Harvard Library Bulletin*.

[16] H. M. Adams, *Catalogue of Books Printed on the Continent of Europe, 1501–1600 in Cambridge Libraries* (Cambridge, 1967).

[17] In his Catalogue No. 19 (ca. 1952), Weil wrote, "I hope to be able to publish one

day a census [of *De revolutionibus*], for which I have collected data for a long time," but no notes for this seem to survive apart from his record book. The most extensive survey prior to my own, listing 70 copies primarily in northern Europe, was published by Ernst Zinner, *Entstehung und Ausbreitung der Coppernicanischen Lehre* (*Sitzungsberichte der physikalisch-medizinischen Sozietät zu Erlangen,* vol. 74) (Erlangen, 1943), pp. 448-55.

[18] Houghton shelf mark B4469.654.5*, illustrated in Plate XI of Owen Gingerich, "Science in the Age of Copernicus," *Harvard Library Bulletin,* vol. 26 (1978), pp. 401–16.

[19] Copy owned by Valentine Engelhart (1516–62?), now in the Sächsische Landesbibliothek, Dresden, Lit Graec B 716^h.

[20] Leipzig Universitätsbibliothek, Libri sep 577r. This copy has been reproduced in facsimile by Edition Leipzig and is currently distributed by the Johnson Reprint Corporation, New York, but the facsimile does not include the paste-in clipping with the price from a different copy. That price was misinterpreted as the absurdly large amount of 28 florins by L. A. Birkenmaier, *Mikolaj Kopernik* (Cracow, 1900), p. 647, and this has unfortunately been repeated elsewhere without challenge.

[21] Stadtbibliothek Schaffhausen, f III 2610 abc; Maestlin's price is written inside the back cover. The Regiomontanus work was printed by Petreius in Nuremberg in 1533.

[22] Georg Erler, *Die Matrikel der Universität Leipzig,* vol. 2 (Leipzig, 1897), p. 671.

[23] Staatsbibliothek Bamberg, MS. no. J. H. Msc. astr. 3, written by Bartholomeus Scultetus in 1562.

[24] Personal communication from Julian Roberts, Bodleian Library; no copy with a Dee provenance has been located.

[25] See Max Caspar and Martha List, *Bibliographia Kepleriana* (Munich, 1968).

[26] Paul F. Grendler, "Venice, Science, and the Index of Prohibited Books," in *The Nature of Scientific Discovery,* ed. by Owen Gingerich (Washington, D. C., 1975), p. 344.

[27] One of the Rheticus presentation copies is mentioned in note 6. Another, presented to Georg Donner, a fellow canon in Copernicus's diocese, is now in the Kopernicus Samlungen of the Uppsala Universitets Bibliotek. The presentation copy to Bishop Tiedemann Giese is no longer found, but notes on it are preserved in a copy at the Biblioteka Jagiellońska in Cracow, Cimelia F. 8288.

[28] Schaffhausen copy; see note 21.

[29] Martha List, "Marginalien zum Handexemplar Keplers von Copernicus: *De revolutionibus orbium coelestium* (Nürnberg, 1543)," *Studia Copernicana,* vol. 16 (1978), pp. 443–60.

[30] Beinecke Library, Yale University, QB41 + C663.

[31] Owen Gingerich, "The Censorship of Copernicus' *De revolutionibus,*" *Annali dell'Istituto e Museo di Storia della Scienza di Firenze,* vol. 6 (1981), fas. 2., pp. 45–61 [reprinted as selection 15 in this anthology].

[32] It is my hope that laser typesetting will soon make *An Annotated Census of Copernicus' 'De revolutionibus' (Nuremberg, 1543 and Basel, 1566)* a published reality, in the Polish Academy of Sciences's series from Brill in Leiden.

The Censorship of Copernicus's De revolutionibus

B y the end of the sixteenth century Copernicus's *De revolution-ibus orbium coelestium* was securely established as an important book, undoubtedly the most significant astronomical treatise since antiquity. The first two editions of the work, published in Nuremberg in 1543 and in Basel in 1566, were by then widely distributed throughout Europe in several hundred copies each; at least one copy had reached America and another would soon arrive in China. A bulky work of 400 pages and well illustrated with 146 diagrams, it was too imposing to be discarded lightly and too formidably technical to be worn out from overuse. As we shall see, this important book received unique treatment from the Inquisition.

Background to the Prohibition of Copernicus's Treatise

The initial reaction of the church, both Catholic and Protestant, was muted. The often-quoted comment from Luther's *Table Talk*, that "this fool would turn the whole art of astronomy upside down" (or in the alternative version, that everyone who would be clever nowadays must come up with something new), is grossly misleading. A casual remark, Luther's off-the-cuff judgment is not at all representative of the actual Lutheran reaction. In fact, Copernicus's book was highly

Selection 15 reprinted from *Annali dell'Istituto e Museo di Storia della Scienza di Firenze*, anno VI, fas. 2 (1971), pp. 44–61.

FIGURE 1. *The 1566* De revolutionibus *in the Biblioteca Statale, Cremona, showing the excision of the entire Chapter 8 of Book I, on the motion of the Earth. Folio six has been sliced out, and the remaining text has been covered over. This is the only known copy where the harsher alternative censoring has been carried out. Note also the paste-overs at the beginning of Chapter 9.*

regarded in Lutheran circles and extensively studied throughout their university system, beginning in the first place with the efforts of Rheticus to have the work published, and especially continuing with the analysis by Erasmus Reinhold, a leading professor at Wittenberg.

The potential ecclesiastical reaction to Copernicus's radical heliocentric cosmology was considerably tempered by the anonymous introduction added in publication by Andreas Osiander, a Lutheran clergyman of Nuremberg hired by the printer Petreius to oversee the final stages of the proofreading. Osiander's "Ad lectorem" took an instrumentalist stance by stating that the hypotheses of the work "need not be true nor even probable," and that their essential requirement was to furnish a model whereby planetary positions could be calculated for any conceivable time. That this was precisely the way the book was received in the Lutheran universities is clearly borne out by the pattern of annotations found in the most heavily marked copies. Particularly noteworthy is the well-worked-over volume owned by Erasmus Reinhold, the astronomer at Wittenberg.[1] Reinhold annotated extensively in Book III of *De revolutionibus*, where Copernicus discussed the technical issues of chronology, precession, and the Earth-Sun relation. When Copernicus proposed several different geometrical arrangements to account for the same celestial motion, Reinhold carefully noted and numbered the alternatives, thereby emphasizing the model-building, hypothesis-as-instrument aspects of Copernicus's treatise. In contrast, Reinhold offered very little marginalia in the cosmological part of Book I.

An Italian echo of the Protestant treatment of *De revolutionibus* is found in the *Commentary on the Sphere of Sacrobosco* written by the Jesuit astronomer Christopher Clavius. In the third edition of this work (1581), Clavius incorporated the remark that "all that can be concluded from Copernicus's assumption is that it is not absolutely certain that the eccentrics and epicycles are arranged as Ptolemy thought."[2]

Meanwhile, this instrumentalist interpretation provided the basis for the teaching of astronomy throughout the northern network of Lutheran universities, and it was within such a framework that Johannes Kepler learned about *De revolutionibus* from his Tübingen teacher, Michael Maestlin. However, the young Kepler was attracted to the heliocentric cosmology not just as geometrical model building, but as a description of physical reality itself. Intensely interested in the Copernican world view, Kepler soon produced his *Mysterium cosmographicum*, the first unabashedly heliocentric treatise to appear since Copernicus had published his *De revolutionibus* over 50 years before. Anxious to have an international hearing for his ideas, Kepler not only sent copies to specific astronomers, but he sent a pair of the books with an emissary to Italy. Apparently the ambassador was already working his way back from Rome when he realized that he had done nothing about Kepler's request. Looking around for a suitable recipient, he handed both copies over to the professor of mathematics in Padua, a man of some local reputation but certainly unknown north of the Alps inasmuch as he had not yet published anything. The Italian wrote a hasty reply, indicating that he, too, had supported Copernicus privately.[3] Kepler mentioned the reply in a letter to his former teacher, Maestlin, bemusedly reporting that there was an Italian astronomer with the same first name as last name: Galileo Galilei.[4] Kepler promptly wrote back to Galileo, urging him to stand forth openly in favor of the Copernican doctrine.[5] There the correspondence temporarily ended, not to be resumed until both men had established brilliant reputations as Copernicans: Kepler as imperial mathematician and author of the *Astronomia nova*, and Galileo as the discoverer of the satellites of Jupiter and the mountains on the moon.

Galileo did not tell Kepler how he had originally learned of the heliocentric theory, but we may infer from the *Dialogo. . . sopra i due Massimi Sistemi del Mondo* (1632) that Galileo first heard the Copernican system seriously discussed in Padua by "a certain foreigner from Rostock whose name I believe was Christian Wursteisen, a supporter of the Copernican opinion." As Stillman Drake has pointed out, a Christopher Wursteisen actually enrolled at Padua in 1595.[6] Christopher Wursteisen could well have been at one of the northern Lutheran universities and perhaps even at Rostock, even though no Wursteisen had formally matriculated at Rostock in the period 1585–94. Thus we again see a possible effect of the network of Protestant schools in propagating Copernicus's ideas. Rostock was one of the important Lutheran universities satellite to Wittenberg. In 1585 Duncan Liddell had matriculated at Rostock, and was soon teaching the Copernican system openly. It was there that he made what is one of the three extant sixteenth-century copies of Copernicus's *Commentariolus*.[7] A few years earlier, in Wroclaw, Liddell had copied into his own *De revolutionibus* marginalia originally derived from Erasmus Reinhold. Perhaps the *De revolutionibus* now at the University of Pisa,[8] which also contains some of Reinhold's annotations, could have come to Italy via one of the itinerant students such as Wursteisen.

In the opening decades of the 1600s both Kepler and Galileo challenged the instrumentalist view of astronomical methodology. Each of them adopted the Copernican doctrine not as a fictional, geometrical hypothesis but as physical reality, and each of them argued that the new interpretation of nature did not violate the inerrancy of Holy Writ. Johannes Kepler, in his introduction to his *Astronomia nova* (1609), wrote,

> But now the Sacred Scriptures, speaking to men on ordinary matters (not intended for instruction) used everyday speech so as to be understood by men. Who is not aware of the poetic allusion in Psalm 19, where the Sun is said to emerge from the tabernacle of the horizon like the bridegroom from his bridal chamber, and like a strong man eager

to run a race? The psalmist knew that the sun did not come from the horizon as if it were coming out of a tabernacle (even if it seems so to his eyes). He thought the Sun moved primarily because it seemed to do so; nevertheless he expresses two ideas since both appear that way to the sight. One ought not to think that he is speaking falsely in either case, but that the psalmist simply uses the visual perception proper to his deeper meaning in representing the course of the Gospel and of the Son of God.

Galileo Galilei wrote along very similar lines in his *Letter to the Grand Duchess Christina* (1615), subtitled "Concerning the Use of Biblical Quotations in Matters of Science":[9]

> I think in the first place that it is very pious to say and prudent to affirm that the Holy Bible can never speak untruth whenever its true meaning is understood. But I believe that nobody will deny that it is often very abstruse, and may say things which are quite different from what its bare words signify. Hence in expounding the Bible if one were always to confine oneself to the unadorned grammatical meaning, one might fall into error.

Galileo quoted from Tertullian, who said "We conclude that God is known first through Nature, and then again more particularly, by his doctrine; by Nature in His works, and by doctrine in His revealed word," and he quoted even more aptly from Cardinal Baronius, that the intention of the Holy Ghost is to teach us how to go to heaven, not how the heavens go.

By arguing for the reconciliation between the new Copernican doctrine and the traditional Biblical text, both Galileo and Kepler hoped to pave the way for an acceptance of the heliocentric system as physical reality and as a true description of the world. Up to this time Truth had been identified with Divine Revelation through Scripture, not through Nature. As long as heliocentrism was regarded as a convenient fiction, its study posed no threat to the churchmen. Their position was succinctly expressed by the influential theologian, Cardinal Bellarmine, who accepted Osiander's instrumentalist stance when he wrote in 1615 in his famous letter to Foscarini:[10]

> It seems to me that your Reverence and Signor Galileo would act prudently were you to content yourselves with

> speaking hypothetically and not absolutely, as I have always
> believed Copernicus spoke.

Unfortunately for the theologians' viewpoint, Copernicus was annoyingly vague concerning whether or not he himself believed in the reality of his system. Furthermore, by this time Kepler had conspicuously announced the true authorship of the anonymous "Ad lectorem" that prefaced *De revolutionibus*, so it could no longer be maintained that Copernicus necessarily subscribed to the instrumentalist view given in that introduction. The entry of such an accomplished polemicist as Galileo into the territory of the theologians was cause for alarm in ecclesiastical circles. It was not so much that they were threatened by the particulars of the heliocentric cosmology; rather, they feared that the Book of Nature might be seen as providing a more direct route to Truth than the Book of Scripture. Heliocentrism as a device or model was not at fault, but in the eyes of the churchmen it was essential for Copernicus's work to be perceived as hypothetical and not as physical reality. Therefore, an obvious place to draw the battle lines was on the *interpretation* of Copernicus's doctrine. Hence, from the vantage point of the Holy Congregation in Rome, the most expeditious course was to suspend Copernicus's book until the appropriate adjustments could be made in the text. Decree XIV of the Holy Congregation of the Index, issued 5 March 1616, placed the *De revolutionibus* on the prohibited list *donec corrigatur,* "until corrected." Similarly, the book on Job by Diego de Zuñiga (Didacus à Stunica) was prohibited until corrected, whereas Foscarini's *Lettera sopra l'opinione de' Pittagorici, e del Copernico, della mobilita della Terra, e stabilita de Sole* (Naples, 1615) was condemned outright, along with five other non-scientific books.

The relevant part of the Decree reads:[11]

> And whereas it has also come to the knowledge of the
> said Congregation, that the Pythagorean doctrine—which is
> false and altogether opposed to Holy Scripture—of the motion of the Earth, and the quiescence of the Sun, which is
> also taught by Nicholas Copernicus in *De revolutionibus orbium coelestium,* and by Diego de Zuñiga in [his book on]
> Job, is now being spread abroad and accepted by many—as
> may be seen from a certain letter of a Carmelite Father,
> entitled, *Letter of the Rev. Father Paolo Antonio Foscarini,
> Carmelite, on the opinion of the Pythagoreans and of Copernicus concerning the motion of the earth, and the stability of*

the sun, and the new Pythagorean system of the world, at Naples, printed by Lazzaro Scorriggio, 1615: wherein the said father attempts to show that the aforesaid doctrine of the quiescence of the Sun in the centre of the world, and of the Earth's motion, is consonant with truth and is not opposed to Holy Scripture. Therefore, in order that this opinion may not insinuate itself any further to the prejudice of Catholic truth, the Holy Congregation has decreed that the said Nicholas Copernicus, *De revolutionibus orbium,* and Diego de Zuñiga, *In Job,* be suspended until they be corrected; but that the book of the Carmelite Father, Paolo Antonio Foscarini, be altogether prohibited and condemned, and that all other works likewise, in which the same is taught, be prohibited, as by this present decree it prohibits, condemns, and suspends them all respectively. In witness whereof the present decree has been signed and sealed with the hands and with the seal of the most eminent and Reverend Lord Cardinal of St. Cecilia, Bishop of Albano, on the 5th day of March, 1616.

The expression *donec corrigatur* belonged to the standard vocabulary of the Inquisition, but in only one case did they ever announce specific corrections. That unique treatment was reserved for Copernicus's *De revolutionibus,* when, four years later, Decree XXI spelled out ten explicit alterations for the book.

The report made to the Congregation of the Index on the proposed censorship of Copernicus's book still survives in the Vatican Library,[12] although it is little known and does not, for example, appear in the Favaro edition of the *Galileo Opere.* Unlike the final decree announcing the corrections, the report in Codex Barberiniano XXXIX gives the background reasons for the action. It was probably written by Bonifacio Caetani, one of the cardinals who had persuaded the Pope to label the heliocentric doctrine as false rather than heretical.[13] A translation of this report follows in the next section.

Codex Barberiniano XXXIX.55

On the Emendation of the Six Books of Nicholas Copernicus's De revolutionibus

To the most illustrious and reverend cardinals of the Congregation of the Index:

There are three things, most illustrious and reverend fathers, that you most illustrious servants of the Lord must diligently consider with regard to the emendation of the six books of Copernicus's *De revolutionibus*. The first consideration is that this book by Copernicus should be entirely preserved and maintained for the sake of the Church because the determination of time, for which the Christian people have great need both on account of celebrating divine and solemn occasions and of conducting business affairs, depends on astronomical calculations, especially of the Sun and Moon and of the precession of the equinoxes, so that from these were brought about the correction of the year instituted by Gregory XIII and which remains his excellent memorial. However, these astronomical calculations are in need of restoration and repair every so many years, since, either because of our ignorance of all the celestial motions or because of the accumulation over time of certain tiny and scarcely discernible motions, they barely show the true locations of the celestial bodies. Truly, this restoration and renewal cannot be made by the astronomers unless they possess the observations of past centuries, as is established by the writings of Ptolemy in the *Almagest* and of Tycho in the *Progymnasmata;* certainly the books of Copernicus are filled with such observations, as is obvious upon reading them, and these should be entirely preserved as being useful to the Church.

The second consideration is that no emendation of Copernicus can be made that posits the immobility of the Earth, according to the truth of the matter and to divine scripture. For since Copernicus assumes as a principle three motions of the Earth and upon it makes all his demonstrations for saving the appearances or the heavenly motions, or arranges his phenomena so as to be supported by this principle, an emendation of Copernicus would not be a correction, but rather the total destruction of his system.

The third consideration is that it is possible to proceed by a middle way so that Copernicus might be preserved in this difficult matter without compromising either the truth of the Holy Writ; this is to be done by emending only certain passages in which he appears to write about the motion of the Earth, not hypothetically, but according to reality; for Copernicus speaks with very few exceptions either according to hypothesis or without asserting the truth of terrestrial motion.

Moreover I say that this emendation can be made without compromising truth or Holy Scripture; because the subject which Copernicus is dealing with is astronomy, whose most distinctive method is to use false and imaginary principles for saving appearances and celestial phenomena, as is established by the epicycles, eccentrics, equants, apogees, and perigees of the ancients. If certain of Copernicus's passages on the motion of the earth are not hypothetical, make them hypothetical; then they will not be against either the truth or the Holy Writ. On the contrary, in a certain sense, they will be in agreement with them, on account of the false nature of suppositions, which the study of astronomy is accustomed to use as its special right.

The specific recommendations that follow in this document are given here in translation and printed in italics, interspersed with remarks both on the fuller context of the corrections and on the several differences between these recommendations and those adopted in the final Decree.

Therefore, with these considerations, let those who have some diligence approach the judgment of this emendation, which is as follows:

In the preface near the end:

I would delete everything from "perhaps" to the words, "my work" and I would insert, "my work and those of others."

The full passage to be deleted reads, in Professor Edward Rosen's new translation:[14]

Perhaps there will be babblers who claim to be judges of astronomy although completely ignorant of the subject and, badly distorting some passage of Scripture to their purpose, will dare to find fault with my undertaking and censure it. I disregard them even to the extent of despising their criticism as unfounded. For it is not unknown that Lactantius, otherwise an illustrious writer but hardly an astronomer, speaks quite childishly about the Earth's shape, when he mocks those who declared that the Earth has the form of a globe. Hence scholars need not be surprised if any such persons will likewise ridicule me. Astronomy is written for astronomers. To them my work too will seem, unless I am mistaken, to make some contribution also to the Church, at the head of which Your Holiness now stands.

Lactantius, the most widely reprinted of the Latin Church Fathers, was hailed by Renaissance humanists as the "Christian Cicero," but it was perhaps not so much for the protection of Lactantius as to preserve the right of theologians to criticize natural science that the principal part of this passage was struck out. It is interesting to note that Galileo had quoted favorably this entire passage in the *Letter to the Grand Duchess Christina.*[15]

[In Chapter 5, folio 3:]

Where it reads, "Nevertheless, if we examine more carefully," I would correct it to "Nevertheless, if we examine the matter more carefully, it makes no difference whether the Earth exists in the middle of the universe, or away from the middle, as long as we judge that the appearances of the heavenly motions are saved. For every, etc."

Copernicus's original sentence reads: "Nevertheless, if we examine the matter more carefully, we will see that this problem has not yet been solved, and is therefore by no means to be disregarded."

In Chapter 8 of the same book:

This whole chapter can be deleted because it admittedly deals with the truth of the Earth's motion, while it discredits the ancient reasons for proving its immobility.

The title of eighth chapter of Book I is "The Inadequacy of the Previous Arguments and a Refutation of Them."

If, however, it would please the most illustrious fathers that this chapter be emended, Copernicus may be made to seem to speak always problematically and from opinion, and it can be arranged as follows. Also, it would better satisfy students, since the sequence and arrangement of the books would remain intact.

The Decree as finally adopted retains the possibility of deleting the entire chapter, but immediately adds "Nevertheless, since it seems to speak problematically, in order to satisfy students and to keep intact the sequence and arrangement of the book, let it be emended as below." (I know of only a single case where the entire Chapter 8 was excised, the second edition in the Biblioteca Statale in Cremona—see Figure 1).

First of all, on folio 6, the phrase must be deleted from "Therefore why" to the words "we sail"; and this passage must be corrected in this manner, "Therefore why can we not concede mobility to it on account of its form rather than attribute a movement to the entire universe, whose limit is unknown and indeed unknowable; what is an appearance in the heavens is so to say, as Virgil's Aeneas *put it. . . ."*

By compressing the text, the emendation removes the polemical bite from Copernicus's original remarks on the relativity of motion:

"Therefore why do we hesitate to concede mobility to it on account of its form, rather than attribute a movement to the entire universe, whose limit is unknown and indeed unknowable; why should we not admit that the daily rotation itself is an appearance of the heavens but real in the Earth? For this is, so to say, as Virgil's *Aeneas* puts it, [we sail forth from the harbor, and the land and cities recede]."

Secondly, on folio 7, the phrase beginning, "Furthermore," can be emended on this manner, "Furthermore, it is no more difficult to ascribe motion to the Earth which is contained and located, than to that which is containing or locating."

This change considerably softens Copernicus's statement: "Furthermore, it would seem rather absurd to ascribe motion to that which is containing or locating instead of that which is contained and located, namely, the Earth."

Thirdly, on the same page, and at the end of the chapter, the phrase beginning, "you see," should be deleted to the end of the chapter.

Copernicus states, "You see then, from all of these arguments, that it is more probable that the Earth moves than that it stays at rest, especially with respect to the daily rotation, which is particularly appropriate to the Earth."

In chapter 9, folio 7:

The beginning of this chapter to the phrase, "for indeed . . .": I would emend it in this manner: "Therefore, since I have assumed the Earth to be moved, I think it must now be considered whether other motions can also be suitable to it; for indeed, etc."

Copernicus has begun the chapter by declaring: "Therefore, since nothing prohibits the mobility of the Earth, I think it must now be considered whether other motions are suitable to it, so that it can be regarded as one of the planets."

In Chapter 10, folio 9:

The phrase beginning "therefore" I would correct thus: "therefore it is no shame to assume"; and a little way below this, where it says, "this is verified better by the mobility of the Earth," I would say: "this is saved better by the mobility of the Earth"; for the truth of the matter is that Copernicus saves the celestial phenomena better with this invention of his than did the ancients.

The first change alters the single world *lateri* to *assumere*, making it "no shame to assume" instead of "to admit." A few lines later *potius* and *verificari* become *melius* and *salvari*, respectively; the context is that any apparent motion of the Sun must preferably be attributed to (or established by) the motion of the Earth (according to Copernicus), whereas the correction indicates that any apparent motion of the Sun is "better explained" by the motion of the Earth.

In the decree of 1620 not only was the compliment to Copernicus eliminated, but the correction was botched, leaving the word *verificari* and substituting *consequenter* for *potius*. Thus the Sun's motion is "appropriately" or "consequently" attributed to the motion of the Earth rather than "preferably," and the notion of hypothetically "saving the phenomena" connoted by *salvari* is lost in the decree.

On folio 10, at the end of the chapter, I would delete these last words: "So vast, without any question, is the divine handiwork of the most excellent Almighty."

In this resounding climax to his cosmological chapter, Copernicus is remarking especially that the universe is so large that the sphere of fixed stars does not exhibit any observable parallactic changes on account of the annual revolution of the Earth—a nuance lost in several of the current translations. The Holy Congregation evidently decided that this closing sentence suggested all too forcefully that the Divine Creator had patterned the real world after a heliocentric blueprint, and therefore struck it out. (See Figure 2.)

In Chapter 11:

The title of the chapter can be adapted in this manner, "On the Hypothesis of the Three-fold Motion of the Earth and its Explication."

The original title reads simply, "On the Explication of the Three-fold Motion of the Earth." Of all the English translations, only the one by J. F. Dobson and S. Brodetsky[16] correctly renders the nuance of *demonstratio* by "explanation." A reading of Chapter 11 clearly shows it gives a description or explanation, not a demonstration in the sense of a geometrical theorem. To translate *demonstratio* as "proof" (as Professor Rosen does) leads to the absurdity that the Holy Congregation rewrote the chapter title as "On the Hypothesis of the Three-fold Motion of the Earth and its Proof," which would be self-contradictory.

In the second and third books, nothing needs to be corrected.

In Book 4, Chapter 20, folio 122:

In the title of the chapter I would delete the words "these three stars," because the Earth is not a star, as Copernicus makes it.

Copernicus's title is "On the Sizes of these three Stars (*siderum*), the Sun, Moon, and Earth, and their Intercomparison."

In the fifth and sixth books, nothing needs to be emended.

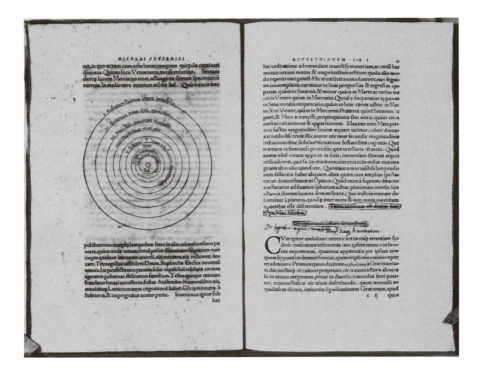

FIGURE 2. *The 1566* De revolutionibus *owned and corrected by Galileo. At the end of the famous cosmological chapter the words "So great, without any question, is the divine handiwork of the most excellent Almighty" have been cancelled, and the heading of Chapter 11 has been altered from "On the Explanation of the Three-fold motion of the Earth" to "On the Hypothesis of the Three-fold Motion of the Earth and its Explanation." Courtesy of the Biblioteca Nazionale Centrale, Florence, shelf mark Palat C. 10.6.26.*

Results of the Survey of Copies of *De revolutionibus*

During the past decade I have attempted to compile a complete census of extant copies of the *De revolutionibus* in both the 1543 Nuremberg and the 1566 Basel editions. In the process I have personally examined nearly every surviving copy—well over 500 books—and because Decree XXI was so explicit with respect to the corrections, it is possible to have a remarkably complete picture concerning the extent of the censorship.[17] The most surprising result is that, although the majority of copies then in Italy were censored, the Decree had comparatively little impact elsewhere, and none whatsoever in Spain or Portugal.

In physical form the censorship was accomplished by one of two methods: the offending text was either inked out or pasted over with slips of paper. Sometimes the ink cancellation was so slight as to highlight the passages in question—for example, in the Lincei copy in Rome or almost so in Galileo's corrected copy in Florence (see Figure 2); in other instances the ink was so heavy that its gallic acid has by now eaten entirely through the paper—for example, in the second edition in the Biblioteca Nazionale in Palermo. With paper cancels, it has sometimes been possible to soak off the slip, so that each copy must be examined carefully to ascertain whether in the seventeenth century it was indeed censored; but alternatively, the glutenous paste has on occasion provided nourishment for bookworms that have destroyed both the offending text and the innocent printing on the opposite side of the page.

Censored copies of the book are now scattered throughout the world, from Tokyo to Stockholm to Chicago. Each of these three specific examples can be traced to Italian provenances. The only censored copy in England, a first edition at Trinity College in Cambridge, came from Italy in the last century. Similarly, many of the uncensored copies in Italy today arrived there after 1650 as part of the general ongoing movement of books. Consequently, in order to draw any conclusions concerning the effectiveness of the Decree and the *Index*, it has been necessary to examine the provenance of each copy in order to establish its locale in the early 1600s. Since earlier ownership marks are sometimes destroyed, either accidentally or deliberately, it is not always possible to pinpoint an earlier venue, but other evidence, such as the style of binding or the foxing and browning characteristic of the copies kept in a Mediterranean climate, serves to build up a representative picture of the seventeenth-century locations of Copernicus's book.

My Copernican census now provides for the first time evidence concerning the effectiveness of the censorship. Of about 400 copies now in Europe, 33 are censored, or about 1 in 12, as shown by the filled symbols in Figure 3. From the analysis of the provenances, it has been possible to reconstruct where the majority of the books were in 1620, and thus I can conclude that about 60 percent of the copies then in Italy were censored, and relatively few elsewhere. For example, in France, where many copies were in Jesuit libraries, there was comparatively little censorship; apparently the Jesuits considered the *Index* primarily a Dominican concern! My census revealed a quite unexpected and initially puzzling situation with respect to Spain and Por-

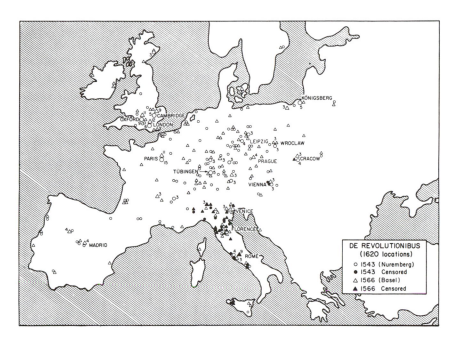

FIGURE 3. *The 1620 locations of* De revolutionibus *deduced from the prove-nances recorded in the census research.*

tugal, where none of the copies is censored. In Madrid I turned up the uncensored copy once owned by Juan de Piñeda, a Spanish theologian active in the early 1600s; subsequent research showed that he had edited the Spanish version of the *Index*. Piñeda's version of the *Index* explicitly mentioned the 1566 Basel edition of Copernicus's book—not to correct it, but to prohibit Rheticus's *Narratio prima* as reprinted at the end of it—and indeed, he sliced these offending pages from his copy. Piñeda could hardly have been unaware of Decree XXI, but the *Index*[18] he edited specified no changes in the *De revolutionibus* text itself, and apparently neither he nor any others on the Iberian penin-sula deleted any part of Copernicus's treatise. Apparently they consid-ered the Copernican censorship to be part of a local Italian imbroglio and of no international concern!

This pattern of Italian censorship and Spanish noncensorship shows up even in the missionary copies: the second edition taken by the Italian Jesuits to China (and today in the National Library in Peking) is censored, but what is apparently the only early copy surviving in Latin America (in the Guadalajara Public Library in Mexico) is not censored.

As for the Italian suspension and censorship of *De revolutionibus*, I suspect its chief effect was to give Copernicus's ideas abundant free publicity. For example, when Kepler's *Epitome of Copernican Astronomy* was placed on the *Index* in 1619, a Venetian correspondent assured him that his work would be read all the more attentively in Italy.[19] And perhaps it was not mere coincidence that a third (and uncorrected) edition of Copernicus's book was published in Amsterdam in 1617, the year after the first Decree.[20]

Galileo's trial and its inhumanity, coming a dozen years after the censorship, was considerably more consequential. It cast a damper on scientific inquiry throughout Catholic Europe and destroyed creative science in Italy for several generations. But the censorship itself had little effect in maintaining the primacy of Scripture over Nature as the path to truth about our physical world. It was as fruitless as King Canute's commanding the tides to stop.

When Galileo had grown old and blind, he was visited in Arcetri by the English poet John Milton. In Milton's later years, with his own eyesight dimmed, he wrote:[21]

> For Heav'n
> Is as the Book of God before thee set
> Wherein to read his wondrous works . . .
> . . . whether Heav'n move or Earth
> Imports not. . . .

Without committing himself on the cosmology, Milton was already conceding the legitimacy of the Book of Nature. Within another generation came Milton's countryman Isaac Newton, who "feigned no hypotheses." Newton read the Book of Nature to establish a physical coherency and thus a persuasive justification for the heliocentric blueprint of the Universe. By 1835 Copernicus's book had finally vanished from the *Index*, but long before, by the time of Newton, the censorship of *De revolutionibus* had run its course—ineffectively.

Notes and References

[1] Owen Gingerich, "The Role of Erasmus Reinhold and the Prutenic Tables in the Dissemination of the Copernican Theory," Studia Copernicana, vol. 6 (1973), esp. pp. 56–58 [reprinted as selection 13 in this anthology].

[2] Christopher Clavius, *In Sphaeram Ioannis de Sacro Bosco commentarius* (Rome, 1581), pp. 436–37; trans. by Edmund Dolan and Chaninal Mischler in Pierre Duhem, *To Save the Phenomena* (Chicago, 1969), pp. 94–95. (It is rather curious to notice that Clavius suppressed the references to Copernicus in subsequent editions of his commentary.)

[3] Galileo to Kepler, 4 August 1597, in *Johannes Kepler Gesammelte Werke* (hereafter called *JKGW*), vol. 13 (1955), pp. 130–31.

[4] Kepler to Michael Maestlin, early October 1597, *JKGW*, vol. 13, p. 143.

[5] Kepler to Galileo, 13 October 1597, *JKGW*, vol. 13, p. 145.

[6] Stillman Drake, *Galileo at Work* (Chicago, 1978), p. 36.

[7] Jerzy Dobrzycki, "The Aberdeen Copy of Copernicus's Commentariolus," *Journal for the History of Astronomy*, vol. 4 (1973), pp. 124–27.

[8] Detailed descriptions of each of the copies will eventually appear in my *An Annotated Census of Copernicus's* De revolutionibus *(Nuremberg 1543 and Basel 1566)*.

[9] Stillman Drake, *Discoveries and Opinions of Galileo Galilei* (Garden City, New York, 1957), p. 181; see also pp. 183, 186.

[10] Robert Bellarmine to P. A. Foscarini, 12 April 1615, translated in Stillman Drake, *op. cit.* (note 8), pp. 162–63.

[11] Karl von Gebler (trans. by Mrs. George Sturge), *Galileo Galilei and the Roman Curia* (London, 1879), pp. 84–85.

[12] Codex Barberiniano XXXIX.55, = XXXIX, fol. 58–60v. Transcribed in Joseph Hilgers, *Der Index der Verbotener Bücher* (Freiburg, 1904), pp. 541–42. I am indebted to a former student, Melvin Tracy, for pointing out this source and for making a preliminary translation.

[13] See Giovanfrancesco Buonamici, 2 May 1663, *Le Opere di Galileo Galilei*, vol. 15 (reprint 1968), p. 111.

[14] Nicolas Copernicus, *On the Revolutions*, ed. by Jerzy Dobrzycki, trans. and commentary by Edward Rosen (Cracow, London, and Baltimore, 1978).

[15] Stillman Drake, *op. cit.* (ref. 9), p. 180.

[16] J. F. Dobson and S. Brodetsky, *Nicolaus Copernicus, De Revolutionibus*, Preface and Book I, *Occasional Notes of the Royal Astronomical Society*, no. 10 (1947).

[17] Owen Gingerich, "The Great Copernicus Chase," *American Scholar*, vol. 49 (1979), pp. 81–88.

[18] *Index novus librum prohibitorum*, ed. by Juan de Piñeda (Seville, 1631).

[19] Vinzenz Bianchi to Kepler, 20 January 1619, *JKGW*, vol. 17 (1955), p. 319.

[20] The suggestion that the 1617 edition took advantage of the publicity generated by the *Index* is made by Elizabeth Eisenstein in *The Printing Press as an Agent of Change* (Cambridge, England, 1979), p. 676.

[21] John Milton, *Paradise Lost*, VIII, lines 66–71 (1667).

Heliocentrism as Model
and as Reality

Near the close of Book One of the autograph manuscript of his great work, Copernicus writes:[1]

> And if we should admit that the course of the Sun and Moon could be demonstrated even if the Earth is fixed, then with respect to the other wandering bodies there is less agreement. It is credible that, for these and similar causes (and not because of the reason of motion, which Aristotle mentions and rejects), Philolaus was aware of the mobility of the Earth, and some even say that Aristarchus of Samos was of the same opinion. But since things were such that they could not be comprehended except by a sharp intellect and continuing diligence, Plato says that generally very few philosophers in that time understood the reason for the sidereal motion.

Before a copy of Copernicus's manuscript was sent to the printer, the work was somewhat reorganized and in the process this passage was struck out. The original first and second books were merged into a single section, and the deleted material was rewritten into the preface to Pope Paul III. Apparently by that time Copernicus had access to

Selection 16 reprinted from *Proceedings of the American Philosophical Society,* vol. 117 (1973), pp. 513–22.

the 1531 Greek edition of Plutarch,[2] and so he chose to use a direct quotation in Greek, which reads in translation:[3]

> Some think that the Earth is at rest, but Philolaus the Pythagorean says that it moves around the fire with an obliquely circular motion, like the Sun and Moon. Herakleides of Pontus and Ekphantus the Pythagorean do not give the Earth any movement of locomotion, but rather a limited movement of rising and setting around its center, like a wheel.

In this way the name of Aristarchus, often called the "Copernicus of Antiquity," was eliminated from the printed edition of *De revolutionibus*. An anniversary such as this, when Copernicus is everywhere apotheosized, inevitably breeds detractors. Among their complaints is the large measure of glory attributed to Copernicus and the silence that attends the speculative suggestions of Aristarchus.

I do not intend to give a judgment here, but rather, I shall first answer with the platitude that nothing succeeds like success. Surely a critical factor is that Copernicus's system has been universally adopted and that of Aristarchus was not. This, then, leads us to the fascinating study of the reception, the near rejection, and the ultimate acceptance of the heliocentric system. By this I do not mean the dramatic story of Galileo and the Inquisition, but a pattern of events that unfolded and reached their denouement before Galileo wrote his *Dialogo* in 1632.

Two of the key figures in the dissemination of the Copernican doctrine were professors of mathematics at the Lutheran University of Wittenberg. About the senior member of the pair, Erasmus Reinhold, few personal facts are known. In 1531 his name is inscribed in the Dean's Book of the University of Wittenberg along with other students, and, in 1536, at age 25, he became professor of higher mathematics, that is, of astronomy. On two occasions he served as dean, and he later became rector of the University of Wittenberg. In 1553, at the peak of his astronomical career, he died of the plague, being only 41 years old.[4]

In the same year that Reinhold became professor of higher mathematics, Georg Joachim Rheticus received the chair of lower mathematics at age 22. Apart from the fact that they both served together for a few years on the Wittenberg faculty, and both played fundamental although different roles in making Copernicus famous, their subsequent lives have little in common. Unlike Reinhold, who became an

establishment figure at Wittenberg, Rheticus became a scholastic itinerant, his interest in Copernicus quickly fading. Nevertheless, his part
in getting Copernicus's work published was memorable, and rather
similar to Halley's role with respect to Newton's *Principia*.

In 1539 the young Rheticus journeyed to Frauenburg (now the
town of Frombork) in remote Polish Prussia to gain first-hand knowledge concerning the astronomical innovations suggested by Copernicus. Although Rheticus came from the hotbed of Lutheranism, the
Catholic Copernicus received him with courage and cordiality. Swept
along by the enthusiasm of his young disciple, Copernicus allowed
Rheticus to publish a first printed report about the heliocentric system.
In a particularly beautiful passage, Rheticus wrote:[5]

> With regard to the apparent motions of the Sun and Moon,
> it is perhaps possible to deny what is said about the motion
> of the Earth. . . . But if anyone desires to look either to the
> principal end of astronomy and the order and harmony of
> the system of the spheres or to ease and elegance and a
> complete explanation of the causes of the phenomena, by no
> other hypotheses will he demonstrate more neatly and cor-
> rectly the apparent motions of the remaining planets. For all
> these phenomena appear to be linked most nobly together,
> as by a golden chain; and each of the planets, by its position
> and order and very inequality of its motion, bears witness
> that the Earth moves and that we who dwell upon the globe
> of the Earth, instead of accepting its changes of position,
> believe that the planets wander in all sorts of motions of
> their own.

His use of the word "hypotheses" is particularly interesting. This
reappears on a subsequent page where Rheticus wrote:

> But my teacher had long been aware that in their own
> right the observations in a certain way required hypotheses
> which would overturn the ideas concerning the order of the
> motions and spheres that had hitherto been discussed and
> promulgated and that were commonly accepted and believed
> to be true; moreover, the required hypotheses would contra-
> dict our senses.

Both of these passages use the word "hypotheses" in a somewhat
different sense from our modern meaning of the word. Rheticus, in
common with most other sixteenth-century astronomical writers, uses

"hypothesis" to mean an arbitrary geometrical device by which the observed celestial motions can be explained. Included within this set of geometrical devices was the grand hypothesis of them all, the heliocentric concept itself. The ultimate nature of the hypotheses, that is to say, whether they were hypothetical models or something real, became a fundamental issue in deciding on the relevance of the heliocentric idea.

As a preface to the next stage in our examination of "hypotheses" in Copernican astronomy, we must note that Rheticus not only gained permission to publish the *Narratio prima*, but he also persuaded Copernicus to allow publication of the magnum opus itself. Consequently, Rheticus obtained a copy of the manuscript, and upon returning to Germany he arranged for the publication of the book in Nuremberg by Johann Petreius, one of the leading scientific publishers of northern Europe. Rheticus temporarily resumed his teaching duties at Wittenberg but then moved to a professorship at Leipzig. Because he was still too far away to oversee the printing, the job fell to a Lutheran theologian, Andreas Osiander, who had previously worked as an editor for Petreius.

When Rheticus received his copies of the printed volume in the spring of 1543, he was annoyed to discover that an anonymous introduction on the nature of hypotheses had been added to the work. On two copies—one in the private collection of Mr. Harrison Horblit in Connecticut [now in a private collection in Italy] and the other preserved in the Uppsala University Library—Rheticus crossed out Osiander's unsigned introduction, in each case with a red pencil or crayon.

Osiander's introduction contains statements that seem quite innocent today, and which must have struck most sixteenth-century readers as eminently reasonable. I cannot believe that his anonymity in the matter stemmed from any malicious mischievousness, but rather simply from a Lutheran reluctance to be associated with a book dedicated to the Pope. He wrote:[6]

> Since the novelty of the hypotheses of this work has already been widely reported, I have no doubt that some learned men have taken serious offense because the book declares that the Earth moves; these men undoubtedly believe that the long established liberal arts should not be thrown into confusion. But if they examine the matter closely, they will find that the author of this work has done

nothing blameworthy. For it is the duty of an astronomer to record celestial motions through careful observation. Then, turning to the causes of these motions he must conceive and devise hypotheses about them, since he cannot in any way attain to the true cause. . . . The present author has performed both these duties excellently. For these hypotheses need not be true nor even probable; if they provide a calculus consistent with the observations, that alone is sufficient. . . . Now when there are offered for the same motion different hypotheses, the astronomer will accept the one which is the easiest to grasp. The philosopher will perhaps rather seek the semblance of the truth. But neither of them will understand or state anything certain, unless it has been divinely revealed to him. . . . So far as hypotheses are concerned, let no one expect anything certain from astronomy, which cannot furnish it, lest he accept as the truth ideas conceived for another purpose, and depart from this study a greater fool than when he entered it. Farewell.

In addition to striking out the introduction in both copies, Rheticus deleted the last two words of the printed title, *De revolutionibus orbium coelestium*. There is an old tradition, further attested to by copies at Yale University and at the Jagiellonian University in Cracow, that Osiander assisted the printer in changing the title from "Concerning the Revolutions" to "Concerning the Revolutions of the Heavenly Spheres." It is difficult to see precisely what Rheticus thought was offensive about the additional words except that, like the introduction, the expression "heavenly spheres" perhaps suggests the idea of model building.

Rheticus's role as midwife in the publication of *De revolutionibus* guarantees his enduring fame. But after his return to Wittenberg, the torch was in effect passed to Erasmus Reinhold. Reinhold himself remains a rather ambiguous figure. In 1551, he published his *Pruten-icae tabulae* in the first of several editions. These were a handy and much expanded form of the Copernican tables in *De revolutionibus*. This widely used reference work became a principal avenue for making Copernicus's name known. In the work Reinhold wrote:[7]

> All posterity will gratefully remember the name of Copernicus, by whose labor and study the doctrine of celestial motions was again restored from its near collapse. Under the light kindled in him by a beneficent God, he found and

explained much which from antiquity till now was either unknown or veiled in darkness.

Though Reinhold's name was closely linked with Copernicus and Copernicanism through these handy tables, his printed writings show a notable lack of commitment with respect to the heliocentric astronomy. For example, his *Prutenic Tables* are carefully framed so that they are essentially independent of the mobility of the Earth.

A number of authors have argued that Reinhold's own philosophical position was very close to that of Osiander.[8] He has left scattered clues throughout his writings, and a few hints in a single long manuscript preserved in Berlin. Although we seem to have less material extant from Reinhold than from Copernicus himself, I was able, by a happy piece of serendipity, to find and identify his personal copy of *De revolutionibus*, now preserved in the Crawford Library of the Royal Observatory in Edinburgh. At the bottom of its title page he wrote in Latin "The axiom of astronomy: celestial motion is circular and uniform or made up of circular and uniform parts." Clearly, what Reinhold saw as important in Copernicus's work was not the heliocentric cosmology, but some of the small technical details—minor hypotheses that were not part of the major cosmological revolution. In particular, he appreciated that Copernicus, in seeking to reform astronomy, had adopted a mechanism to eliminate the so-called equant of Ptolemaic astronomy, thereby returning the description of celestial motion to a pure combination of circles. Reinhold's *Prutenic Tables* are strictly based on this technical scheme, as I have demonstrated by a modern recomputation.

I should now like to describe some new material that shows the influence of this attitude on the teaching of astronomy at Reinhold's university 25 years after the publication of *De revolutionibus*. Two years ago, when the Smithsonian Institution enabled me to spend part of a sabbatical year in England, I made a systematic search of the manuscript astronomical tables in the Cambridge colleges. In the course of this investigation, I came upon a manuscript that proved to be a set of notes for the astronomy lectures at the University of Wittenberg in the late 1560s, roughly two decades after the death of Reinhold. Because no comparable material has ever been described in the literature, I should like to present some details, especially to show in what connection Copernicus's name came up in the lectures.

At that time, the introductory astronomy course was based on the late medieval text of John of Hollywood, better known as Sacrobosco.

Sacrobosco's *Sphere* was a very low-level treatment of spherical astronomy that scarcely mentioned planetary motion or the sophistication of the Ptolemaic theory. A new feature of the teaching at Wittenberg, however, was the recent availability of cheap printed textbooks. It almost seemed as if each astronomy teacher had printed, or was organizing his notes for the printing of, a new commentary on Sacrobosco.

The manuscript is No. 387 in the Gonville and Caius College Library; it contains about 200 leaves, written in two different hands.[9] The first three quires of eight leaves each appear to have been written by Laurentius Rankghe of Colberg in 1564,[10] and constitute a Latin commentary on Sacrobosco's *Sphere*. The commentary goes up to the definitions of circles including the zodiac, and then stops in midstream. Rankghe's writing ends on the first page of the fourth quire, thus suggesting that the entire volume with its vellum binding was bound together originally as a blank notebook.

The rest of the manuscript has apparently been written by Johannes Balduinus between 27 May 1566, and sometime in 1570. All the dates given are consecutive, and sometimes record weekly progress through the astronomy lectures. Balduinus became dean in the autumn of 1569 and, therefore, he may have been taking an official record of the lectures.[11] This could perhaps account for the fact that elementary material is covered repeatedly. He begins by recording the "Erotemata in Questiones Sphaerae" (which might be roughly translated "questions on the questions of the sphere") of Sebastian Theodoricus Winshemius. Theodoricus was professor of mathematics at Wittenberg at that time, and his textbook on this subject was printed at least seven times in Wittenberg beginning in 1564.[12] The manuscript notes approximately parallel the printed textbook. Of particular interest are the references to Copernicus, who is first mentioned in both the manuscript and the printed text in connection with the size of the Earth.[13] A little later, in a discussion of precession found in the notes but not in the printed textbook, Copernicus is cited for his numerical values, along with Reinhold's *Prutenic Tables*.[14] A few pages later Copernicus's name appears again in a discussion of the Moon.[15]

Both the manuscript and textbook then move on to the question "does the Earth move?" The discussion proceeds through the standard arguments of the preceding centuries, and there is no hint that Copernicus had proposed the mobility of the Earth.

The next group of notes in the manuscript probably comes from lectures given in the 1567 winter term by Bartholomew Schönborn, a medical doctor who published some small astronomical works during that decade.[16] The material follows in part the Wittenberg astronomer Casper Peucer's *Elementa doctrinae de circulis coelestibus*.[17] Needless to say this book does not espouse heliocentrism, but it does give Copernicus, as well as Reinhold, a certain prominence in the chronological section that opens the book. In the manuscript Copernicus is mentioned along with Regiomontanus and Apianus for his trigonometric tables.[18] The manuscript notes then turn to a second book by Peucer, *De dimensione terrae*, where another reference to Copernicus's trigonometric tables occurs.[19] The section ends with calculations and a poem for the eclipse of 8 April 1567, by Sebastian Theodoricus.

In May, Schönborn lectured on still another work of Peucer, *Novae questiones sphaerae*, another of the seemingly endless commentaries on Sacrobosco, but one not actually printed until 1573. Here we find a more interesting and more technical citation of Copernicus, in connection with the motion of the solar apogee. The words "Etsi aut Copernici hypothesis ut absurdas" jump out from the page, but the reference turns out be a technical point on the motion of the apogee, and not on the mobility of the Earth itself.[20] However, a few pages later, the numerical information for Mars is quoted with a book and chapter reference to *De revolutionibus*. Several pages later, after a section of rough calculations, Copernicus's name appears again, in a discussion on the measurement of star positions. The same topic reappears in more detail again with Copernicus's name, in the next section, in which yet another commentary on the sphere becomes the subject of the lectures.[21] This time the book is apparently *Epitome doctrinae de primo motu* of Vitorin Strigel,[22] a former student of Caspar Peucer's who was at that time professor of theology at Leipzig and just about to be silenced because of suspected Calvinism. Once more Copernicus's sine table is mentioned, and a few pages later his value of the obliquity of ecliptic is contrasted with that of Ptolemy. But when the notes discuss the possible mobility of the Earth, once more the standard rebuttals appear, and Copernicus is nowhere in sight.

The concluding section, the largest single section of the manuscript, deals with a slightly different work, Caspar Peucer's *Hypotheses astronomicae*. It is not clear to me whether these lectures were given by Sebastian Theodoricus or by Peucer himself. Peucer, the son-in-law of the Lutheran theologian Melanchthon, held considerable authority in the University at that time, although in 1576 he lost out in a faculty power struggle and was jailed, ostensibly for theological errors. In any event, the first part of the lectures mostly parallels a work called

Hypotyposes orbium coelestium, published anonymously in Strasbourg in 1568, but republished in 1571 (that is, a year or two after the date of the lectures) under the title *Hypotyposes astronomicae*, with Caspar Peucer as author. At the beginning, Peucer declared that the Strasbourg edition had been pirated from him. His preface to this rare printed text is an extraordinary commentary on the state of the Copernican hypothesis as taught in Wittenberg. Peucer complains of the "offensive absurdity so alien to the truth, of the Copernican theories."[23] The proper solution, he contends, is the Ptolemaic model made consistent with recent observations. This is implied in the full title of his book, which in English reads "Astronomical Hypotheses or the Theory of the Planets, from Ptolemy and other old doctrines, accommodated to the observations of Nicholas Copernicus and the canon of motion based on them."

Interestingly enough, the manuscript notes themselves are not so specific in their rejection of Copernican cosmology, but nonetheless this topic is given the treatment of silence. The manuscript contains numerous numerical comparisons between the tables of Johann Schöner and Reinhold's *Prutenic Tables*, but entirely divorced from any questions of the Earth's motion. Finally the manuscript ends with a horoscope and calculations for the eclipse of August 15, 1570, these apparently being the ultimate product of an astronomical education at Wittenberg on the eve of the first centenary of Copernicus's birth.

The document shows clearly that Copernicus was well known and esteemed as a mathematician and astronomer. Nevertheless, at that great academic center—the home of the *Prutenic Tables* and the spot from which Rheticus had gone to Poland to encourage the publication of *De revolutionibus*—the students were fully protected from possible confusion by Copernicus's absurd cosmology. These lecture notes show vividly the remarkable silence that seemed almost everywhere to shroud the Copernican system in the sixteenth century.

In fact, this is not news to any attentive reader of the astronomical literature between 1550 and 1600. Thus, Copernicus's name appears often in print, but his heliocentric system is virtually never discussed. A nice example, worth noting only because it is comparatively early (1556), is the *Tractatus brevis et utilis, de erigendis figuris coeli* of Johannes Garcaeus. Primarily an astrological work, it cites Copernicus ten times and uses his numbers to get celestial positions of the planets. Another, more interesting, example is Michael Maestlin's *Epitome astronomiae*, first printed in 1582 and then issued six more times, the last

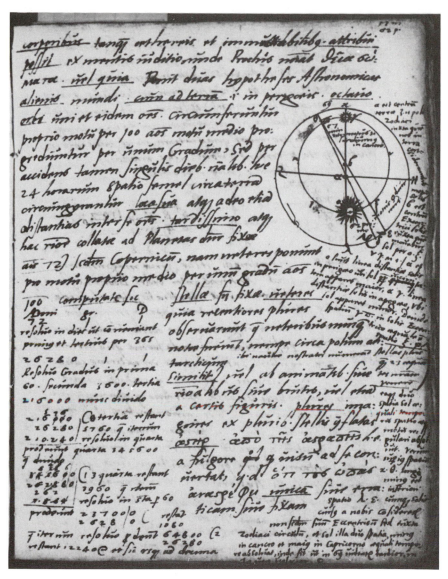

FIGURE 1. *Copernicus's ideas on the motion of the solar apogee are mentioned near the center of the page in these astronomy lecture notes from the University of Wittenberg in 1566. Gonville and Caius MS 387, Sec. 4, fol. 9r.*

in 1624. In this textbook Maestlin mentioned the name of Copernicus several times, but never once did he breathe a hint of the heliocentric cosmology.

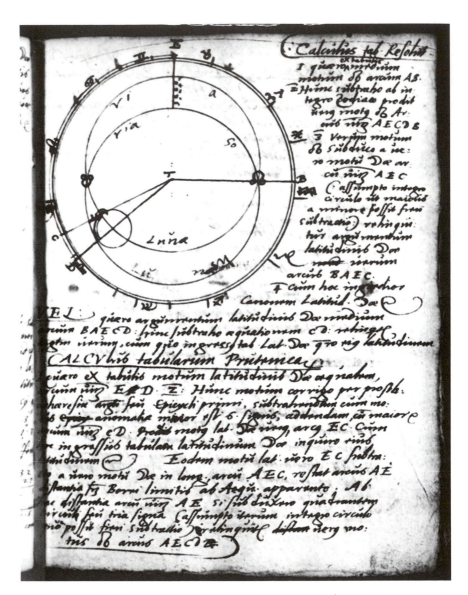

FIGURE 2. *Calculation of lunar positions according to both Schöner's* Tabulae resolutae *and Reinhold's* Prutenicae tabulae, *in astronomy lecture notes from 1569. Gonville and Caius MS 387, Sec. 12.*

In spite of the silence accorded the Copernican system in printed works and in academic lectures, I am convinced that Copernicus's arguments were rather widely known. Arthur Koestler, in his *Sleep-*

walkers, has called *De revolutionibus* "the book that nobody read" and "all-time worst seller." In fact, my examination of over 100 copies of the first edition of the book has convinced me that the contrary is the case. Sixteenth-century astronomers read with pen in hand, and their tracks persuade me that the book had a fair readership. So far I have found locations for approximately 180 copies of the original 1543 edition of the book. It is difficult to estimate the rate of attrition, but the original edition must have totaled at least 400 copies, a substantial number for a Renaissance science book. Nevertheless, by 1566 a second edition of a comparable size had become economically feasible. It would seem, then, that anyone seriously interested in astronomy would not have had much difficulty in encountering Copernicus's ideas.

This then leads us back to the theme of this paper—from model to reality. I can well imagine that the majority of sixteenth-century readers found Copernicus's ideas profoundly unsettling. As long as people could view the heliocentric idea as just another geometrical hypothesis for saving the phenomena, there was no need to get particularly upset. One could always hope for another alternative, such as the geocentric model later developed by Tycho Brahe. The matter is very nicely put around 1555 in a letter from Gemma Frisius that was published in several editions of Stadius's *Ephemerides*. In one of the rather few printed references in that century to the heliocentric hypothesis, Gemma allowed that the Copernican system gave a better understanding of planetary distances as well as certain features of retrograde motion. He added, however, that those who objected to the ephemerides because of the underlying hypothesis understood neither causes nor the use of hypotheses. "For these are not posited by the authors as if this must exist this way and no other." He further remarks: "Nay, even if someone wished to refer to the sky those motions that Copernicus assigns to the Earth, he could do so and according to the very canons of calculation."[24]

This position is even more clearly confirmed by the censorship imposed by the Inquisition when *De revolutionibus* was placed on the *Index* "until corrected," and by the dozen corrections issued in 1620. For example, the title of Chapter 11 was changed from "On the demonstration of the triple motion of the Earth" to "On the hypothesis of the triple motion of the Earth and its demonstration." Most of the other corrections have a similar nature.

Nowadays this Osianderian view of hypotheses strikes a sympathetic chord. Hypothetical model building is once more a familiar procedure, not only for astronomers and physicists, but for biologists and sociologists. To this extent our world view finds kinship with the astronomers of the late 1500s. Thus the "progress" from model to realism in the sixteenth century and the profound philosophical revolution of the early seventeenth century concerning the knowability of physical reality now takes on a bittersweet poignancy.

How did the view of heliocentrism change from a mere model to physical reality? Two men played the leading roles in the transformation of the prevailing opinions: Johannes Kepler, who found that the aesthetic arrangement of the Copernican system led to a coherent mathematical description of the motions, and Galileo Galilei, whose telescopic observations helped convince people that the Copernican system was not so absurd after all.

Kepler's own account of becoming a Copernican appears in the introduction to his *Mysterium cosmographicum*, where he mentions hearing about Copernicus in Michael Maestlin's astronomy lectures at Tübingen; Kepler was so delighted that he began to collect all the advantages that Copernicus had over Ptolemy, and he initiated a quest for the mathematical relationships between the number, the dimensions, and the motions of the planets. "At last on a quite trifling occasion I came near the truth," he wrote. "I believe Divine Providence intervened so that by chance I found what I could never obtain by my own efforts. I believe this all the more because I have constantly prayed to God that I might succeed if what Copernicus had said was true."[25]

What Kepler found was that the spacing of the planets could be closely approximated by an appropriately arranged nesting of the five regular polyhedra between spheres for the six planets of the Copernican system. Quixotic or chimeral as Kepler's polyhedra may appear today, we must remember that the *Mysterium cosmographicum* was essentially the first Copernican treatise of any significance since *De revolutionibus* itself. Without a Sun-centered universe, the entire rationale of his book would have collapsed.

Furthermore, Kepler recognized that, although in the Copernican system the Sun was near the center, it played no physical role. Kepler argued that the Sun's centrality was crucial, for the Sun itself must provide the driving force to keep the planets in motion, and he set out for the first time to show this connection mathematically.

Kepler knew that the more distant a planet was from the Sun, the

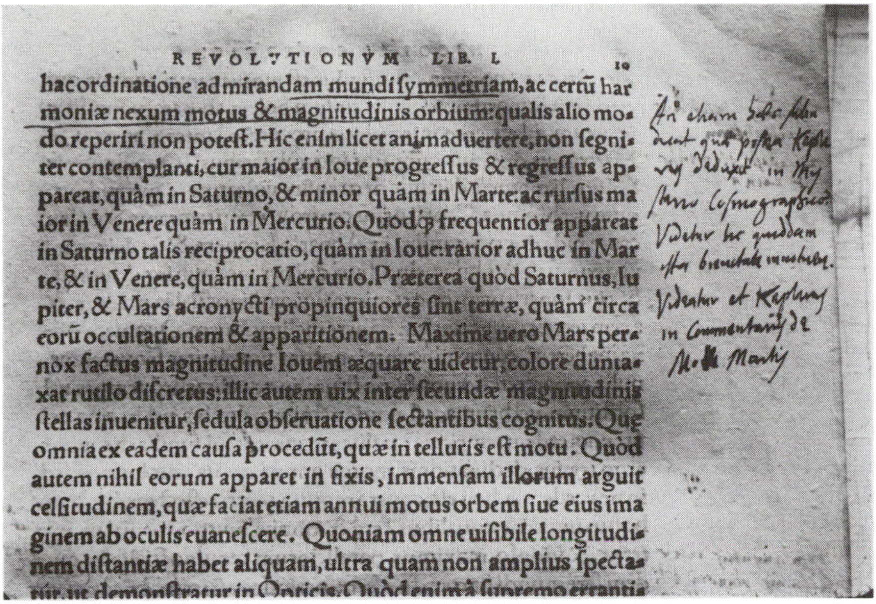
REVOLVTIONVM LIB. I. 10

hac ordinatione admirandam mundi symmetriam, ac certũ har
moniæ nexum motus & magnitudinis orbium: qualis alio mo-
do reperiri non potest. Hic enim licet animaduertere, non segni-
ter contemplanti, cur maior in Ioue progressus & regressus appareat, quàm in Saturno, & minor quàm in Marte: ac rursus ma
ior in Venere quàm in Mercurio. Quodq; frequentior appareat
in Saturno talis reciprocatio, quàm in Ioue: rarior adhuc in Mar
te, & in Venere, quàm in Mercurio. Præterea quòd Saturnus, Iu
piter, & Mars acronycti propinquiores sint terræ, quàm circa
eorũ occultationem & apparitionem. Maxime uero Mars per-
nox factus magnitudine Iouem æquare uidetur, colore dunta-
xat rutilo discretus: illic autem uix inter secundæ magnitudinis
stellas inuenitur, sedula obseruatione sectantibus cognitus. Quę
omnia ex eadem causa procedũt, quæ in telluris est motu. Quòd
autem nihil eorum apparet in fixis, immensam illorum arguit
celsitudinem, quæ faciat etiam annui motus orbem siue eius ima
ginem ab oculis euanescere. Quoniam omne uisibile longitudinem distantiæ habet aliquam, ultra quam non amplius specta-
tur, ut demonstratur in Opticis. Quòd enim à supremo errantia

FIGURE 3. *Broscius's annotation in his copy of Copernicus's* De revolutionibus *(1566) in the Cracow Observatory Library; a transcription appears in note 27.*

longer its period—indeed, this was one of the most important regularities of the heliocentric system, already noted by Copernicus, who wrote:[26]

> In this arrangement, therefore, we discover a marvelous symmetry of the universe, and an established harmonious linkage between the motion of the spheres and their size, such as can be found in no other way.

Undoubtedly Kepler himself had been inspired by this passage in *De revolutionibus*. And it is fascinating to notice that at least one of Kepler's contemporaries recognized this connection. Johannes Broscius, professor at Cracow, underscored those lines in his own copy of Copernicus's book, and in the margin wrote in Latin (see Figure 3):[27]

> Was perhaps this underlined part what Kepler afterwards deduced in his *Mysterium cosmographicum?* It seems by the brevity that something more is involved. See also Kepler in his *Commentary on the Motion of Mars.*

For Kepler, there was an essential physical difference between a geocentric and a heliocentric universe; only in the latter case would the Sun provide the central motive power for the planetary system. Hence, Kepler believed firmly in the reality of the Copernican system. Armed with this conviction, he realized that, if the orbit had a physical reality, the same orbit must yield latitudes as well as longitudes. This may be obvious today but, in Kepler's age, this was a novel idea that became a fundamental tool for his attack on the problem of Mars and an important link in the chain that led to the discovery of the elliptical orbit of Mars. Thus, for Kepler's work, belief in the heliocentric system really mattered, and made a vital difference in his approach to the subject.

In 1609, when Kepler published the results of his researches on Mars and his *Astronomia nova*, he placed on the back of the title page an indignant notice revealing in print for the first time that Osiander was the author of the anonymous preface to Copernicus's book. He wrote, "It is a most absurd fiction, I admit, that the phenomena of nature can be demonstrated by false causes. But this fiction is not in Copernicus . . . as evidence, I offer this work."

Accompanying Kepler's bold proclamation was a second remarkable paragraph. Petrus Ramus, professor of philosophy and rhetoric in Paris during the middle of the sixteenth century, had offered his chair to anyone who could produce an "astronomy without hypotheses," and Kepler declared that if Ramus were still alive he would have claimed the reward. Clearly Kepler believed that his recourse to physics had freed astronomy from the arbitrary geometrical devices that were still present in the work of Copernicus. Fundamental to Kepler's "astronomy without hypotheses" was the concept that one special physical object, the Sun, was physically and mathematically linked to planetary motions. In essence this is the central power of the Copernican idea and the essential stepping stone to Newton's law of gravitation. It is, of course, in this context that Copernicus, rather than Aristarchus, is being celebrated in 1973.

Notes and References

[1] Translated from the transcription given in *Nikolaus Kopernicus Gesamtausgabe*, ed. by F. Zeller and C. Zeller, vol. 2 (Munich, 1949), p. 30. I wish to thank Miss Joanne Phillips for preparing an initial translation.

[2] Πλουταρχου Χαερωνεως περι των αρεσκοντων τοις φιλοσοφοις [*de placitis philosophorum*] (Joan. Hervagius, Basel, 1531), book 3, chap. 13. The location of the copy Copernicus used is unknown, although his copy of the 1516 Strasbourg

Latin edition is preserved in Uppsala. I have compared the Copernicus text against the British Museum copy of the Basel edition, 524. g. 18, where it is cataloged under "Supposititous Works." In spite of a few minor differences, it seems likely that Copernicus used the Basel edition for his quotation.

[3] N. Copernicus, *Revolutions of the Heavenly Spheres*, in *Great Books of the Western World*, vol. 16 (Chicago, 1952), p. 508.

[4] See my "Reinhold" in the *Dictionary of Scientific Biography*; also see Karl Heinz Burmeister, *Georg Joachim Rheticus*, three volumes (Wiesbaden, 1967–68).

[5] Translations of the *Narratio prima* slightly modified from E. Rosen, *Three Copernican Treatises* (New York, 1971), pp. 165, 192.

[6] The greatly abridged text printed here is based on the translation of E. Rosen, *op. cit.*, pp. 24–25.

[7] E. Reinhold, *Prutenicae tabulae* (Tübingen, 1551), part 1, fol. 35 in Sec. 21, "Praeceptum. De Calculo adparentis magnitudinis tropici anni ad datum tempus." This quotation was inserted as an advertisement in the second edition of Copernicus, *De revolutionibus* (Basel, 1566).

[8] O. Gingerich, "The Role of Erasmus Reinhold and the Prutenic Tables in the Dissemination of the Copernican Theory," *Studia Copernicana*, vol. 6 (1973): pp. 43–62; this article cites previous authors including P. Duhem, L. A. Birkenmajer, E. Zinner, and A. Birkenmajer [reprinted as selection 13 in this anthology].

[9] See M. R. James, *A Descriptive Catalogue of the Manuscripts in the Library of Gonville and Caius College* (Cambridge, 1907–08). Under 387, p. 447, in line 3 read "Vuinshemii" in place of "Avinstemii" and in lines 9 and 10, read "Peuceri" in place of "Pruerii."

[10] Originally I stated that I had not been able to locate Rankghe in the student lists in *Album Academiae Vitebergensis ab A. Ch. MDII Usque ad A. MDCII*, vol. 2 (Halle, 1894); I was under the false impression that no name index existed. In fact, Fourentius Rancke Colbergen is indexed; he matriculated in January, 1566.

[11] Balduinus's handwriting in this manuscript agrees with the more formal specimen in the Wittenberg Dean's Book, which is now preserved at the Archives of the Martin Luther University in Halle. Balduinus was from Witternberg and matriculated at the University in October of 1542. In 1574 Balduinus published *Vorhersage für 1574* (Wittenberg), Zinner 2664, but I have found no other trace of him.

[12] Sebastian Theodoricus Winshemius, *Novae questiones spherae, hoc est, de circulis coelestis, primo mobile, in gratiam studiosae iuuentutis scriptae* (Wittenberg, 1564, 1567, 1570, 1578, 1583, 1591, 1605). Theodoricus served as dean at Wittenberg in the spring of 1568.

[13] MS Sec. 4, fol. 2r. The printed text reads on p. 90: "Terra maior est centies sexagies sexies. Est enim proportio Diametrorum secundum Ptolemaeum, quintupla sesquialtera, que est 11 ad 2. Secundum Copernicum vero quintupla superpartiens novem vicesimas, quae est 5 inteq. & 27 scrup ad unum."

[14] MS Sec. 4, fol. 64; also fol. 7r. On fol. 7v, "vide Reinholdum in tabulis Prutenicis."

[15] *Ibid.*, fol. 8v.

[16] Schönborn authored *Computus astronomicus* (Wittenberg, 1567, 1579) and *Oratio de studiis astronomices astronomices* (Wittenberg, 1564). He was dean at Wittenberg in 1564.

[17] Caspar Peucer, *Elementa doctrinae de circulis coelestibus* (Wittenberg, 1551, 1553, 1558, 1563, 1569, 1576, 1587); *De dimensione terrae* (Wittenberg, 1550, 1554, 1579); *Novae questione sphaerae* (Wittenberg, 1573).

[18] MS Sec. 6, fol. 3v.

[19] MS Sec. 7, fol. 10v.

[20] MS Sec. 9, fol. 9v; also fol. 9r.

[21] MS Sec. 11, fol. 16r; also section 10, fol. 5r.

[22] Victorin Strigelius, *Epitome doctrinae de primo motu, aliquot demonstrationibus illustrata* (Leipzig, 1564; Wittenberg, 1565). MS Sec. 11, fol. 22r; fol. 24v.

[23] Caspar Peucer, *Hypotyposes astronomicae, seu theorias planetarium. Ex Ptolemaei et aliorum veterum doctrina ad observationes Nicolai Copernici & canones motuum ab eo condilos accomodatae* (Wittenberg, 1571).

[24] In Johannes Stadius, *Ephemerides novae* (Cologne, 1556, 1559, 1560, 1570). Translation by Joanne Phillips.

[25] Johannes Kepler, *Mysterium cosmographicum* (Tübingen, 1596), p. 6; trans. by Owen Gingerich in "Kepler," *Dictionary of Scientific Biography*, vol. 7 (New York, 1973), pp. 289–312.

[26] *Nicholas Copernicus Complete Works II. On the Revolutions*, trans. by Edward Rosen, translator (London-Warsaw-Cracow, 1973), book 1, chap. 10.

[27] "An etiam haec subindicat quam postea Keplerus deduxit in Mysterio Cosmographico. Videtur hic quiddam ista brevitate involvere. Videatur et Keplerus in Commentariis de Motibus Martis." Broscius's copy of the 1566 edition is preserved at the Observatory in Cracow. I wish to thank Professor E. Rybka for arranging for me to see and photograph this book. My transcription differs slightly from the one given by L. A. Birkenmajer, *Mikolaj Kopernik* (Cracow, 1900), p. 657.

KEPLER AND
THE NEW
ASTRONOMY

Johannes Kepler and the New Astronomy

Johannes Kepler was conceived on 16 May 1571 at 4:37 A.M. and born on 27 December at 2:30 P.M. We therefore see that 1971 was the four-hundredth anniversary not only of Kepler's birth but also of his conception. The existence of such accurate dates reminds us that Kepler lived in an age when astronomer still meant astrologer and when the word "scientist" had not yet been invented. Kepler wrote down these dates when he was 25 years old and much fascinated by astrology. Like many of the world's greatest scientists, he had a profound feeling for the harmony of the heavens; Kepler believed in a powerful concord between the cosmos and the individual, although he rejected many of the traditional details of astrology.

From our own scientific and philosophical vantage point far removed from the turn of the seventeenth century, any assessment of this man's genius must be incomplete and imperfect. Nevertheless, our twentieth-century perspective can offer insights overlooked by the interpreters of previous generations. If Kepler could have chosen from our twentieth-century words, I suppose that he would have called himself a cosmologist. I should like to argue that we can accurately call Kepler the first astrophysicist.

Kepler stands at a junction in the history of astronomy when the old Earth-centered universe was giving way to the new Sun-centered sys-

Selection 17 reprinted from *Quarterly Journal of the Royal Astronomical Society*, vol. 13 (1972), pp. 346–60.

tem. Yet the heliocentric system as presented by Copernicus contained many vestiges of the old astronomy. Kepler's greatest book was the *Astronomia nova*. Published in 1609, it broke the two-millennium spell of perfect circles and uniform angular momentum—it was truly the New Astronomy. It is this work, which Kepler called his "warfare on Mars," that will form the focus of my remarks.

There was little in Kepler's youth to indicate that he would become one of the foremost astronomers of all time. Although Tycho's supernova of 1572 burst forth when Kepler was a mere infant, the Great Comet of 1577 made a lasting impression. Kepler was a weak and sickly child, but intelligent, and after the elementary Latin school he easily won a scholarship to the nearby Tübingen University so that he could study to become a Lutheran clergyman. There he produced a straight-A record—but the grade records preserved at Tübingen show that nearly everyone was an A student in 1589. In recommending him for a scholarship renewal, the University Senate noted that Kepler had "such a superior and magnificent mind that something special may be expected of him."

Yet Kepler himself wrote that, although he had done well in the prescribed mathematical studies, nothing indicated to him a special talent for astronomy. Hence, he was surprised and distressed when, midway through his third and last year as a theology student at Tübingen, he was summoned to Graz, far away in southern Austria, to become an astronomy teacher and the provincial mathematician.

At the Protestant high school in Graz, Kepler turned out not to be a very good teacher. In the first year he had only a few students, in the second none at all. Needless to say, this gave him more time to pursue his own research! Nevertheless, it was in one of his class lectures that Kepler hit upon what he believed to be the secret key to the construction of the universe.

This key hung upon a crucial thread: at Tübingen, Kepler had become a Copernican. The astronomy teacher at the University, Michael Maestlin, was remarkably knowledgeable about Copernicus's *De revolutionibus*. Yet, strangely enough, his popular and often reprinted textbook, *Epitome astronomiae*, never even hinted at the heliocentric cosmology. Nevertheless, in his lectures at Tübingen, Maestlin included a discussion of the new Copernican system. He explained how this system accounted for the retrogradations in a most natural way, and how the planets were laid out in a very harmonic fashion, both with respect to their spacing from the Sun and with respect to their periods.

It was undoubtedly the beautiful harmonic regularities "so pleasing to the mind" that appealed strongly to Kepler's sense of the aesthetic and induced him to become such an enthusiastic Copernican—as opposed to Maestlin, the timid Copernican. To Kepler the theologian, such regularities revealed the glory of God. When he finally hit upon that secret key to the universe, he attributed it to Divine Providence. "I believe this," he wrote, "because I have constantly prayed to God that I might succeed in what Copernicus had said was true."[1] Later, in writing to his teacher Maestlin, he said, "For a long time I wanted to become a theologian; for a long time I was restless. Now, however, behold how through my effort God is being celebrated in astronomy."[2]

Because of his preoccupation with the Copernican system, Kepler began to ask himself three unusual questions: Why are the planets spaced this way? Why do they move with these regularities? Why are there just six planets? All these questions are very Copernican, the last one particularly so because a traditional geocentrist would have counted both the Sun and the Moon, but not the Earth, thereby listing seven planets.

Figure 1 recalls the circumstances under which Kepler hit upon his secret key to the universe. In a lecture to his class, he had drawn the ecliptic circle and he was illustrating how the great conjunctions of Jupiter and Saturn, which take place every 20 years, fall almost one-third of the way around the sky in successive approaches. As he connected the successive conjunctions by quasi-triangles, the envelope of lines outlined a circle with a radius half as large as that of the outer ecliptic circle. The proportion between the circles struck Kepler's eye as almost identical with the proportions between the orbits of Saturn and Jupiter. Immediately, he began a search for a similar geometrical relation to account for the spacing of Mars and the other planets, but his quest was in vain.

"And then again it struck me," he wrote. "Why have plane figures among three-dimensional orbits? Behold, reader, the invention and the whole substance of this little book!"[3] He knew that there were five regular polyhedra—that is, solid figures each with faces all the same kind of regular polygon. By inscribing and circumscribing these figures with spheres (all nested in the proper order), he found that the positions of the spheres closely approximated the spacings of the planets (Figure 2). Since there are five and only five of these regular or Platonic polyhedra, Kepler thought that he had explained the reason why there were precisely six planets in the solar system.

Kepler published this scheme in 1596 in his *Mysterium cosmo-*

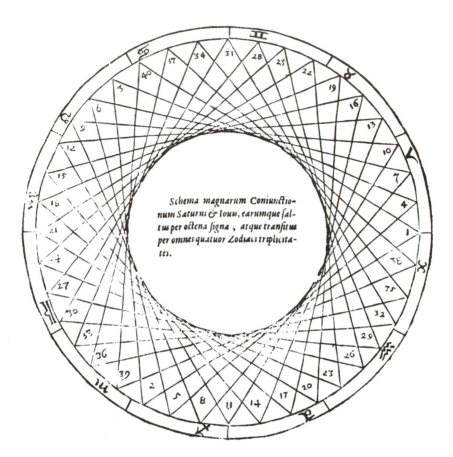

FIGURE 1. *The pattern of successive conjunctions of Jupiter and Saturn, from the* Mysterium cosmographicum.

graphicum, the "Cosmographic Secret." It was the first new and enthusiastic Copernican treatise in over 50 years, since *De revolutionibus* itself. Without a Sun-centered universe, the entire rationale of the book would have collapsed.

Kepler also realized that the center of the Copernican system was the center of the Earth's orbit, not the Sun. Although the Sun was nearby, it played no physical role. But Kepler argued that the Sun's centrality was essential and that the Sun itself must supply the driving force to keep the planets in motion. Not only did he propose this very significant physical idea, but he attempted to describe mathematically how the Sun's driving force diminished with distance. Again, his result

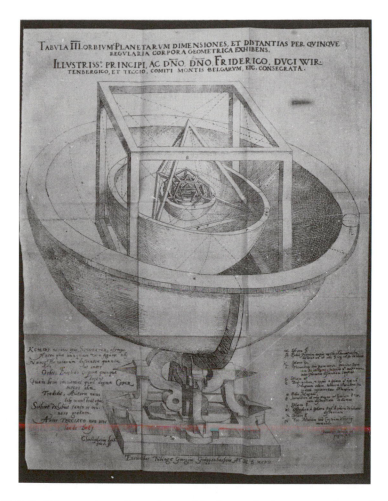

FIGURE 2. *The nesting of regular polyhedra and planetary spheres that accounted for the spacing and number of planets according to Kepler. From his* Mysterium cosmographicum *(1596).*

was only approximate, but at least the important physical-mathematical step had been taken. This idea, which was to be much further developed in the *Astronomia nova*, establishes Kepler as the first scientist to demand physical explanations for celestial phenomena. Although the principal idea of the *Mysterium cosmographicum* was erroneous, never in history has a book so wrong been so seminal in directing the future course of science.

Kepler sent a copy of his remarkable book to the most famous astronomer of the day, Tycho Brahe. Unknown to Kepler, the renowned Danish astronomer was in the process of leaving his homeland. He had boasted that his magnificent Uraniborg Observatory had cost the king more than a ton of gold. Now, however, fearing the loss of royal support, Tycho had decided to join the court of Rudolf II in Prague. Emperor Rudolf was a moody, eccentric man whose love of the occult made him more than willing to support a distinguished astronomer-astrologer.

Kepler describes this sequence of events in the *Astronomia nova* itself. The Danish astronomer had been impressed by the *Mysterium cosmographicum*, though he was unwilling to accept all its strange arguments; then, Kepler writes,[4]

> Tycho Brahe, himself an important part in my destiny, continually urged me to come to visit him. But since the distance of the two places would have deterred me, I ascribe it to Divine Providence that he came to Bohemia. I arrived there just before the beginning of the year 1600 with the hope of obtaining the correct eccentricities of the planetary orbits. Now at that time Longomontanus had taken up the theory of Mars, which was placed in his hands so that he might study the Martian opposition with the Sun in 9° of Leo [that is, Mars near perihelion]. Had he been occupied with another planet, I would have started with that same one. That is why I again consider it an effect of Divine Providence that I arrived in Prague at the time when he was studying Mars; because for us to arrive at the secret knowledge of astronomy, it is absolutely necessary to use the motion of Mars; otherwise that knowledge would remain eternally hidden.

Kepler's *Astronomia nova* was not to be published until nine years later. Never had there been a book like it. Both Ptolemy in the *Almagest* and Copernicus in *De revolutionibus* had carefully dismantled the scaffolding by which they had erected their mathematical models. Although Kepler's book is well organized, it is nearly an order of magnitude more complete and complex than anything that had gone before; our astronomer himself admits that he might have been too prolix.

In the first great battle in his warfare on Mars, Kepler describes the so-called vicarious hypothesis. This was an attempt to represent the

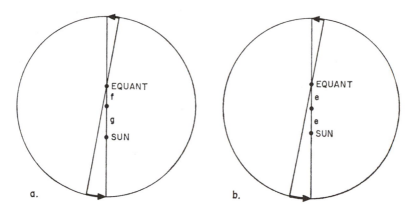

FIGURE 3. *The orbit of Mars with an equant, the seat of uniform angular motion. In the vicarious hypothesis (a), accurate longitudes are obtained by setting g/f=5/3. The quasi-Ptolemaic scheme (b), with its equal-and-opposite equant, satisfies the Sun-Mars distances but errs in longitude by 8′ in the octants.*

motions of Mars on an eccentric circle driven by uniform angular motion about a point called an equant—essentially a traditional model cast into the new heliocentric pattern. Kepler achieved the great accuracy in the longitudes by allowing the equant to fall at an arbitrary position, as shown in Figure 3(a). In this scheme, which he was to call the "vicarious hypothesis," the true anomaly is (neglecting terms in e^3):

$$v = T - (f+g)\sin T + g\frac{(f+g)}{2}\sin 2T + \ldots, \qquad (1)$$

where g is the eccentricity from the Sun to the center of the circle, and f is the eccentricity from the equant to the center. For comparison, the motion in an ellipse with the law of areas is

$$v = T - 2e \sin T + \tfrac{5}{4}e^2 \sin 2T + \ldots .$$

Hence, if $g = \tfrac{5}{4}e$ and $f = \tfrac{3}{4}e$, the vicarious hypothesis satisfies the equations to second order and we can show that the remaining error is approximately $\tfrac{1}{4}e^3$, which, in the case of Mars with its eccentricity of nearly 0.1, amounts to about 1′. Thus, in predicting the longitudes, Kepler succeeded brilliantly, with an accuracy almost two orders of magnitude better than that of either Ptolemy or Copernicus.

There exists among Kepler's manuscript pages still preserved in Leningrad a remarkable sheet showing a diagram of the vicarious orbit (Figure 4).[5] It is very carefully laid out in a publishable form as one of the first few pages of a book on Mars, and it includes the opening lines of the poetic tribute to Tycho that ultimately appeared in the *Astronomia nova*. The diagram, with its unequally spaced equant in the ratio 5:3, can be seen at the bottom of the page. Kepler was always very eager to publish, and elsewhere in the manuscript material we see the titles for chapters in a book that he was organizing before he even knew that Mars had a noncircular orbit. Apparently, this page comes from about the same period—evidently at one point he was prepared to publish his vicarious orbit as the solution of the riddle of Mars. Fortunately, Divine Fate prevented him from publishing his commentary on Mars until it indeed was truly the New Astronomy.

Although Kepler's scheme had achieved a great triumph with respect to the longitudes, it failed with respect to distances. In observational astronomy, longitudes can be determined directly with great precision, but in general the distances must be deduced by other methods. Here Kepler very cleverly used the latitudes of Mars to deduce the distances—but alas, this led to an absurdity and showed that his orbit could not, in fact, be the real one. Hence, he named it the vicarious orbit in contradistinction to the real or "physical" hypothesis that he was seeking.

Ptolemy, in his orbit for Mars, had constrained the equant to fall directly opposite the center of the orbit from the Earth and equally distant from it. Kepler now realized that such an equal-and-opposite equant more closely approximated the real orbit than did his vicarious orbit, which satisfied the longitude so well. This case is represented by equation (1) when $e = f = g$, or

$$v = T - 2e \sin T + e^2 \sin 2T + \ldots,$$

so that the error in heliocentric longitude is $\frac{1}{4}e^2 \sin 2T$; in the case of Mars, this gives 8' in the octants, an error easily detectable with Tycho's data, which Kepler believed were generally accurate to about 2'. In a celebrated passage, Kepler wrote: "God's goodness has granted us such a diligent observer in Tycho Brahe that his observations convicted the Ptolemaic calculation of an error of 8' of arc. It is therefore right that we should with a grateful mind make use of this gift to find the true celestial motions."[6]

Figure 3(b) depicts the eccentric orbit with its equal-and-opposite equant. Because the angular motion is uniform about the equant, the

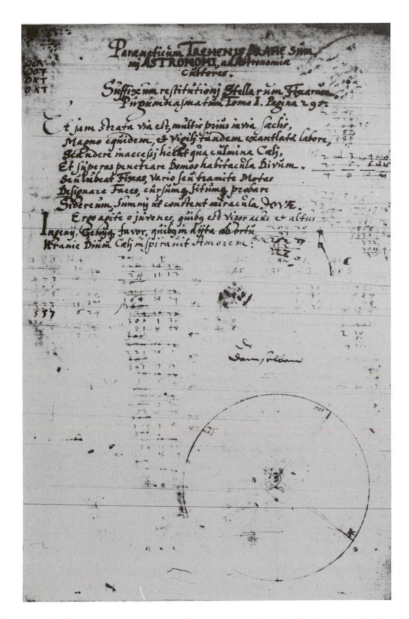

FIGURE 4. *Kepler's manuscript for a proposed introductory leaf of a Mars ephemeris. Note the vicarious orbit near the bottom. After he abandoned this publication, Kepler used the page for other calculations. Archives of the Academy of Sciences of the USSR, Leningrad Kepler manuscripts, XIV, fol. 372.*

opposite angles are equal, and the orbital motion at the aphelion is much less than at the perihelion. This was precisely the kind of motion that Kepler the astrophysicist desired: the planet's speed is inversely proportional to its distance from the Sun, a quite reasonable hypothesis if we assume that some physical emanation from the Sun is responsible for propelling the planet in its orbit. For Kepler this was a very fundamental idea; we can call it his distance law.

Although the outer planets had an equal-and-opposite equant in the Ptolemaic system, Kepler knew that the Sun-Earth orbit did not. In order for the Copernican system to be a real physical one, Kepler recognized that the same mechanism must apply for the orbit of the Earth as for that of Mars. The varying speed in longitude of the apparent Sun throughout the year required a certain definite spacing between the Sun and the point of uniform angular motion (traditionally taken as the center of the orbital circle). The same spacing can be preserved, however, if we retain an equant but recenter the circle midway between the old center and the position of the Sun. Such a model will predict virtually the same longitudinal motion, but with different Earth-Sun distances, and would, of course, provide a physical mechanism similar to that for Mars.

But how to find the varying distance of the Sun? One way would be to measure the apparent diameter of the Sun at different times throughout the year. And so let me digress here, just as Kepler did.

When Kepler arrived in Prague, he bet Longomontanus that he could solve the theory of Mars within a week. He lost the bet, of course—it actually took five years, but, as he apologized in the *Astronomia nova*, he took one year out for optics. His resultant work, the *Astronomiae pars optica*, lays the foundation for modern geometrical optics. In it he explains, for the first time, how an inverted image is formed on the retina of the eye, and he clearly defines the light ray. Also, he investigates the effects of apertures of various sizes and shapes on the formation of an image.[7] Such considerations were of fundamental importance in observing the solar diameter, because the variations were rather small, but unfortunately the results were not conclusive.

Thus, Kepler turned his attention to another exceedingly ingenious way to locate the position of the Earth's orbit. He knew that Mars returned to the same point in space every 687 days, but that the Earth would be at two different points in its orbit since in that time it would not yet have completed its second full revolution. Kepler's manuscripts for the first two years of his work on Mars are apparently almost completely intact, and in Figure 5 we see the very first time when

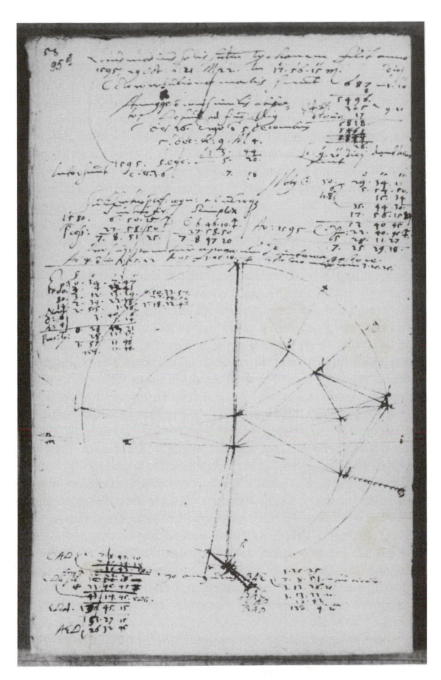

FIGURE 5. *Kepler's earliest Earth-Mars triangulation attempt, page 58 in his Mars workbook of 1600. Archives of the Academy of Sciences of the USSR, Leningrad Kepler manuscripts, XIV, fol. 95v.*

Kepler tried such a triangulation. These results were ultimately very important, for they showed that his physical intuition was correct and that the Earth's orbit had to be moved to a new center. Hence, it could have a physical mechanism and a distance law, just as did the other planets.

Kepler, in fact, already had rather definite ideas about the physical mechanism involved. Through Johannes Taisner's book on the magnet (1562) and, later, William Gilbert's, he convinced himself that the planetary driving force emanating from the Sun must be magnetic. He believed that both the Sun and the magnetic emanation were necessarily rotating in order to impart a continuous motion to the planets. From the distance law, he deduced that the strength of the emanation decreased in inverse proportion to distance, and he therefore concluded that the emanation spread out in a thin plane—unlike light, which filled space and decreased as $1/r^2$.

When he applied the distance law to the Earth's orbit, a difficult quadrature resulted that he could handle only by laborious numerical calculations. Then Kepler had the fortunate inspiration to replace the sums of the radius vectors required by the distance law with the areas within the orbit. Thus, the radius vector swept out equal areas in equal times. Kepler recognized that this was mathematically objectionable, but, like a miracle, it provided an accurate approximation to the orbital motion predicted by the distance law. In Figure 3(b) it is easily seen that the equant theory represents the law of areas only if the equant is placed directly opposite the Sun and at an equal distance from the center; the distance law and the law of areas are then rigorously equivalent at aphelion and perihelion.

At this point, Kepler had (1) an accurate but physically inadmissible scheme for calculating longitudes (the vicarious hypothesis) and (2) an intuitively satisfactory physical principle (the distance law) that was applicable to the Earth as well as to the other planets but which left an unacceptable 8' error in predicting the heliocentric longitudes of Mars. In order to preserve simultaneously both his accurate longitude predictions and the properly centered circular orbit, Kepler next added a small epicycle to his circle. This was a time-honored device, used not only by Ptolemy but by Copernicus and Brahe as well. The earliest pages of Kepler's Mars notebook from the first few weeks with Tycho Brahe in Prague show numerous experiments with epicycles. It is fascinating to see that, although Kepler is here exploring very new ground, he can still adapt his tools from traditional astronomy. Nevertheless, he was distressed by having to introduce such an absurd

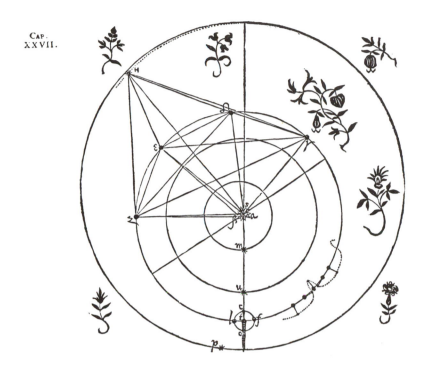

FIGURE 6. *The triangulation that revealed the noncircular orbit of Mars, from the* Astronomia nova *(1609), Chapter 27.*

device. He argued that, just as sailors cannot know from the sea alone how much water they have traversed, since their route is not distinguished by any markers, so the mind of the planet will have no control over its motion in an imaginary epicycle except possibly by watching the apparent diameter of the Sun.

Kepler had difficulty in preserving the circular motion when he adopted an epicycle; it is therefore not surprising to find that our astronomer next turned to a closer examination of the shape of Mars's path. Having established the proper position of the Earth's orbit by triangulation of Mars, he was able to turn the procedure around and to investigate a few points in the orbit of Mars itself. The results are shown schematically in Figure 6, taken from the *Astronomia nova*. I say schematically because this method did not yield an exact quantitative position. Instead it showed only qualitatively that the orbit was noncircular. Kepler recognized that observational errors

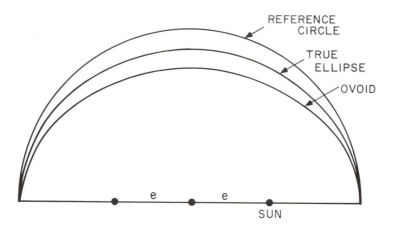

FIGURE 7. *Kepler's ovoid orbit compared with the final ellipse: the eccentricity is greatly exaggerated.*

prevented him from getting precise distances to the orbit. Because of this scatter, he had to use, as he picturesquely described it, a method of "votes and ballots."

Armed with these results, Kepler found in the epicycle the convenient means for generating a simple noncircular path. The resultant curve is shown in Figure 7. On this scale, it differs imperceptibly from an ellipse, although actually the curve is slightly egg-shaped, with the fat end toward the Sun.

Kepler required that the motion with the generating epicycle should satisfy his law of distances, which could be approximated by the law of areas; some details of the construction are found in the extended caption to Figure 8. If Kepler had had access to the integral calculus, he would have found that the egg-shaped or ovoid curve has a very elementary equation, but this he did not know. We must remember that even the equals sign had been invented only in the preceding generation, and Descartes' analytic geometry was still in the coming generation.

In working with the ovoid, Kepler got himself into a very messy quadrature problem that could best be tackled with the help of an approximating ellipse. Most popular accounts of his warfare on Mars leave the reader puzzled as to why Kepler did not immediately abandon the ovoid and adopt the ellipse. As Figure 8 demonstrates, the approximating ellipse has an effective eccentricity of $\sqrt{2}e$, where e is the

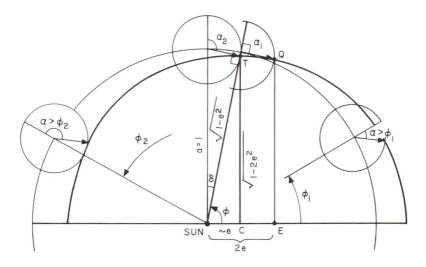

FIGURE 8. *The epicyclic construction of Kepler's ovoid (the thicker curve). The epicycle has radius e. Angle α moves uniformly with time, whereas φ moves nonuniformly in order to satisfy the area law, so that $\int_0^\phi r^2 d\phi = c\alpha$. If this construction had an equant, it would fall 2e from the Sun at E, and Mars would reach Q in a quarter period; hence, α_1 in the epicycle must be very close to 90°. As the epicycle center moves through angle δ, the epicycle vector will also advance by δ since dφ is very near its mean rate in this part of the orbit (so $d\alpha \approx d\phi$). Thus, $\alpha_2 = a_1 + \delta$, and the angle at T is a right angle. Then the line SUN-T $= \sqrt{1 - e^2}$ and TC $= \sqrt{1 - 2e^2}$. Since the semiminor axis of an ellipse is $a\sqrt{1 - e^2}$, the approximating ellipse to Kepler's ovoid has eccentricity $\sqrt{2}e$. Kepler called angle δ the "optical equation." He finally realized that an ellipse of eccentricity of e gave the required path when he noticed that sec δ($= 1/\sqrt{1 - e^2} = 1 + \frac{1}{2}e^2$) exceeded unity by precisely the width of the lunula between the circle and the noncircular orbit. "It was as if I had awakened from a sleep," wrote Kepler.*

distance from the center of the circle to the Sun. The diagrams show how inaccurately the triangulation method determined the points on the orbit of Mars. If these points had been well determined, Kepler would have seen immediately that the ovoid departed from the circular orbit by twice as much as it should have. His real hold on the problem came through the predicted longitudes, not the distances, and here again he found the 8′ discrepancy in the longitudes at the octants. Kepler wrote the previously quoted celebrated passage about the 8′ in connection with the errors of a circle, but it is quite possible that he first discovered it in examining the ovoid. The symmetry of the situation shows that, if there is an 8′ error in a circular Ptolemaic orbit, there ought also to be about an 8′ error in the ovoid, which deviates equally on the other side of the correct ellipse.

Luckily, in the end Kepler abandoned the epicycle and adopted the ellipse that lay halfway between his ovoid and the circle. But the process was not simply the method of "cut and try" so often imputed to Kepler by popular accounts. He was still seeking a single physical mechanism to explain not merely the varying speed of Mars in its orbit but also the varying distances. His answer came in an extension of the magnetic effects that propel the planets in their orbits. Kepler drew a very charming analogy to a boatman in an amusement park. Apparently, a cable was stretched across a stream and the small boat attached to the cable. The oarsman, by directing the rudder, could use the flow of the stream to propel the craft back and forth from one side to another.

From Gilbert's book, Kepler knew about the magnetic axis of the Earth. Such a magnetic axis, he proposed, could act as the rudder in the Sun's magnetic emanation, guiding the planet first near and then far from the Sun. If the magnetic axis is fixed in space, then its projection as seen from the Sun will be $\cos \theta$. Such a cosine term appears in the polar equation for the ellipse:

$$r = a(1 - e^2)/(1 + e \cos \theta).$$

To the first order in eccentricity, the ellipse satisfies this physical picture of the magnetic axis governing the advance and retreat of the planet. For Kepler, who did not work with the polar equation, the real hurdle was to find the geometrical equivalents between the librating magnetic axis and the ellipse. "I was almost driven to madness in considering and calculating this matter," he wrote. "I could not find out why the planet would rather go on an elliptical orbit. Oh, ridiculous me! As if the libration on the diameter could not also be the way to the ellipse. So this notion brought me up short, that the ellipse exists because of the libration. With reasoning derived from physical principles agreeing with experience, there is no figure left for the orbit of the planet except a perfect ellipse."[8]

Indeed, Kepler was luckier than he knew. Just as there is an approximating ellipse to the oval he originally tried, so there is an approximating oval indistinguishable (with Tycho's data) from this final ellipse, the so-called *via buccosa*. But when Kepler found that the ellipse satisfied the observations, he must have hastily assumed that no other curve would do; thus, driven by his persistent physical intuition, he had continued until he almost accidentally hit upon the right curve.

With justifiable pride he could call his book *The New Astronomy*; its subtitle emphasizes its repeated theme: "Based on Causes, or Celestial Physics, Brought Out by a Commentary on the Motion of the Planet Mars." Today, Kepler is primarily remembered for his laws of planetary motion. Although his magnetic forces have fallen by the wayside, his requirement for a celestial physics based on causes has deeply molded science as we know it today. It was, in effect, the mechanization and the cleansing of the Copernican system, setting it into motion like clockwork and sweeping away the vestiges of Ptolemaic astronomy. The results can very appropriately be called the Keplerian system. In the preface to Kepler's long-awaited *Rudolphine Tables*, finally published in 1627, he felt compelled to excuse the extended delay. He says that "the novelty of my discoveries and the unexpected transfer of the whole of astronomy from fictitious circles to natural causes were most profound to investigate, difficult to explain, and difficult to calculate, since mine was the first attempt."[9] Kepler's *Astronomia nova* might have been forgotten had it not been for the brilliant success of the *Rudolphine Tables*, whose predictions were nearly two orders of magnitude more accurate than previous methods. Today, with the clarity of hindsight, we see not only that the *Astronomia nova* was truly "the new astronomy" but that Johannes Kepler himself deserves to be remembered as the first astrophysicist.

Notes and References

[1] J. Kepler, *Mysterium cosmographicum* (1596). Reprinted in the standard Kepler edition *Johannes Kepler Gesammelte Werke* (*JKGW*, Munich, 1937–), vol. 1, p. 11.

[2] Kepler to Maestlin, 3 October 1595; *JKGW*, vol. 13, p. 40.

[3] J. Kepler, *Mysterium cosmographicum* (1596); *JKGW*, vol. 1, p. 13.

[4] J. Kepler, *Astronomia nova* (1609); *JKGW*, vol. 3, p. 109. It is scandalous that this work has never been published in English; the passage quoted has been prepared by Ann W. Brinkley and myself.

[5] After Kepler's death in 1630, his manuscripts changed hands repeatedly—they were once owned by Hevelius and a list of them appears in the *Philosophical Transactions*, vol. 9 (1674), 29–31, but they were finally bought by Catherine the Great for the Academy of Sciences in St. Petersburg, and to this day they are preserved there. I wish to thank Dr. P. G. Kulikovsky and Dr. V. L. Chenakal for providing me with an excellent microfilm of Volume XIV, which contains the principal papers relating to the analysis of Mars. The greater part of this volume comprises a notebook of nearly 600 pages, which represents virtually all the work for 1600 and 1601. There are several draft chapters for the *Astronomia nova* and material for an ephemeris. It

was not until 1605 that he found the ellipse, and unfortunately virtually no manuscripts remain from that part of the work.

[6] J. Kepler (1609), *op. cit.*, p. 178.

[7] 1971 was not only the four-hundredth anniversary of Kepler, but also the five hundredth anniversary of Albrecht Dürer. It is interesting to note that Kepler knew Dürer's great book on art theory, the *Underweysung der Messung* (1525), which illustrates not only the Platonic solids and a construction for an erroneously egg-shaped ellipse but also a device that uses threads to assist in drawing an object in perspective. Kepler employed a rather similar procedure of threads to investigate the formation of an image, and Stephen Straker, in a doctoral dissertation written at the University of Indiana, suggests that Kepler may have gotten the ideas from Dürer.

[8] J. Kepler (1609), *op. cit.*, p. 366.

[9] J. Kepler, *Tabulae Rudolphinae* (1627); *JKGW*, vol. 10, pp. 42–43.

Kepler as a Copernican

Two of the most fascinating questions in the history of Copernican astronomy are: Why did Johannes Kepler become a Copernican? and, Why did Galileo Galilei, somewhat later, also become a Copernican? These questions become all the more intriguing when we notice the lack of enthusiasm for the heliocentric system among other astronomers at the turn of the seventeenth century. Ironically, Copernicus himself was esteemed as an astronomer and mathematician, but his cosmology was virtually discarded. The erosion of the Copernican philosophy had already begun with the addition of Osiander's anonymous preface to Copernicus's *De revolutionibus*, which stated that "these hypotheses need not be true nor even probable." Erasmus Reinhold, the cleverest mathematical astronomer of the mid-sixteenth century and author of the Copernican-based *Prutenicae tabulae*, saw at once that the Copernican system was equivalent to the geocentric "homocentric epicycle," and the great Tycho Brahe envisioned a similar Earth-centered system as the foundation for the reform of astronomy.

Against this background the work of Kepler appears all the more extraordinary. In retrospect, the pivotal but complementary roles of Kepler and of Galileo appear to have turned the course of astronomy even more than the work of Copernicus himself.

Why Kepler became a Copernican no one can ever explain with certainty. Nevertheless, we can examine the state of astronomy in

Selection 18 reprinted from *Johannes Kepler, Werk und Leistung, Katalog des Oberösterreiches Landesmuseums,* vol. 74 (Linz, 1971), pp. 109–14.

Kepler's formative student years, and we can identify some of the motivations that appear in his work.

Nowadays the student of modern science views Copernicus primarily as a cosmological innovator. This was not the case in Kepler's student days. In the late sixteenth century Copernicus was highly respected not for his heliocentric theory but for his competence as a mathematician.

In recent years Copernicus has on occasion been severely criticized for various inadequacies in the technical apparatus of *De revolutionibus*. Without doubt Kepler was by far the better mathematician and physicist. Nevertheless, in the details of mathematical astronomy, Copernicus stood head and shoulders above his contemporaries. One very specialized point can illustrate clearly his abilities as a technician. When Ptolemy set up the tables for planetary longitudes in the *Almagest*, he ingeniously avoided a cumbersome double-entry table by using instead the product of two single-variable functions. During late antiquity one of these sophisticated functions was replaced by an elementary sine function; the fact that this procedure alone could introduce deviations of over 1° in the positions of Mars simply illustrates the decadence of pre-Copernican astronomy, which could scarcely be bothered by errors of this magnitude in the predictions versus observations. On a technical level, Copernicus had the understanding to restore a correct interpolating function. This point, not unimportant in the prediction of accurate planetary positions, was too subtle to be appreciated by nearly every astronomer of the next few generations; however, these astronomers did recognize the value of many other numerical changes that Copernicus introduced.

The acceptance of Copernicus's numerical innovations and rejection of his cosmology is clearly shown in a set of manuscript astronomy lecture notes written at the University of Wittenberg in the late 1560s, only a few years before Kepler's birth. These notes, now preserved at Gonville and Caius College in Cambridge, contain repeated references to Copernicus, but they never mention the heliocentric system.[1] The manuscripts include the traditional arguments against the motion of the earth, without citing Copernicus in this connection.

A similar situation must have prevailed when Kepler matriculated at the University of Tübingen in 1589. Michael Maestlin, the astronomy professor there, knew Copernican astronomy well; he edited the 1571 edition of the *Prutenicae tabulae*, and he used these to compute his own *Ephemerides*. Maestlin's *Epitome astronomiae* (which went through seven editions between 1582 and 1624) mentioned Copernicus several times, but the references always cite technical, not cosmological points. When Kepler tells us that "I have by degrees—partly out of hearing Maestlin, partly by myself—collected all the advantages that Copernicus has over Ptolemy,"[2] we must suppose that he is referring principally to the technical, numerical aspects that Maestlin freely conceded, especially because Kepler may not yet have had access to *De revolutionibus* itself. However, it was the cosmological aspects of Copernicus's teachings that seized Kepler's imagination, and there can be no doubt that he was converted to heliocentrism in his student days.

Maestlin, perhaps even unwittingly, planted the seed that later blossomed into a full Copernicanism. The ground was fertile; Kepler's quarterly grades at the university, still preserved, show him as a straight-A student, and when he applied for a scholarship renewal at Tübingen, the senate noted that he had "such a superior and magnificent mind that something special may be expected of him."[3] Yet as Kepler himself wrote concerning the science and mathematics of his university curriculum, "these were the prescribed studies, and nothing indicated to me a particular bent for astronomy."[4]

Nevertheless, within just a few years, we find the 25-year-old Kepler publishing the first unabashedly Copernican book since *De revolutionibus* itself. A thin volume of youthful exuberance, the *Mysterium cosmographicum* nonetheless contained hints of much that would be developed in Kepler's later work. Maestlin, for the one time in his entire life, found himself involved in a fully Copernican production as he computed for Kepler the thickness of the sphere in the Copernican system.

The contrast between Maestlin's and Kepler's views of Copernicus could scarcely have been more striking. For Kepler, the heliocentric system was *true*; without a Sun-centered universe, the entire rationale of his book would collapse. Its leading idea was a nesting of regular solids and spheres that accounted not only for the number but also the spacing of the planets, and this scheme required the Sun to be in the middle of the planetary orbits. Furthermore, the rhythmic ordering of the planetary motions, which appealed so forcibly to Kepler's sense of order in the cosmos, came about only with a heliocentric arrangement. For Maestlin, on the other hand, the Copernican system was a challenging mathematical exercise, but hardly more than one possible hypothesis. Maestlin would undoubtedly have agreed with Matthias Hafenreffer, the rector of the Tübingen senate, who urged Kepler to "proceed in the presentation of such hypotheses clearly only as a mathematician, who does not have to bother himself about the question whether these theories correspond to existing things or not."[5]

Kepler's personal introduction to the Copernican system bordered on a religious experience. Its loveliness filled him "with unbelievable rapture" when he contemplated it. While working on the *Mysterium cosmographicum*, he wrote to Maestlin, "I wanted to become a theologian; for a long time I was restless: Now, however, observe how through my effort God is being celebrated in astronomy."[6] And many years later, at the close of his *Harmonice mundi*, he exclaimed, "Praise Him, ye celestial harmonies, praise Him, ye judges of the harmonies revealed, and thou my soul, praise the Lord thy Creator, as long as I shall be!"[7] In Book IV of the *Epitome astronomiae Copernicanae*, Kepler's most mature work, he sees the Copernican system as the embodiment of the Trinity: "For in the sphere, which is the image of God the Creator and Archetype of the world, there are three regions, symbols of the three persons of the Holy Trinity—the center, a symbol of the Father; the surface, of the Son; and the intermediate space, of the Holy Spirit."[8]

Yet even in his more naively conceived *Mysterium cosmographicum* we find that his plan is no less than to discover the pattern of Creation of God the Supreme Architect. Already he expresses his theological approach: "And there were three things above all for which I sought the causes as to why it was this way and not another—the number, the dimensions, and the motions of the orbs. I have dared to carry out this search because of the beautiful correspondence of the immobile Sun, fixed stars, and intermediate space with God the Father, the Son, and the Holy Spirit."[9]

In Kepler's theological assimilation of the heliocentric system we can find important clues about which aspects of Copernicus's work attracted him. Clearly it was the grand patterns, the aesthetic appeal of celestial regularities. Unlike Maestlin, the skilled calculator of Copernican-based ephemerides, Kepler demonstrated little concern for the technical accomplishments of Copernicus. It is surely ironical that Kepler the cosmographer should in the end become the computer-of-ephemerides *par excellence*, and should through his massive technical achievements give a major impetus to the acceptance of the Copernican world view.

The standard textbook version of the Copernican revolution has rewritten the history of science into mythology. By the time of Copernicus, so the myth runs, the Ptolemaic system had become so patched up that it was ready to collapse under the weight of its complexities, and furthermore, the discrepancies between theory and observation had brought astronomy to a crisis. Copernicus, in a brilliant house-cleaning, swept away the epicycles, restoring astronomy with a system so simple that it had to be right.

In actual fact, an unembellished Ptolemaic system still served the astronomers of Copernicus's day; the only new feature was a more adequate reckoning of precession, which added no great complexity. To be sure, the tables erred in predicting the rare but astrologically important conjunctions of Jupiter and Saturn, but beyond this so few precise observations were available that scarcely anyone realized how bad the ephemerides really were. The textbook mythology ignores the fact that Copernicus's planetary predictions were by and large little better than his predecessors—indeed, at about the same time that Maestlin was computing his ephemerides from the *Prutenicae tabulae*, Tycho Brahe was complaining in his observing book that by then the Prutenic positions were sometimes *worse* than the Ptolemaic *Alfonsine Tables*. Kepler, in the Preface to his *Rudolphine Tables*, cites specific dates when the *Prutenic Tables* gave errors of 4° and 5° for Mars.[10] Moreover, the textbook mythology ignores the fact that Copernicus's use of small epicycles actually produced a system more complex in technical detail than what had gone before.

But in a sense the textbook version is right, for those large-scale simplifications and regularities that the Copernican system achieved were precisely those aspects that kindled Kepler's imagination. When the Earth was hurled into orbit, its 365-day period fell logically between the 225 days for Venus and the 687 days for Mars. For Kepler, this was the sort of harmony that mattered.

Kepler writes in the *Mysterium cosmographicum*, "Copernicus alone gives an explanation to those things that provoke astonishment among other astronomers, thus destroying the source of astonishment, which lies in the ignorance of the causes."[11] By this Kepler means the facts that Mercury, Venus, and the Sun have the same period of revolution on their deferents, that five planets (but not the Sun and Moon) exhibit retrograde motion, that the relative sizes of the Ptolemaic epicycles form an unexplained progression, and that the epicycle motions of Mars, Jupiter, and Saturn always operate in concert. Copernicus himself had pointed to the same phenomena in the most majestic passage of *De revolutionibus*, Chapter 10 of Book I. In a sense, Kepler was offering nothing new. But, in another sense, the speculative flights of the *Mysterium cosmographicum* dramatically reminded astronomers of the underlying orderliness and harmony of the Copernican system. It was, ultimately, an appeal to aesthetics.

It was not a crisis in astronomy that impelled Kepler to become a revolutionary. Rather, it was the powerful feeling of rightness and coherence offered by the new view of the universe. The fact that planetary predictions were scandalously bad, that astronomy *should* have been in a state of crisis, or that Kepler himself ultimately took major steps toward effecting the reform, have little or nothing to do with Kepler's becoming a Copernican. Kepler's thought was permeated by the deep belief that the underlying structure of the universe resides in rational relationships of great beauty, and for Kepler the heliocentric system revealed a glimpse of this loveliness. Such a keen appreciation of the aesthetic qualities of physical theory is not unique with Kepler; it has guided many men of genius. Albert Einstein, for example, constructed the theory of relativity almost exclusively from aesthetic motivations; the presumed "crisis" in physics produced by the Michelson-Morley experiment appears to be primarily a pedagogical device introduced afterward.

Kepler's well-developed sense of the aesthetic took varied forms. In the *Harmonice mundi* we find him at work on the theory of musical harmony to build a foundation for the planetary harmonies—and, when mathematical motivation failed, his ear served as guide. Or again, when the clumsy typeface of the title page for the *Tabulae Rudolphinae* printed in Prague failed to please him, he had struck off in Ulm another version, differing only in typography.

To argue further, that Kepler's profound commitment to the use of physical causes (as opposed to "fictitious hypotheses") is another manifestation of his powerful aesthetic sense, lies beyond the scope of this brief inquiry into the roots of Kepler's Copernicanism. Kepler believed that the Sun was the seat of the force that drove the planets in their orbits. This physical insight, already touched on in the *Mysterium cosmographicum*, was tightly coupled to his espousal of a heliocentric structure. This idea received a penetrating development in the *Astronomia nova* (published about a decade later), where it motivated his researches and where it guided him safely past the trap of building his theory of motions on the traditional epicyclic devices.

Again, Kepler's approach divided him from Maestlin, who wrote in a letter of 1 October 1616: "Concerning the motion of the Moon, you write that you have traced all the inequalities to physical causes; I do not quite understand this. I think rather that one should leave physical causes out of account, and should explain astronomical matters only according to astronomical method with the aid of astronomical, not physical, causes and hypotheses. That is, the calculation demands astronomical bases in the field of geometry and arithmetic"[12]—that is to say, the circles, epicycles, and equants that already in his *Astronomia nova* Kepler had abandoned.

The *Astronomia nova* is a truly remarkable book, where for the first time in history we can see a scientist grappling with a redundancy of raw data. In the face of the ensuing contradictions, Kepler would surely have faltered without the guidance of physical causes and without the driving motivation to discover the large cosmological patterns of the universe. In some respects, the *Astronomia nova* was a digression from his primary quest for an understanding of the cosmic harmonies, yet ultimately it proved to be Kepler's greatest contribution toward the acceptance of the Copernican system.

Kepler wrote prolifically, but a much greater audience awaited the cosmological writings of a far more gifted polemicist, Galileo Galilei. Kepler's intensely personal cosmology did not carry a broad appeal to the coming age of rationalism. It was Galileo, through the brilliant clarity and enthusiasm of his *Dialogo* and his *Discorsi*, who became the persuasive purveyor of the new cosmology.

Kepler was an astronomer's astronomer. It was the astronomers who recognized the immense superiority of the *Tabulae Rudolphinae*, finally published in 1627, which were based on the discoveries of the *Astronomia nova*. They knew that, before Kepler, the best predictions for Mars erred by several degrees, whereas the new tables gave positions within a few minutes of arc. For the professionals, this improvement was a forceful testimony to the efficacy of the Copernican system.

Today Kepler is remembered primarily for his three laws of planetary motion, two of them reported in the *Astronomia nova*, the third in the *Harmonice mundi*. But, to Kepler himself, these laws were only a few representatives of the harmonies he saw, and the tables were ancillary considerations; he was primarily a cosmographer enraptured by the aesthetic delights of God's creation. His prayer in the *Harmonice mundi* summarizes his approach to his life's work[13]: "If I have been allured into rashness by the wonderful beauty of Thy works, or if I have loved my own glory among men, while I am advancing in the work destined for Thy glory, be gentle and merciful and pardon me: and finally deign graciously to effect that these demonstrations give way to Thy glory and the salvation of souls and nowhere be an obstacle to that."

Notes and References

[1] See Owen Gingerich, "From Copernicus to Kepler: Heliocentrisim as Model and as Reality," *Proceedings of the American Philosophical Society*, vol. 117 (1973), pp. 513–22 [reprinted as selection 16 in this anthology].

[2] Kepler, *Mysterium cosmographicum, JKGW*, vol. 1, p. 9: 19–21. *JKGW* refers to the standard modern Kepler edition, *Johannes Kepler Gesammelte Werke* (Munich, 1937–); numbers after the colon refer to the lines on the page or within a letter.

[3] *JKGW*, vol. 13, letter no. 3:1055.

[4] Kepler, *Astronomia nova, JKGW*, vol. 3, p. 108:13–15.

[5] Hafenreffer to Kepler, 12/22 April 1598, *JKGW*, vol. 13, no. 93:46–48.

[6] Kepler to Maestlin, 3 October 1595, *JKGW*, vol. 13, no. 23:256–257.

[7] Kepler, *Harmonice mundi, JKGW*, vol. 6, p. 368:17–19.

[8] Kepler, *Epitome astronomiae Copernicanae, JKGW*, vol. 7, p. 258:22–31.

[9] Kepler, *Mysterium cosmographicum, JKGW*, vol. 1, p. 9:33–34.

[10] Kepler, *Tabulae Rudolphinae, JKGW*, vol. 10, pp. 40:47–41:1.

[11] Kepler, *Mysterium cosmographicum, JKGW*, vol. 1, pp. 14:35–15:1.

[12] Maestlin to Kepler, 1 October 1616, *JKGW*, vol. 17, no. 744:24–29.

[13] Kepler, *Harmonice mundi, JKGW*, vol. 6, p. 363:9–13. The translation, by Charles Glenn Wallis, is found in *Great Books of the Western World*, vol. 16 (Chicago, 1952), p. 1080.

Kepler's Place in Astronomy

Any complete assessment of Kepler's place in astronomy would necessarily notice his *De stella nova* (1606), with its discussion of the supernova of 1604, still called Kepler's nova; his *De cometis libelli tres* (1619), which resulted in one of the few mentions of his name in Newton's *Principia*; and his continuing but unfinished work on the lunar theory, including his anticipation of Tycho Brahe in the discovery of the annual equation. It would include a consideration of his attempt to establish an absolute distance scale for the solar system; his thoughtful analysis but rejection of the idea of an infinite universe; and a description of his remarkable Copernican polemic in disguise, that is, his *Somnium* (1634) or dream trip to the Moon. Surely such an assessment would take cognizance of his work on optics, especially of the new form of telescope described in his *Dioptrice* (1611). It might also discuss Kepler's little-recognized contributions as an observer. In his *Astronomiae pars optica* (1604) he showed that the Sun's apparent diameter varied throughout the year. In 1607 he caught between clouds a fleeting glimpse of a sunspot, which he unfortunately mistook for a transit of Mercury (*Phaenomenon singulare, seu Mercurius in sole*, 1609). And some of his own observations of Mars play a role in the *Astronomia nova* (1609).

Nevertheless, in the space at my disposal I cannot aspire to a *complete* evaluation of Kepler as an astronomer; instead, I shall try to identify his single most important contribution to astronomy. I shall

Selection 19 reprinted from *Vistas in Astronomy*, ed. by A. Beer and P. Beer, vol. 18 (1975), pp. 261–78.

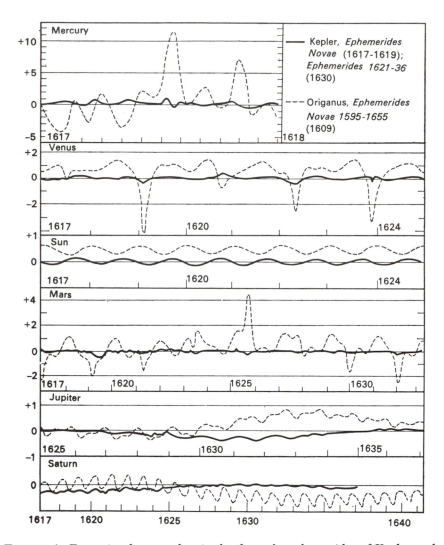

FIGURE 1. *Errors in planetary longitudes from the ephemerides of Kepler and of David Origanus compared with B. Tuckerman's* Planetary, Lunar and Solar Positions A.D. 2 to A.D. 1649, *American Philosophical Society Memoirs, vol. 59 (Philadelphia, 1964). Origanus based his calculations on Reinhold's* Prutenicae tabulae *(Tübingen, 1551). The displacement for his solar positions is according to Tycho's data, which agree closely with Kepler's. These graphs were produced by Barbara L. Welther.*

argue that Kepler's most consequential achievement was the mechanizing and perfecting of the world system. By the *mechanization* of the solar system, I mean his insistence on "a new astronomy based on causes, or the celestial physics," as he tells us in the title of his great book. By the *perfection* of the planetary system, I mean the fantastic improvement of nearly two orders of magnitude in the prediction of planetary positions (Figure 1).

I should remind you that this differs from the traditional assessment of Kepler's achievements, which remembers him primarily for his discovery of the three laws of planetary motion. Although these are essential elements in Kepler's work, I think it fair to claim more. Copernicus gave the world a revolutionary helio*static* system, but Kepler made it into a helio*centric* system. In Kepler's universe, the Sun has a fundamental physically motivated centrality that is essentially lacking in *De revolutionibus*. We have grown so accustomed to calling this the Copernican system that we usually forget that many of its attributes could better be called the Keplerian system.

Some of these distinctions lie in the technical intricacies of the Copernican planetary mechanism, and as such they are generally glossed over in elementary discussions. I hope that I shall not be considered anti-Copernican if I discuss three of these antique vestiges in the formulation of Copernicus. These are, first, his use of the center of the Earth's orbit and not the Sun as the central reference point of the solar system; second, his anomalous model for the motion of Mercury; and finally, his epicyclets.

Copernicus's use of the center of the Earth's orbit as the reference point for his system is a natural consequence of his transformation of the geocentric Ptolemaic theory. In Ptolemy's epicyclic scheme, the epicycle of a superior planet is equivalent to the Earth's heliocentric orbit. This, however, requires two important approximations. First, the rules call for uniform motion and no eccentricity of the epicycle itself; thus, the epicycle generates the average uniform motion, that is, the so-called mean motion of the Sun. Second, the eccentricities assigned by Ptolemy to each of the planets are essentially vector sums of the Earth's and the planet's eccentricities, with the result that in each case the line of apsides is slightly wrong. When Copernicus transformed the Ptolemaic scheme into a heliocentric system, the major epicycles coalesced into a single circular orbit for the Earth. Copernicus did not bother to sort out the individual eccentricities, so each planetary eccentricity implicitly combines the Earth's. In addition, his apsidal lines are drawn through the mean Sun, that is, through the fictitious Sun in the middle of the Earth's orbit.

When Kepler, in his *Mysterium cosmographicum* (1596), worked out the highly imaginative set of nested spheres and Platonic polyhedra that he believed to be God's blueprint for establishing the planetary spacings, he realized that the Copernican use of the mean Sun gave the planet Earth an unnecessarily privileged role. Kepler's own account of the matter is given in the *Astronomia nova*:[1]

> In 1597 I wrote to Tycho Brahe asking him what he thought of my little work, and when in answer he mentioned among other things his observations, he fired me with an enormous desire to see them. From then on Tycho Brahe, an important part of destiny, continually urged that I should come to visit him. But since the distance of the two places would have deterred me, I ascribe it to Divine Providence that he came to Bohemia.
>
> I thus arrived there at the beginning of 1600, with the hope of obtaining the correct eccentricities of the planetary orbits. When in the first week I learned that he himself along with Ptolemy and Copernicus employed the mean motion of the Sun, although in fact the apparent motion agreed more with my little book, I was authorized to use the observations in my manner.

Kepler's workbook from this period still exists in the Archives of the USSR Academy of Sciences in Leningrad,[2] the proper diagram and calculations appear on the first few pages (Figure 2), and on page 5 (Figure 3) some Latin text stating:[3]

> However, what must be used is not the mean but the true opposition of Mars and the Sun because such is the place where all the parallax is secured. Moreover, it is on Mars' line of apsides. Therefore, if the apsides depend on the mean place of the Sun, the Earth must be stationed between the Sun's mean place and Mars. But because I now assume that the apsides depend on the Sun's true place, it is necessary to use the true opposition.

In retrospect, this appears as the logical and necessary first move in the rectification of the Copernican system, but the observational proof is far from simple. In fact, not until Chapter 52 of the *Astronomia nova* is Kepler fully prepared to show that the planetary apsidal lines, and hence the planetary planes, must pass through the Sun.

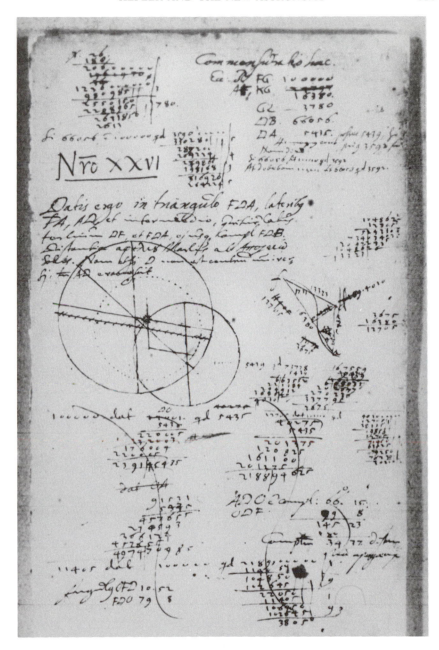

FIGURE 2. *The opening page of Kepler's workbook on Mars (Leningrad XIV, fol. 68). Written early in 1600, the diagram shows apsidal lines for Mars drawn through both the true Sun and the mean Sun (center of the dotted terrestrial orbit). The triangle near the center of the circles with the Sun as its apex is redrawn to a larger scale at right.*

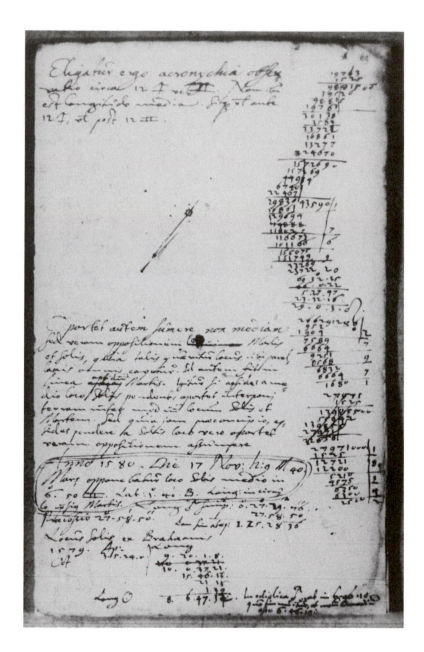

FIGURE 3. *Page 5 of Kepler's workbook shows the Sun, Earth, and Mars in the configuration for the true opposition (Leningrad XIV, fol. 69). The Latin text is transcribed in note 3. Notice how Kepler has doubly circled one of the rare Mars observations entrusted to him by Tycho Brahe at this initial stage of his research.*

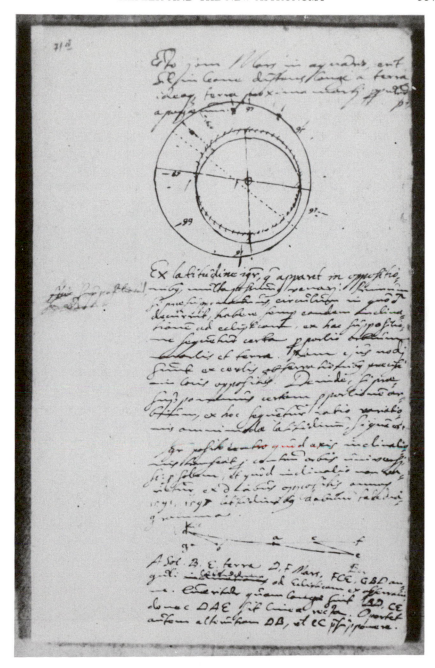

FIGURE 4. *Page 8 of Kepler's workbook shows that he considered the latitudes at a very early stage (Leningrad XIV, 71v). The diagram near the bottom of the page displays a cross section of the Earth's orbit,* gaf, *and one of Mars' orbit,* dae, *corresponding to the Martian oppositions of 1591 and 1597.*

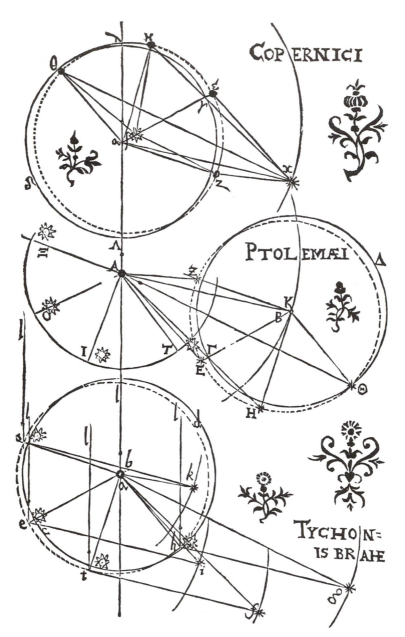

FIGURE 5. *The Mars triangulation method forces a displacement in the traditionally accepted position of the Earth's orbit, regardless of the cosmological system, as shown in Chapter 24 of Kepler's* Astronomia nova.

The idea is so important that we should perhaps call it Kepler's zeroth law. It is a sign of Kepler's genius that he so quickly recognized this as the crucial first move in the reform of astronomy.

Kepler's own account in the *Astronomia nova* continues:

> Now at that time, Tycho's personal assistant Christian Severinus [Longomontanus] had taken up the theory of Mars so that he might study the opposition of Mars with the Sun in 9° of Leo [that is, Mars near perihelion]. Had Christian been studying another planet, I would have started with that same one also. That is why I again consider it an effect of Divine Providence that I arrived in Prague when he was studying Mars; because for us to discover the secret knowledge of astronomy, it is absolutely necessary to use Mars; otherwise, that knowledge would remain eternally hidden.

This passage has customarily been understood to refer to the comparatively large eccentricity of Mars, which produces clearly observable effects in longitude, but Kepler also realized that only Mars comes close enough to the Earth for its latitudes to provide a powerful leverage for establishing its orbit in space. In his manuscript workbook, the second topic he takes up is indeed the problem of latitudes (Figure 4). Once again, Kepler's keen insight into the problem manifests itself, for previous astronomers always segregated the discussions of longitude and latitude into two separate parts. Kepler realized from the beginning the importance of the three-dimensional nature of his problem and formulated his attack accordingly. Quite possibly his earlier struggles with the nested polyhedra and spheres had prepared his mind for a three-dimensional approach to the Martian orbit in space.

I have mentioned the privileged status of the Earth that arose in the Copernican system because the planets were referred to the center of the Earth's orbit. The terrestrial orbit was unique in another way as well; it did not have the equant mechanism (or its epicyclet equivalent) employed traditionally for Venus, Mars, Jupiter, and Saturn. Not only did Kepler feel that the Earth should have the same clockwork as the other planets, but his desire for physical causes demanded just such a device, which drove a planet more quickly when it approached closer to the Sun (Figure 5). Kepler called the Sun "the fountain of strength" of the solar system. He envisioned it as a great magnetic body, whose rotation also turned the magnetic emanations that propelled the planets in space.

To demonstrate that the Earth had the same physical mechanism as the other planets, Kepler essentially had to show that the entire circle of the Earth's orbit had been falsely placed by an amount equal to half its traditional eccentricity. This he succeeded in establishing by a clever and now rather well known triangulation to the planet Mars. Because Mars returns to the same spot in its orbit every 687 days, but in that interval the Earth has not yet completed its second full revolution, he could obtain, from Tycho's observations over many years, several Earth-Mars sight-lines. These showed that the Earth's orbit required a different displacement from the Sun, one compatible with the equant device. In this fashion, described in Book III of the *Astronomia nova*, Kepler brought the Earth's orbit into physical uniformity with those of Venus and the outer planets. "Consider whether I have made a step toward establishing a physical astronomy without hypotheses, or rather, fictions," wrote Kepler to Fabricius in 1602. "The force is fixed in the Sun, and the ascent and descent of the planets are likewise fixed according to the greater or lesser apparent emanation from the Sun. These, therefore, are not hypotheses (or as Ramus calls them, figments), but the very truth, as the stars themselves; indeed, I assume nothing except this."[4]

Not only the Earth but also Mercury had a unique mechanism in the Copernican system. I suspect that Copernicus was only too aware of this anomaly but was frustrated from altering it by the paucity of observations at his disposal. The Copernican Mercury model was essentially a direct heliocentric transformation of Ptolemy's model.[5] Ptolemy employed a central crank to pull in the deferent circle slightly on the sides, because his erroneous interpretation of the observations had indicated that Mercury made its closest approaches to the Earth in two places, not just one. This mechanism produces an ovoid figure that differs only slightly from an ellipse. Figure 6 shows the shape of the curve when the eccentricity is greatly exaggerated. Note that the period of motion in this figure is one year, so the ovoid is essentially a distorted reflex of the Earth's orbit, a curve that has nothing at all to do with the elliptical orbit of Mercury. Copernicus retained an equivalent mechanism whose annual period was suspiciously anomalous in the Mercury model. Why should one of the Mercury circles happen to have a period of exactly one year?

Kepler was in a much better position to reconstruct the Mercury model than Copernicus had been, for he had 85 Mercury observations from Tycho Brahe. Nevertheless, I believe that with his characteristic boldness and physical acumen he would in any event have adopted the same model for Mercury as for the other planets. Writing to Maestlin

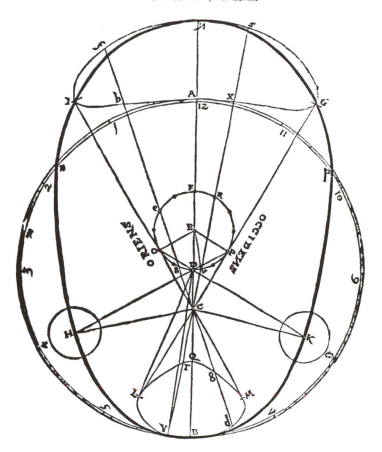

THEORICA OMNEM FERE VA-
rietatem motus centri epicycli & apogij eccen-
trici Mercurij ostendens.

Collocanda est hæc figura ante fol. 69.

FIGURE 6. *In the Ptolemaic Mercury theory, a small interior circle combined with the larger deferent produces an ovoid curve. Kepler never illustrates the ovoid, but specifically directs the reader to Reinhold's edition of Peurbach's* Theoricae novae planetarum. *Reproduced here is the Mercury diagram from the author's copy of the Paris, 1558, printing.*

already in 1601, Kepler stated, "I have been consumed with the hope that the other planets will be reduced to the same laws, even Mercury itself."[6]

The traditional Mercury model is admittedly esoteric, and the reader may think it is idiosyncratic to include it here. But I mention it for two particular reasons. In the first place, the predictions of Mercury provided one of the greatest triumphs for Kepler's theories. At the top of Figure 1, note the errors in the longitude as predicted for Mercury by Copernicus; these errors become especially large at the times of inferior conjunction, when the planet is normally visible. The same figure shows how greatly reduced the errors become when the positions are predicted by Kepler.[7]

For the first time, an adequate method was at hand to predict a transit of Mercury across the face of the Sun, on 7 November 1631, but, alas, Kepler did not live to see his prediction fulfilled. Fortunately, the transit was observed on schedule in Paris by Pierre Gassendi. Since tables based on Ptolemy, Copernicus, and others missed by 5° in predicting this event, compared to Kepler's 10′ of arc, the evidence for the new system was overwhelming.[8]

The second reason for mentioning the Mercury model lies in the inspiration it provided for Kepler's initial planetary models, before he finally adopted the ellipse. This in turn is related to a third antique vestige in Copernicus's planetary theory, namely, his continued use of epicycles (Figure 6).

Copernicus felt very strongly that the use of the equant violated the rule of uniform motion in perfect circles.[9] He substituted for the objectionable equant a very small epicycle, or epicyclet. Although this produced the same nonuniform angular motion, Kepler was quick to notice that the resulting trajectory was not a circle but a curve bowed out in the opposite direction compared to an ellipse.

Perhaps it was Copernicus's use of epicyclets that induced Kepler to try them, for his notebook shows that, within a few weeks after joining Tycho in Prague, he was already exploring the effects of small epicycles. This is not the occasion for a detailed analysis of the "warfare on Mars" that resulted in Kepler's greatest book, the *Astronomia nova*.[10] Suffice it to say that in the opening stages he discovered a scheme for predicting the longitudes of Mars to within 2′—better than ten times as accurately as any predecessor—but it failed completely with respect to the distances as revealed by his three-dimensional latitude analysis. It was then that he again turned to epicyclic mechanisms, always demanding some construction directed to the true Sun itself. Such constructions always lead to an ovoid curve; there is nothing metaphysical about this shape—it is simply the result of using a single epicycle on a deferent with a noncentral point for generating the motion.[11] The resulting ovoid curve was closely related to the traditional Mercury model, as Kepler points out in a short marginal note.

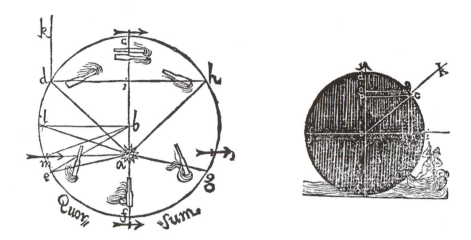

FIGURE 7. *A pair of figures from Chapter 57 of the* Astronomia nova. *At right is the sailor navigating his boat; at left, the magnetic axis, represented by the arrow, is analogous to the oar in the water for directing the planet near and far from the Sun.*

Kepler was distressed by the introduction of such a nonphysical device as an epicycle. He argues that, just as sailors cannot know from the sea alone how much water they have traversed, since their route is not distinguished by any markers, so the mind of the planet will have no control over its motion except possibly by watching the apparent diameters of the Sun.[12] In the end, it was once more his trust in physical simplicity that steered him past this shoal. His goal, as he now wrote to Herwart, was "to show that the celestial machine is not so much a divine organism but rather a clock-work. . .inasmuch as all the variety of motions carried out by means of a single very simple magnetic force of the body, just as in a clock all the motions arise from a very simple weight."[13]

For Kepler was still seeking a single physical explanation not only for the varying speed of Mars in its orbit but also for the varying distances. His answer came in an extension of the magnetic effects that propel the planets in their orbits. Kepler draws a very charming analogy to a boatman in an amusement park[14] (Figure 7):

> If a rope or cord is hung across a river, firmly affixed to
> the banks, and from a pulley running along the rope another
> cord constrains a small boat riding to and fro in the stream,

and if the passenger fastens a rudder or oar behind in the right way, holding everything else still, the boat is carried transversely by the simple downward force of the stream, and is transported from one bank to the other by the pulley running on the rope above. Indeed, on a wider river they can drive the boats in circles, going here and there, playing a thousand games without ever touching the bottom or banks, but with the help of the oar alone converting the very simple downward flow of the stream to their own ends.

Kepler had read with interest the work of William Gilbert and his hypothesis of the magnetic axis of the Earth. Such a magnetic axis, Kepler proposed, could act as the rudder guiding the planet first near and then far from the Sun. This rudder is depicted in the left part of Figure 7. If we keep the planet's magnetic axis fixed in space, as shown by the arrow, it will alternatively attract and repel the planet to and from the Sun's unipolar magnetism. The projection of the planet's magnetic axis as seen from the Sun will be $\cos \theta$. To the first order, the ellipse satisfies this physical picture of the magnetic axis governing the advance and retreat of the planet. Kepler did not work with the polar equation of the ellipse, however, and the real hurdle for him was to find the geometrical equivalents between the librating magnetic axis and the ellipse. "I was almost driven to madness in considering and calculating this matter," writes Kepler.[15]

> I could not find out why the planet would rather go on an elliptical orbit. Oh, ridiculous me? As if the libration on the diameter could not also be the way to the ellipse. So this notion brought me up short, that the ellipse exists because of the libration. . . . With reasoning derived from physical principles agreeing with experience, there is no figure left for the orbit of the planet except a perfect ellipse.

With the discovery of that perfect ellipse, Kepler could simultaneously satisfy his self-imposed requirement for a physical explanation and also represent the accurate Mars observations of Tycho Brahe. He had at last swept away the remnants of antiquity, breaking the two-millennium spell of perfect circles and of uniform angular motion. His *Astronomia nova* was truly "The New Astronomy, Based on Physical Causes, or the Celestial Physics." This requirement, for a celestial physics, has deeply molded science as we know it today. Kepler was the first and, until Descartes, the only scientist to demand this sort of physical explanation.[16] This essential step in the mechanization of the world picture provided the motivation to finally throw out the anomalous geometric kinematic devices of the past.

Concerning the *Astronomia nova*, J. L. E. Dreyer has written, "In the history of astronomy there are only two other works of equal importance, the book *De revolutionibus* of Copernicus and the *Principia* of Newton."[17] Nevertheless, Kepler's *Astronomia nova* has recently been called his most impenetrable work, and it has long been accused of involved, mystical digressions. Careful inspection shows that most of these so-called mystical passages are precisely the places where Kepler is wrestling with the concept of magnetic forces, so crucial to his line of reasoning.[18] His earlier work, the *Mysterium cosmographicum* (1596), with its fantastic "mystical" nesting of spheres and polyhedra, nonetheless contained much of his celestial physics in an embryonic form. And his third or harmonic law, reported in the *Harmonice mundi* (1619), was embedded with a cosmological context of musical harmony, astrology, and Platonic solids. To modern readers, it is certainly strange, perhaps even mystical. Yet we cannot divorce the context and the motivations in any of Kepler's work.

It is appropriate to end with the concluding remarks made in an earlier Kepler anniversary symposium, in 1930, also cosponsored by the American Association for the Advancement of Science, where Carl Rufus elegantly enunciated this same thesis:[19]

> As an astronomer he was thinking God's thoughts. In Kepler science and religion were united. His mystical speculations have sometimes been decried. Some biographers, with kind intent, with fans and sieves of their own fabrication, have attempted to winnow away what to them was straw and chaff and to preserve a few kernels of scientific truth to represent the abundant harvest of his fertile imagination and prodigious labors. That method is as disastrous to the real Kepler as plucking the petals from a wild rose and displaying the bare stamens and pistils. . . .
>
> In a certain sense science and mathematics have made mystics of us all, tho we try to remain with our feet on the solid ground. Eddington says: "we have found that where science has progressed the farthest, the mind has but regained from nature that which the mind has put into nature. We have found a strange foot-print on the shores of the unknown. We have devised profound theories, one after an-

other, to account for its origin. At last, we have succeeded in reconstructing the creature that made the foot-print. And Lo! it is our own." That is mysticism in a supersense. Who put law and order into the universe? There are two answers. The first is—the human mind. Kepler, in his message to the twentieth century, answers—God.

Notes and References

[1] *Astronomia nova*, 1609; page citations will be taken from the standard modern edition, *Johannes Kepler Gesammelte Werke* (*JKGW*) vol. 3 (Munich, 1937) chap. 7, p. 109.

[2] I wish to thank Dr. P. G. Kulikovsky and Dr. V. L. Chenakal for providing me with an excellent microfilm of Volume XIV of the Leningrad Kepler material.

[3] Leningrad Kepler MS XIV, fol. 69: "Oportet autem sumere non mediam sed veram oppositionem Martis et Solis, quia talis quaeritur locus ubi parallaxis omnis cavitur. It autem fuit in linea apsidum Martis. Igitur si apsides a medio loco Solis pendent oportet interponi terram inter medium locum Solis et Martem. Sed quia jam praeconcipio apsidas pendere a Solis loco vero oportet veram oppositionem assumere."

[4] Letter no. 226, to David Fabricius, 1 October 1602, in *JKGW*, vol. 14, p. 226. Petrus Ramus (1515–72), Professor of Philosophy and Rhetoric in Paris, had offered his chair to anyone who could produce an "astronomy without hypotheses," and Kepler, on the back of the title page of the *Astronomia nova*, declared that, if Ramus had still been alive, he would have claimed the reward.

[5] See Owen Gingerich, "The Theory of Mercury from Antiquity to Kepler," *Actes du XIIe Congrès International d'Histoire des Sciences, 1968*, vol. 3A (Paris, 1971), pp. 57–64 [reprinted as selection 23 in this anthology].

[6] Letter no. 203, to Michael Maestlin, 10/20 December 1601, in *JKGW*, vol. 14, p. 204.

[7] J. Kepler, *Ephemerides novae motuum coelestium, ab anno vulgaris aerae MDCXVII* (Linz, 1617–19). These ephemerides actually antedated the *Tabulae Rudolphinae*, and, as can be seen in the Jupiter and Saturn errors in Figure 1, Kepler's accuracy improves in the later predictions.

[8] The impact of the successful Mercury theory is clearly seen in the work of a leading almanac maker and French astrologer, Noël Durret. Durret's *Richeliene Tables* (Paris, 1635), were based primarily on Lansberg's tables, as were the years 1637–42 in his *Novae motuum coelestium ephemerides Richelianae* (Paris, 1641). In 1639, however, he issued his *Supplementum tabularum Richelienarum cum brevi planetarum theoria ex Kepleri*, which advised the reader: "Four years ago my new theory of the planets with the *Richeliene Tables* for Paris were favorably received by his excellency. . . .But experience has shown that the astronomical tables of Lansberg do not agree so exactly with the observations of Mercury as those of Kepler, as one can see in the observations of that excellent observer M. Gassendi, principally those made in the years 1631, 1633, and 1634." In a similar text on folio aiij of his *Ephemerides*, he specifically compares predictions for the 1631 transit. Thus, beginning with the ephemerides for 1643, he switches to the Keplerian theory.

[9] Apparently, Erasmus Reinhold, who calculated the *Prutenic Tables* from the works of Copernicus, got this message from *De revolutionibus*. I was very fortunate to find and identify Reinhold's personal annotated copy of *De revolutionibus* in Edinburgh; and, on the title page, Reinhold wrote: "Axioma Astronomicum. Motus coelestis aequalis est et circularis. vel ex aequalibus et circularibus compositus," almost a direct quotation from the title of Book I, Chapter 4.

[10] Alexandre Koyré, *The Astronomical Revolution* (Ithaca, 1973), pp. 172–279; Curtis Wilson, "Kepler's Derivation of the Elliptical Path," *Isis*, vol. 59 (1968), pp. 5–25; E. J. Aiton, "Kepler's Second Law of Planetary Motion," *Isis*, vol. 60 (1969) pp. 75–90; Owen Gingerich, "Johannes Kepler and the New Astronomy," *Quarterly Journal of the Royal Astronomical Society*, vol. 13 (1972), pp. 346–60 [reprinted as selection 17 in this anthology].

[11] See K. Fladt, "Das Keplerische Ei," *Elemente der Mathematik*, vol. 17 (1962) pp. 73–78 (1962), Fladt's remark that the ovoid and the ellipse are the only two elementary functions for planetary theory, that is, simple algebraic relations employing an elementary transcendental function, is interesting but irrelevant, because Kepler did not have the equation for either figure, and the "egg" was not chosen for algebraic reasons. Similarly, the reasons about the necessity of choosing a one-focus curve in N. R. Hanson's *Patterns of Discovery* (Cambridge, 1958), p. 78 are equally irrelevant.

[12] J. Kepler, *Astronomia nova*, chap. 39, p. 260.

[13] Letter no. 325, to Herwart von Hohenburg, 10 February 1605, in *JKGW*, vol. 15, p. 146; the significance of this letter was first stressed by Gerald Holton, "Johannes Kepler's Universe, Its Physics and Metaphysics," *American Journal of Physics*, vol. 24 (1956), pp. 340–51.

[14] J. Kepler, *Astronomia nova*, chap. 38, p. 255. I take this opportunity to thank Ann Wegner Brinkley for assistance with the translation of the entire *Astronomia nova*, and also the American Philosophical Society for supporting grants 4074 and 4286 from their Penrose Fund.

[15] *Ibid.*, chap. 58, p. 366.

[16] With this claim I am in agreement with Koyré, *op. cit.*, p. 121.

[17] J. L. E. Dreyer, *History of the Planetary Systems from Thales to Kepler* (Cambridge, 1906), p. 410 (reprinted New York, 1953).

[18] Kepler's magnetic forces were considered so disreputable by Robert Small in 1804 that he skipped right over Chapter 57 in his otherwise detailed analysis of the *Astronomia nova* in his *An Account of the Astronomical Discoveries of Kepler* (London, 1804, reprinted Madison, 1963).

[19] W. Carl Rufus, "Kepler as an Astronomer," in *Johann Kepler, 1571–1630, A Tercentenary Commemoration. . .prepared under the auspices of the History of Science Society in collaboration with the American Association for the Advancement of Science (Section A, D, and L)* (Baltimore, 1931), pp. 1–38.

The Origins of Kepler's Third Law

Kepler's Third or Harmonic Law appears to have sprung up full grown, like Minerva from the brow of Zeus. It is first recorded in the fifth and astronomical book of the *Harmonice mundi*, already set out in italics in a precise form: the ratio that exists between the periodic times of any two planets is precisely the ratio of the 3/2th power of the mean distances. If you want the exact time, Kepler says, "it was conceived on March 8 of this year, 1618, but unfelicitously submitted to calculation and rejected as false, and recalled only on May 15, when by a new onset it overcame by storm the darkness of my mind with such full agreement between this idea and my labor of seventeen years on Brahe's observations that at first I believed I was dreaming and had presupposed my result in the first assumptions."

Apart from the candid charm of his statement, we are struck mostly by its terseness. Kepler, the lover of tables and computations, does not bother to show how accurate the relation really is, either here or in one of the few other references to it later in his work. Needless to say, Kepler neither called this a law nor selected and assigned numbers to three of his discoveries. Though later scientists singled out the importance of this relation and named it the Harmonic Law, it was to Kepler simply one of many celestial relationships; probably Kepler felt that it

Selection 20 reprinted from *Vistas in Astronomy*, ed. by A. Beer and P. Beer, vol. 18 (1975), pp. 595–601.

was not so fundamental an insight into the nature of the cosmos as was his nest of spheres and regular polyhedra, which for him offered the "archetypal" explanation for the number and spacing of the planets.

Nevertheless, to Kepler it was also a particularly beautiful relation, one that filled him with ecstasy. The *Harmonice mundi* was sent to press on 17 May 1618, only two days after the discovery of the harmonic law, but Kepler had enough time to add these rhapsodic lines to the introduction to Book V:[2]

> Now, since the dawn eight months ago, since the broad daylight three months ago, and since a few days ago, when the full Sun illuminated my wonderful speculations, nothing holds me back. I yield freely to the sacred frenzy; I dare frankly to confess that I have stolen the golden vessels of the Egyptians to build a tabernacle for my God far from the bounds of Egypt. If you pardon me, I shall rejoice; if you reproach me, I shall endure. The die is cast, and I am writing the book—to be read either now or by posterity, it matters not. It can wait a century for a reader, as God Himself has waited six thousand years for a witness.

Why was Kepler so ecstatic about this harmonic relation? In his youthful *Mysterium cosmographicum* of 1596 he thought he had found the secret key to the cosmos: that God had used a nesting of five regular Platonic solids among the planetary spheres to establish the spacings of the planetary orbits around the Sun. The scheme fits within 5 percent except for Jupiter, quite satisfactory for the young astronomer. But by 1618, with the exhausting rigor of his warfare on Mars behind him, the 5 percent precision of the nested polyhedra and sphere would no longer suffice. Convinced that God had somehow adjusted this master pattern for the distances, he sought and found an answer in the harmonic ratios of the velocities of the various planets in the aphelia and perihelia of their orbits. Thus the harmonic law served as a bridge between the planetary distances and their velocities or periods. This empirically correct relation seemed to strengthen and fortify the a priori premises of the *Mysterium* and the *Harmonice mundi*. This was his triumph and joy.[3]

Nevertheless, Kepler's presentation of the harmonic law itself stands in stark contrast to what have now become known as his first and second laws. Kepler had found the elliptical orbit of Mars 13 years earlier, in 1605, after a laborious analysis resting on an involved physical argument that only a very few of his contemporaries had the patience to follow. The harmonic law, on the other hand, was so simple that any astronomer could confirm it for himself with the data readily available in Copernicus's *De revolutionibus*.

The question therefore arises: Why did it take Kepler so long to find this relation between the periodic times and the mean heliocentric distances? The answer, I believe, sheds considerable light on Kepler's approach to physical problems and on his methodology. We have evidence in Kepler's very first scientific work, the *Mysterium cosmographicum*, that he was already seeking a relation of this sort. The fact that he did not find the harmonic law while he was still in his twenties indicates that he was being guided, or rather misguided, by theoretical considerations, and not merely engaging in the numerology or number juggling that he is often accused of in popular textbooks. In the *Mysterium cosmographicum*, Kepler gives the first evidence of the physical reasoning that will become highly developed in his more mature *Epitome astronomiae Copernicanae*.

After announcing his celebrated nest of spheres and regular solids, which seemed to explain the spacing of the planets, he turns to search for the basis of the regularities in the periods. Indeed (as Kepler notes), Copernicus himself, at the end of the glorious passage that climaxes the cosmological arguments of *De revolutionibus*, states that "therefore in this ordering we find that the universe has an admirable proportion and that there is a sure bond of harmony in the motion and magnitude of the orbits that cannot be found in any other way."[4]

In the *Mysterium cosmographicum*, Kepler already believes that the Sun furnishes the driving force to propel the planets in their orbits.[5] By examining the planetary periods and distances, he discovers that there is proportionally more driving strength the closer a planet lies to the Sun. He asks: Is P_2/P_1 proportional to r_2/r_1? And he finds that P_2/P_1 is too large, or that

$$\frac{P_2}{P_1}\frac{r_1}{r_2} > 1.$$

Kepler then phrases his inquiry in terms of *increments* rather than ratios, essentially arguing that the excess driving strength is proportional to the incremental distance between successive orbits:

$$\frac{P_2}{P_1}\frac{r_2}{r_1} - 1 = \frac{\Delta r_1}{r_2}.$$

There is a remarkable parallel between the approach to this problem and Kepler's attempt a few years later to find a law of refraction. In his *Paralipomena in Vitellionem* in 1604, Kepler also sought a relation expressing an incremental quantity, in that case the difference in the incident and refracted light rays; although he could fit the data tolerably well, this approach effectively prevented him from finding the correct functional relationship between incidence and refraction.[6]

A similar situation applies here; the relation above yields a rather satisfactory agreement for the predicted r_2/r_1, which Kepler exhibits:

	Predicted	Copernicus
Jupiter/Saturn	574	572
Mars/Jupiter	274	290
Earth/Mars	694	658
Venus/Earth	762	719
Mercury/Venus	563	500

Kepler's relation is nearly equivalent to $P_1/P_2 = (r_2/r_1)^2$, instead of the correct $(r_2/r_1)^{3/2}$.[7] (In the second edition of the *Mysterium cosmographicum*, in 1621, Kepler simply calls attention to the corrected form in a footnote.[8])

Kepler's embryonic but tenacious belief in the quantifiable driving power of the Sun was considerably sharpened in his *Astronomia nova* (1609), in which a lengthy mathematical analysis of the motion of Mars led to the ellipse and area laws. In this work, the quantitative driving force of the Sun, which he takes as $1/r$, plays a prominent role. In chapter 39, he writes: "Assuming the same planet maintains two alternating distances from the Sun during one revolution, the periodic times would be proportional to the square of the distances or sizes of the circles." In other words,

$$P = \frac{\text{Length of path}}{\text{Strength of driving force}} = \frac{r}{1/r} = r^2.$$

We may suppose that Kepler was temporarily satisfied with this solution, which to him represented not a fundamental astronomical law but an interesting consequence of the $1/r$ driving force.

By 1618, Kepler had returned to the problems of planetary harmonies; the *Harmonice mundi* was to his mind a delayed sequel to his youthful *Mysterium cosmographicum*. In his earlier work of 1596, Kepler was satisfied with a 5 percent error in the predicted planetary spacing and somewhat more in the periods. But now, greatly matured

in his scientific outlook and imbued with a powerful respect for the
efficacy of observation after his encounter with Tycho's data, he carefully approached the problem of discrepancies in his nest of spheres
and polyhedra. By this time, he must have been willing to concede that
the $P = r^2$ correlation was less than satisfactory, and now that he was
once more attuned to this problem, the answer could not have been far
off. Using Kepler's own data (*Harmonice mundi*, V,4), we can calculate the table he failed to exhibit:

	P (days)	Mean r	P^2 (years)	r^3
Saturn	10759	9510	860.08	867.69
Jupiter	4333	5200	140.61	140.73
Mars	687	1524	3.540	3.538
Earth	365	1000	1.000	1.000
Venus	225	724	0.3795	0.3795
Mercury	88	388	0.0584	0.0580

Kepler viewed this not as a fundamental astronomical law but simply
as an observable consequence of underlying principles. This attitude is
quite evident in Book IV of the *Epitome astronomiae Copernicanae*,
which was completed soon after the *Harmonice mundi*. Books I–III of
the *Epitome*, a comparatively straightforward exposition of spherical
astronomy, had already been published, and the last books were awaiting a publisher, apparently in a sufficiently finished form for them to be
referenced in the *Harmonice mundi*. Kepler now turned to lunar theory, which he hoped to place on a more physical basis, a research that
met with success in April 1620. Consequently, Book IV, essentially
later in its conception than were Books V–VII, was written to epito-
mize both the *Harmonice mundi* and the new lunar theory.

In the *Epitome*, Kepler takes the harmonic law for granted and does
not attempt to show any empirical evidence when the law is extended
to Jupiter's satellites. Kepler gives neither table nor calculation to
confirm his claim.[9] But Kepler's harmonic law requires no elaborate
proof. Unlike the ellipse and area laws, which were not at all self-
evident and which were well-nigh impossible to test without Tycho's
observations, the third law stands with a stark numerical simplicity.

The *Epitome* takes the form of questions and answers. The harmonic law is introduced in answer to a query in Book IV, part 2,
section 3:

Q. *By what reasons are you led to make the Sun the source of move-*
ment for the planets?

A. *1. Because it is apparent that, insofar as any planet is more distant from the Sun than the rest, it moves more slowly—so that the ratio of the periodic times is the ratio of the 3/2 powers of the distances from the Sun. Therefore, we reason from this that the Sun is the source of movement.*

Section 4 of Book IV, part 2, opens with an inquiry into the cause of the 3/2 power law. Kepler's answer, in modern form, is

$$\text{Period} \propto \frac{\text{Length of path} \times \text{Amount of matter in planet}}{\text{Strength of motor virtue} \times \text{Volume of matter in planet}}.$$

Clearly, the longer the path, the greater the period; the greater the magnetic effluvia from the Sun that provides the push, the shorter the period. The matter in the planet itself provides a resistance to continued motion: the more matter, the more inertia, the more time required. Finally, with a greater bulk of volume of matter, the "motor virtue" can be soaked up more greedily, and the period proportionately shortened. To achieve $P = r^{3/2}$ requires that the amount of matter per volume of matter (which is the density) be proportional to $1/\sqrt{r}$, since the first part of the expression equals r^3 as before. In other words, the amount of matter and volume of each planet must depend monotonically on its distance from the Sun (or at least their ratio must). That this was so appeared entirely reasonable to Kepler, and even a century later Newton (in *Principia*, Book III, Proposition VIII, Corrollary 4) was to state that "The smaller the planets are, they are *caeteris paribus*, of so much the greater density. . . so Jupiter is more dense than Saturn, and the Earth than Jupiter. For the Planets were to be placed at different distances from the Sun, that according to their degrees of density, they might enjoy a greater or less proportion of the Sun's heat."

But how can Kepler independently establish these dependences? In the absence of other information, Kepler always turns to "archetypal reasons"—the blueprint of God that can be found from geometrical or numerological reasoning.

The distance of a planet from the Sun, he says, can be proportional to the size, or to the surface, or to the volume. And here we see a most remarkable interaction between theory and observation:[10]

> Of these three modes, the first is refuted beyond controversy both by archetypal reasons and now also by the observations of the diameters made with the Dutch telescope; up to now I have approved the second mode, and Remus

Quietanus the third. On my side the better reasons, the archetypal, seemed to stand; on Remus's side the observations stand; but in such a delicate question I was afraid that the observations were not certain enough not to be taken exception to. Nevertheless, I yield the place to Remus and his observations.

In the *Epitome*, Kepler gives diameters for Saturn, Jupiter, and Mars; for the outer two, the agreement is excellent. For Mars, no amount of manipulation can conceal the complete lack of agreement, but Kepler marches on without further explanation. With the volume proportional to r, it is necessary to have the amount of matter proportional to $\sqrt{r}$. Kepler bravely argues that this must be so from archetypal reasons, but it is perfectly obvious that he would never have concluded so without knowing the answer. Yet he does make use of data in his *Messekunst Archimedis* (1616) to show how reasonably the $1/\sqrt{r}$ density relation fits (which he considers more fundamental than the individual patterns for amount and volume of matter):[11]

Saturn	324	The hardest precious stones
Jupiter	438	The lodestone
Mars	810	Iron
Earth	1000	Silver
Venus	1175	Lead
Mercury	1605	Quicksilver
Sun	1800	Gold

(At least for the Sun and Mercury, the archetypal principles can be discerned!) Armed with these "results," we have

$$P=\frac{r}{1/r}\frac{\sqrt{r}}{r}=r^{3/2}$$

and the theoretical derivation is complete.

Although the discovery comes comparatively late in Kepler's career and plays no particularly prominent role in his work (especially when compared to the law of ellipses or the law of areas), the pattern of physical reasoning that led to this law appears in his earliest writings. A study of the origins of the third law illuminates Kepler's innovative physical approach to astronomical problems. The harmonic relation was seen by Kepler not as a fundamental law in itself but simply as a clear and accurate manifestation of the more fundamental principles underlying the cosmos—both physical and archetypal. Thus, although

Kepler gives the harmonic law very little emphasis, it represents the culmination of a lifelong search and illustrates Kepler's imaginative approach to the mysteries of the cosmos.

Notes and References

[1] J. Kepler, *Harmonice mundi* (1618), in *Johannes Kepler Gesammelte Werke (JKGW)*, vol. 6 (Munich, 1940), p. 302.

[2] *Ibid.*, p. 290.

[3] I am here greatly indebted to the extensive analysis of the *Harmonice mundi* in Max Caspar's *Kepler*, trans. by Doris Hellman, esp. (New York, 1959), pp. 285–87.

[4] N. Copernicus, *De revolutionibus orbium coelestium*, Book I, Chapter 10 (Nuremberg, 1543.

[5] J. Kepler, *Mysterium cosmographicum* (1596), in *JKGW*, vol. 1, chap. 20. Kepler explores this relation with the following table:

Now if	10759 ♄	is taken as 1000	♃ 403	and if the mean	♃ 572
for the	4332 ♃	for the higher	♂ 159	distance of the	♂ 290
period	686 ♂	of each pair,	T 532	superior one is	T 658
(here	365 T	the quantity	♀ 615	1000, the inferior	♀ 719
in days)	224 ♀	of motion is	☿ 392	is according to	☿ 500
				Copernicus	

[6] See G. Buchdahl, "Methodological Aspects of Kepler's Theory of Refraction," *Internationales Kepler-Symposium, Weil der Stadt* (1971), pp. 141–167.

[7] The relation

$$\frac{r_2 - r_1}{r_2} = \frac{P_2 \, r_1}{P_1 r_2} - 1$$

easily reduces to

$$\frac{r_2}{r_1} = \frac{(P_1 + P_2)/2}{P_1} .$$

If we replace the arithmetic mean $(P_1 + P_2)/2$ by the geometrical mean $\sqrt{P_1 P_2}$, we find

$$\frac{P_2}{P_1} = \left(\frac{r_2}{r_1}\right)^2 .$$

[8] J. Kepler, *Mysterium cosmographicum* (1621), in *JKGW*, vol. 8, p. 113.

[9] J. Kepler, *Epitome astronomiae Copernicanae* (1620), in *JKGW*, vol. 7, bk. IV, part 2, sec. 6, pp. 318–19.

[10] *Ibid.*, bk. IV, Part 1, sec. 4, p. 282. Johannes Remus Quietanus, the physician to the Emperor, was an enthusiastic amateur astronomer. The extensive correspondence between him and Kepler during these years must be a pale reflection of their personal discussions. When on 4 October 1619 Remus expresses delight with the proportions of sizes of the planets with distance, Kepler responds (October 1619): "I am surprised that you say you just recently discovered the proportion of the celestial

bodies to the distances, since a year ago or more you told me of this discovery when you came through Linz, and you asked me to examine them, for I would find a beautiful proportion between the bodies of the Earth and Moon and their orbits and periods, and also you showed the quantities of the diameters of Jupiter, Venus, and Mars, which you had proven with a telescope." (*JKGW*, vol. 17, letter no. 859, p. 406.)

[11] J. Kepler, *loc. cit.*, *JKGW*, vol. 7, p. 284.

The Computer versus Kepler

We often hear, in discussions of modern high-speed computers, how an electronic machine can calculate more in a day than a man can calculate in a lifetime by older methods. It occurred to me that a concrete demonstration of some properly chosen specific case would not only be intrinsically interesting, but might shed some light on the historical situation in question, and might also provide a dramatic example of the application of computers in the history of science.

An especially appropriate example is found in the work of Kepler on the orbit of Mars, since he gives some indication of the computational time involved. In *Astronomia nova*, Kepler describes in detail his attempt to fit a circular orbit to a series of observations of Mars at opposition. Since he wished to investigate a somewhat more general orbit than had been adopted classically, he was led to a thorny trigonometric problem that can be solved only iteratively.

Concerning this involved procedure, Kepler implores his reader: "If you are wearied by this tedious method, take pity on me, who carried out at least 70 trials of it, with the loss of much time, and don't be surprised that this already is the fifth year since I have attacked Mars, although the year 1603 was almost entirely spent on optical investigations."[1]

(To this, the French astronomer Delambre replied: "Kepler was sustained by his desire to have a case against Tycho, Copernicus, Ptolemy, and all the astronomers in the world; he has tasted this

Selection 21 reprinted from *American Scientist*, vol. 52 (1964), pp. 218–26.

satisfaction, and I don't believe he deserves our pity for making all these calculations.[2])

The implication that this problem required four years must be taken with a grain of salt, but we do get a rough idea of the time involved.

It is this tedious, time-consuming procedure that I have programmed for the IBM-7094 at the Harvard Computing Center. Before describing my quite unexpected results, let me outline Kepler's problem in somewhat greater detail.

When Kepler started his investigation on the motion of Mars, in 1601, he was already a convinced Copernican, and therefore he assumed a heliostatic orbit. Nevertheless, at the beginning, he accepted the classical idea of using circles to represent the motion, and not until two years later did he work out the elliptical form of the orbit. The "vicarious orbit" that caused Kepler so much anguish and loss of time was a circle, and in the end was completely abandoned.

Kepler had in hand a dozen observations of Mars at opposition— ten from Tycho Brahe and, later, two of his own[3] When Mars is at opposition, the Sun, Earth, and Mars lie in a straight line, so the heliocentric longitude of Mars is immediately known. Figure 1, reproduced from Delambre's *Histoire de l'astronomie moderne*, shows us the basic diagram for this problem. In the diagram, the Sun is at A, and four observations of Mars, carefully chosen for a reasonably uniform distribution, are laid out from it. Note that the Earth does not enter into this discussion. Now the correct elliptical orbit of Mars does not differ very much from a circle, except that the Sun is at one focus and reasonably far displaced from the center. In this circular approximation, the Sun lies off the center of the circle, which is at B.

We know that Mars moves most quickly when nearest the Sun and slowest when at aphelion (that is, when farthest from the Sun), a fact later expressed in the law of areas. Kepler believed this must be so from physical reasons, and therefore he was already convinced that the seat of uniform angular motion in the orbit must lie on the line through A and B—that is, on the line of apsides. In the analogous case, Ptolemy had placed this seat of uniform angular motion, or equant, equally spaced opposite A from the center of the circle. We now know that such a configuration produces the best possible approximation to an ellipse, and when we have the equant at the empty focus of the ellipse, the resulting errors in fitting the observed longitudes reach a maximum of 8' of arc. This is the figure later found by Kepler, which, for him, proved to be such a large discrepancy from Tycho's observations that he felt obliged to abandon the circular orbits.

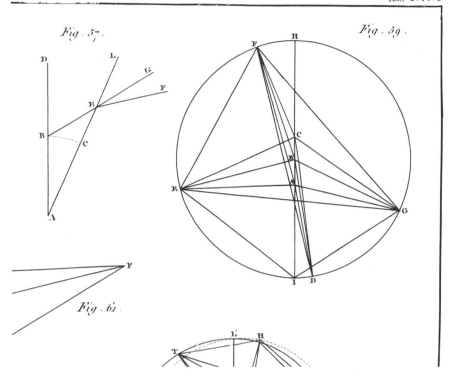

FIGURE 1. *The geometry of Kepler's vicarious orbit solution as given by Delambre.*

Kepler, however, wished to keep the spacing of *A* and *C* along the line of apsides as an unknown quantity to be determined. Also, he knew the direction of the aphelion fairly well, but he wished to improve its position. Kepler was therefore obliged to use four observations to determine all these quantities. Nowadays, we would try to use all twelve observations, combining them into a least-squares solution. This technique was, of course, unavailable to Kepler. Note that the angles from *A* are all determined by observation. The angles from *C* are known relative to one another, because the motion about this point is uniform in time and the times of observations are known. The zero point of this system is to be determined, and also the direction of the aphelion *AH*.

Kepler starts by assuming these two quantities and solves trigonometrically the various angles of this inscribed quadrilateral. The result

NOTES. 335

$\left(\dfrac{AG+AE}{AG-AE}\right)$ = tan. 19° 57' 40'' $\left(\dfrac{103021}{1541}\right)$ = tan.18'11'';

and AEG = 19° 53' 51''; as also EG = $\dfrac{AG.\sin. EAG}{\sin. AEG}$ =

$\dfrac{57282.63271}{34088}$ = 97041.

3. Since the base EG of the isosceles triangle EBG, and the vertical angle EBG are thus found, the angle BEG at the base is given, and = 25° 46' 53''; and, therefore, BE = $\dfrac{EG.\sin. BEG}{\sin. EBG}$ = $\dfrac{97041.43494}{78327}$ = 53860.

4. In the triangle BEA, the angle BEA is given, for it is = BEG — AEG = 5° 51' 2'', and also its ½ suppl. = 87° 4' 47''. Therefore tan. ½ (BAE — ABE) = tan. ½ suppl. BEA $\left(\dfrac{BE-AE}{bE+AE}\right)$ = $\dfrac{1957200.5121}{104599}$ = 58402 = tan. 30° 17' 8''; so that BAE = 117° 21' 37''.

But, since in the second operation the aphelion H was found to be too far advanced in longitude, let it now, in consequence of the last correction, be considered as advanced no more than 3' 8'', instead of 3' 29'' beyond the longitude first assumed. Then, since AH is in 4s. 28° 47' 8'', and AE in 8s. 26° 39' 23'', CAE or HAE will be = 117° 52' 5''; that is, greater by 30' 28'' than BAE; and B is not situated in AC, but on side of it towards E. The suppositions therefore for FAH and FCH, must, one of them, or perhaps both, be false.

But these angles of anomaly cannot be varied by the mere variation of the assumed longitude of the aphelion; because no other position of it will permit the points D, F, F, G, to be situated in the circumference of the same circle; and before it can be farther varied, the mean longitudes, or the position of the lines FC, FE, &c. must be varied. This, therefore, was the next step of Kepler's procedure; and he tells us, it was not till after a great variety of unsuccessful trials, that he found his purpose would be nearly accomplished by the addition of 2' more to the longitude of the aphelion, and of 30'' at the same time to the mean longitudes. By these additions the mean anomalies FCH, ECH, &c. are all diminished 1' 30'' each; and we have FCH = 32° 3' 36''; KCE = 53° 7' 2''; KCD = 11° 0' 44''; and KCG = 68° 18' 1''. The angles again of equation will become AFC = 5° 8' 29''; AEC = 9° 4' 41''; ADC = 2° 17' 10''; and AGC = 10° 13' 45''; being increased 50'' in the first semi-circle of anomaly, and as much diminished in the second: consequently, the

```
      EGA(4*N   )=SECF(AFC(N))
1CC0 CCNCINUE
      N1CNT=N1CNT+1
      WRITE CUTPUT TAPE 6,101,N1CNT,N2CNT,ADDS,BCDATE,(CHS,N=1,4),HMEAN
      WANCM,BAPP,EGN.(AF(N),N=1,4),TAN2,SUM1S,SUM2S
      GC TC K1,(220,230)
C-----ACC ARBITRARY INCREMENT IN FIRST ITERATION.
 220  ACC=RACF(0.,0.,5.,0.C1)
      ASSIGN 23C TC K1
 225  SM1=SUM1
      SM2=SUM2
      SC=SUMC
      CFCLC=CF
      CF=CF+ACC
      GC TC 190
C-----ACC PRCPORTICNAL INCREMENTS IN REMAINING ITERATIONS.
 230  IF(ACC-PACF(C.,0.,10.))235,235,234
 234  ALC=AUC/(SC-SUMC)*SUMC
      IF(N1CNT-2C) ? 5,300,3C0
C
C-----BEGIN CLTER ITERATICN.
 235  EBG= ThCPI-FAF-FAE(4) -SUM1-SUM2
      EAC=FAE(2)+FAF(3)
      AECGAGE=2.*ATANF(TNHSUPF(EAG)*ABAHF(AF(4),AF(2)) )
      AEG=(PI+AFCAGF-EAG)/2.
      EG=AF(4)*SINF(EAG)/SINF(AEG)
      BEC=(PI-EPC)/2.
      BL=EG*SINF(BEG)/SINF(EBG)
      CA=1./PE
      BEA=BEG-AEG
      BAEABE=2.*ATANF(TNHSUPF(BEA)*ABABF(RE,AF(2)) )
      PAF=(PI+BAEABE-BEA)/2.
```

FIGURE 2. *The comparison of the Robert Small commentary with a portion of the FORTRAN program shows how closely the notations agree.*

tells him whether or not the points lie on a circle. In the first instance they do not, so the direction *AH* is altered and the solution made again. A comparison of the results of these trials suggests a better position for *AH*, and the calculation is again repeated. This process I shall call the inner iteration. When it has finally converged, Kepler solves this triangle *EGB* to find if the center of circle *B* lies on the line *CA* between the sun and the equant. Again in this first instance it does not. This time, the zero point of the mean angles at *C* is altered, and the inner iteration is repeated. Eventually, the outer iteration also succeeds, and the points *A*, *B*, and *C* are found to lie on a straight line. I am sure Kepler is counting the inner iterations when he tells us that 70 trials were required.

The programming followed Kepler's procedure almost exactly. I was greatly helped by a book by Robert Small,[4] which was recently reprinted through the efforts of William Stahlman. Figure 2 shows how closely the FORTRAN programming followed his notation. The principal difference in my approach is that, when Kepler got close to the solution, he jumped to the answer using small corrections made by proportional parts, whereas I found it easier simply to repeat the entire calculation. Also, the program used accuracy criteria somewhat more rigid than Kepler's.

After I had set up and "debugged" this program, I found that the machine could polish off the entire problem in a little less than eight seconds! This is not too surprising, when we realize that only about 25 trigonometric functions are required in each trial. Unlike Kepler, the computer does not need to look up and laboriously interpolate each of these. Instead, it computes them from scratch as needed, at the rate of 3000 per second!

At least some readers will want to know how long it took *me* to set up the program. When Kepler first arrived at Tycho's establishment, he made a bet that he would have the Mars orbit all cleaned up within eight days. When I agreed to report on this project, I too hoped to finish the calculations very quickly. But I procrastinated, and finally only eight days remained before the Christmas meeting. Thus, circumstances forced me to carry out these computations within that time span. In all, I had nine tries on the computer for this work. In the first two, the computer system detected errors of typography and nomenclature, so those trials "went up in smoke," as Kepler might say[5] (see Figure 3).

This was followed by a series of runs in which other logical flaws were detected—for example, there turned out to be an error in the Robert Small book, which I had blindly followed. By the sixth try, I already had in hand one very interesting result, after a total of eight minutes of computer time. In the ensuing runs, I corrected several more errors and also computed with different initial conditions, as I shall explain. Altogether, I used 12.4 minutes of IBM-7094 time. Now that the program has been written and "debugged," additional cases require only the eight seconds quoted above. Figure 4 illustrates an example of the output.

The results I have just quoted sound more like a publicity release for electronic computers than a serious paper in the history of science. However, one quite remarkable fact turned up in this investigation. Instead of requiring 70 trials as Kepler did, the computer program, using identical methods, took only nine trials! In fact, we might have anticipated this result without doing any calculations at all, from the following considerations. Suppose the aphelion and the zero point of the mean longitudes are originally known to 1° (actually they were much better known than this). Suppose we wish to get these to 30" of arc, that is, an improvement by a factor of 120. Since 2^7 is 128, seven inner iterations should be required in each of seven outer iterations, if the error is halved each time. This total number of iterations, about 50, should probably be halved because the inner and outer iterations are

```
KEPLER'S VICARICUS ORBIT, OR, 'THE COMPUTER VERSUS KEPLER'          12/18/63          PAGE 6

                    709/7090 FORTRAN DIAGNOSTIC PROGRAM RESULTS


          SUM1=ATANF(TAN2(4)-ATANF(TAN2(1))

03111     TCC MANY LEFT PARENTHESIS.


          SUM2= ATANF(TAN2(3)-ATANF(TAN2(2))

03111     TCC MANY LEFT PARENTHESIS.


1C2       FCRMAT(18H4OUTER ITERATION =I3,24X4HADM=3X,3F4.0,F5.1,8X,23HFINAL COMPARISON ANGLES/14X,4HEBG=3X,3F4.0,F5.1,
          8X,4HEAG=3X,3F4.0,F5.1, 8X,4HBAE=3X,3F4.0,F5.1/14X,4HAEG=3X,3F4.0,F5.1,8X,4HBEG=3X,3F4.0,F5.1,8X,4HHAE=3X,5
          F4.0,F5.1/15X,3HEG=3X,F11.8,3HBE=3X,F11.8/     15X,3HBA=3X,F11.8,14X,3HCA=3X,F11.8)

04C32     FORMAT STATEMENT IS INCORRECTLY WRITTEN.
                    END OF DIAGNOSTIC PROGRAM RESULTS.

          SCURCE PROGRAM ERROR.  NO COMPILATION.

EXECUTICN DELETED.
```

FIGURE 3. *FORTRAN diagnostic. A decimal has been mispunched in place of a comma in the format statement.*

```
N=  1   ITERATION=  1              ADD=  0. -0. -0. -0.

                         1587 MAR  6          1591 JUN  8          1593 AUG 25          1595 OCT 31
APHELICN               4. 28. 44.  0.0      4. 28. 44.  0.0     4. 28. 44.  0.0      4. 28. 44.  0.0
MEAN LCNGITUDE         6.  0. 47. 40.0      9.  5. 40. 19.1    11.  9. 49. 35.8      1.  7.  6. 50.3
MEAN ANOMALY           1.  2.  6. 56.0      4.  6. 56. 19.1     6. 11.  5. 35.8      8.  8. 22. 50.3
APPARENT LONGITUDE     5. 25. 43. -0.       8. 26. 39. 24.2    11. 12. 10. 31.8      1. 17. 24. 21.3
EQUATICN OF CENTER     0.  5.  7. 56.0      0.  9.  4. 11.0    11. 27. 42. 20.0     11. 19. 45. 45.0
RADIUS VECTOR          5.94300860           5.07040501          4.80596131           5.23072946
TAN(HALF DIFFERENCES)  -0.07794160         -0.03456920          0.06613767           0.03087720

           SUM1  0.  6. 13. 31.0
           SUM2  0.  5. 45. 49.6


N=  2   ITERATION=  1              ADD=  0.  0.  5.  0.0

                         1587 MAR  6          1591 JUN  8          1593 AUG 25          1595 OCT 31
APHELION               4. 28. 49.  0.0      4. 28. 49.  0.0     4. 28. 49.  0.0      4. 28. 49.  0.0
MEAN LCNGITUDE         6.  0. 47. 40.0      9.  5. 40. 19.1    11.  9. 49. 35.8      1.  7.  6. 50.3
MEAN ANOMALY           1.  2.  1. 56.0      4.  6. 51. 19.1     6. 11.  0. 35.8      8.  8. 17. 50.3
APPARENT LONGITUDE     5. 25. 43. -0.       8. 26. 39. 24.2    11. 12. 10. 31.8      1. 17. 24. 21.3
EQUATICN OF CENTER     0.  5.  7. 56.0      0.  9.  4. 11.0    11. 27. 42. 20.0     11. 19. 45. 45.0
RADIUS VECTOR          5.92923105           5.07594460          4.77030510           5.22770876
TAN(HALF DIFFERENCES)  -0.07627323         -0.04007670          0.07149462           0.03045668

           SUM1  0.  6.  6. 22.3
           SUM2  0.  6. 23.  3.8


N=  3   ITERATION=  1              ADD= -0.  0.  1. 52.8

                         1587 MAR  6          1591 JUN  8          1593 AUG 25          1595 OCT 31
APHELICN               4. 28. 47.  7.2      4. 28. 47.  7.2     11.  9. 49. 35.8      4. 28. 47.  7.2
MEAN LCNGITUDE         6.  0. 47. 40.0      9.  5. 40. 19.1    11.  9. 49. 35.8      1.  7.  7. 50.3
MEAN ANOMALY           1.  2.  3. 48.8      4.  6. 53. 12.0     6. 11.  2. 28.6      8.  8. 19. 43.1
APPARENT LONGITUDE     5. 25. 43. -0.       8. 26. 39. 24.2    11. 12. 10. 31.8      1. 17. 24. 21.3
EQUATICN OF CENTER     0.  5.  7. 56.0      0.  9.  4. 11.0    11. 27. 42. 20.0     11. 19. 45. 45.0
RADIUS VECTOR          5.93441421           5.07386243          4.78371644           5.22884607
TAN(HALF DIFFERENCES)  -0.07690100         -0.03800140          0.06947503           0.03061479

           SUM1  0.  6.  9.  3.6
           SUM2  0.  6.  9.  1.8


OUTER ITERATICN =  1                   ADM=  0. -0. -0. -0.      FINAL COMPARISON ANGLES
          EBG=  4.  8. 26. 51.6        EAG=  4. 20. 44. 57.1     BAE=  3. 27. 34. 13.1
          AEG=  0. 19. 55. 57.8        BEG=  0. 25. 46. 34.2     HAE=  3. 27. 52. 17.0
          EG=  9.70434964              BE=  5.38831323
          BA=  0.11485431              CA=  0.18558683
```

FIGURE 4. *Intermediate computer printout. The inner iterations are carried out until SUM1 and SUM2 agree; in the outer iteration angles BAE and HAE are compared, and the process repeated until they match.*

not independent and, as the outer iteration converges, the inner set will require fewer than seven tries each time. Furthermore, since the problem turns out to be fairly linear, we can use proportional parts to speed the convergence, and hence we might again halve the number of iterations, making about twelve. On the other hand, we make an initial try, then a try with an arbitary displacement, and finally a try with proportional parts based on the first results. Thus, three tries in each inner iteration, and three outer iterations, give a minimum of nine trials by this method, precisely the number used by the computer.

Why, then, did Kepler require 70 trials? Since Kepler already started with an arbitrary correction to Tycho's zero point on the mean longitudes, we suspect that he may have used many trials to reach the starting point shown in *Astronomia nova*. Therefore, the calculations were repeated, starting directly from Tycho's figures. Now, 13 iterations are required, still a very small number.

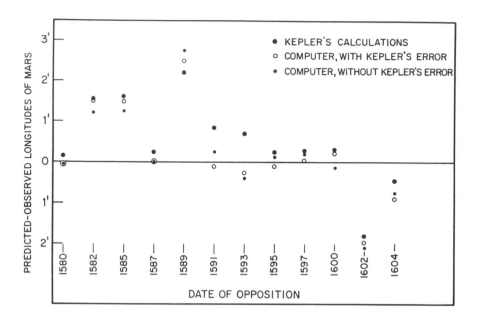

FIGURE 5. *Comparison of Kepler's calculations and computer's calculations with respect to predicted-observed longitudes of Mars and date of opposition.*

I can only conclude that Kepler was horribly plagued by numerical errors, that his trials accidentally diverged nearly as often as they converged. No wonder he was so frustrated in his attempt to solve this problem, which was apparently just at the limit of his computational ability! Do we have any evidence for this conclusion? Yes. At the very beginning of his calculation, Kepler makes numerical errors in three of his eight starting angles—errors of the same order of magnitude as the corrections he was seeking. These errors were noted both by Small and by Delambre. I therefore programmed the computer to solve the problem both with and without this initial error. The final solution appears comparatively insensitive to these errors, but it is curious to note that Kepler gets about the same answer with the errors that the machine computes *without*!

After Kepler completed his solution with four of the twelve oppositions, he carefully calculated the predicted positions for all twelve observations.[6] The results, shown in Figure 5, exhibit several interesting features.

First of all, since the solution was carried out exactly for the oppositions of 1587, 1591, 1593, and 1595, the same observed positions

FIGURE 6. *Kepler's original manuscripts, including the 900 pages of Mars calculations, are still preserved in Leningrad. "Deo et Publico" is the motto of Catherine the Great, who purchased the volumes for the Russian Academy of Sciences in 1773. Photograph courtesy Harvard College Observatory.*

ought to be predicted by the theory. But here, Kepler has taken a very curious step: he corrects each of the positions for the advance of nodes of Mars—a curious step because the correction is made *after* the main calculation instead of *before*![7] Thus, only the pivotal 1587 opposition must predict exactly the observed position; yet, as the graph indicates, Kepler has made a small computational error of 15″. Given a uniform motion of the nodes, the 1591, 1593, and 1595 observations should show increasing errors, yet again this is not the case. Compared to the machine calculations, Kepler's results for 1591 and 1593 show computational errors as large as 1′. One final comment: note from the graph how Kepler's errors generally increase the deviations between observation and prediction, *except* for the most discordant cases!

The best possible solution with this type of model, as stated previously, leaves errors up to 8′ of arc. We see here that Kepler was incredibly lucky in his particular choice of observations—or perhaps we should say unlucky, because, with larger errors, he would probably have recognized the inadequacy of this construction earlier. As a test,

I chose other well-distributed sets of four oppositions as the basis of the solution, and I indeed found larger errors, up to 8' of arc.

I hope this study has shed some light on the difficulties encountered by Kepler, and perhaps on his computational ability. My thesis, that his calculations were incredibly loaded with numerical errors, has already been observed in another section of *Astronomia nova* by O. Neugebauer.[8] Perhaps it will someday be further confirmed by a full analysis of the 900 pages of original manuscript computations, still extant in Leningrad.[9] I do not wish, however, to detract in any way from the magnitude of Kepler's scientific achievement. Perhaps the most appropriate conclusion would be a further quotation from *Astronomia nova*:[10]

> There will be some clever geometers such as Vieta who will think it is something great to demonstrate the inelegance of this method. (As a matter of fact, Vieta has already made this charge against Ptolemy, Copernicus, and Regiomontanus.) Well, let them go solve this scheme themselves by geometry, and they will for me be a great Apollo. For me it suffices to draw four or five conclusions from one argument (in which there are included four observations and two hypotheses), and to have shown by the light of geometry an inelegant thread for finding the way out of the labyrinth. If this method is difficult to grasp, how much more difficult it is to investigate things without any method.

Notes and References

[1] J. Kepler, *Astronomia nova* (1609) in *Johannes Kepler Gesammelte Werke*, vol. 3, ed. by M. Caspar (Munich, 1937) chap. 16, p. 156.

[2] J-B. J. Delambre, *Histoire de l'astronomie moderne*, (Paris, 1821), p. 417.

[3] J. Kepler, *op. cit.*, chap. 16, p. 150.

[4] R. Small, *An Account of the Astronomical Discoveries of Kepler* (1804) (reprinted by the University of Wisconsin Press, Madison, 1963).

[5] "Itaque causae Physicae cap XLV in fumos abeunt." J. Kepler, *op. cit.*, chap. 55, p. 345.

[6] J. Kepler, *ibid.*, chap. 18, pp. 172–73.

[7] *Ibid.*, chap. 18, pp. 169–71.

[8] O. Neugebauer, "Notes on Kepler," *Comments on Pure and Applied Mathematics*, vol. 14, (1961) pp. 593–597.

[9] M. Caspar, "Nachbericht" to *Johannes Kepler Gesammelte Werke*, vol. 3 (Munich, 1927), p. 445.

[10] J. Kepler, *op. cit.*, chap. 16, p. 156.

The Computer versus Kepler Revisited

In Kepler's *Astronomia nova* we can see, for the first time in the history of science, how an astronomer struggles to establish a theoretical model from a conflicting redundancy of data. The great observational heritage of Tycho Brahe provided Kepler with planetary positions consistently more accurate and far more numerous than previously known. Yet, as he pressed toward greater precision within a quasi-traditional framework, he soon reached the level where the scatter in the data led to ambiguities in the results.

Kepler was fully cognizant of the possible accidental observational errors, but he had no error theory to smooth his path or to guide him in handling the ostensibly redundant material. The *Astronomia nova* informs us about some of his procedures, but at least one intriguing mystery remains: Why did he require so many trials to establish his initial model, the so-called "vicarious orbit"? To answer this we shall turn to an essentially unexamined source, his manuscripts, which cast new light on the first round of his battle against Mars.

Ptolemy and Copernicus, the great innovators in planetary astronomy before Kepler, had in their treatises restricted themselves to the minimum number of observations required to derive the orbital parameters, and hence the problem of observational error or redundant conflicting data is not explicitly considered.[1] In contrast, Kepler's en-

Selection 22 reprinted from *Proceedings of the Internationals Kepler-Symposium, Weil der Stadt, 1971,* ed. by F. Krafft, K. Meyer, and B. Sticker (Hildesheim, 1973), pp. 307–14.

tire *Astronomia nova* is permeated by considerations of observational error, for he is obliged to use the multiplicity of data to control the accuracy in the subtle phenomena under investigation. This is rarely shown so clearly as in the procedures of Chapters 16–18.[2] Here he presents the basis of his computational scheme for generating very accurate Martian longitudes—a scheme that cannot be physical or true because it did not simultaneously give the correct distances for Mars. Hence he called it the vicarious hypothesis, as opposed to the physical hypothesis that he was still seeking. Though physically false, Kepler's vicarious hypothesis was a most important device, for it permitted him to specify accurately the Martian longitudes without further recourse to observations beyond those he had used to establish it. Thus the vicarious hypothesis played a significant role in the building of his planetary theory, like an essential but temporary scaffolding.

The vicarious hypothesis is a heliocentric orbit arranged with a quasi-traditional mechanism. In essence, it consists of an eccentric circular orbit with a classical equant, but modified so that the equant could fall at an arbitrary place along the excenter, or apsidal, line. In Chapter 18, Kepler shows that the vicarious hypothesis satisfies 12 acronychal (that is, opposition) observations to within about 2′. With a series approximation, we can show that this is about the expected accuracy using the optimum ratio of 3/5 for the spacing Sun-to-center/Sun-to-equant; Kepler's final ratio of 11,332/18,564 is close to this. (In Kepler's notation, this is the ratio AB/AC.)

Four parameters specify the vicarious orbit: the eccentricity AB and the equant offset AC (both given in terms of the size of the orbit), the direction of the apsidal line AH (which Kepler always calls the apogee, a curious geocentric vestige), and the time when Mars was at the aphelion point. For this Kepler required four observations; these were taken at opposition so that the geometry of the Earth's orbit presumably did not enter. To the four observed angles there correspond four observed times, which in turn specify four angles about the equant point proportional to the time intervals. As may be seen in Figure 1, this fixes four sets of triangles; in the final solution, their apices must lie on a circle whose center falls on the line AC.[3] Since Kepler could find no explicit solution for this geometry, he attacked the problem iteratively, guessing the direction AH and computing a pair of opposite angles in the quadrilateral $DEFG$; these must of course total 180° if the corners of the quadrilateral lie on a circle. This we shall call the inner iteration. Kepler then had to find whether the center of the circle fell on the line AC; if it did not, a slight adjustment to the time of aphelion

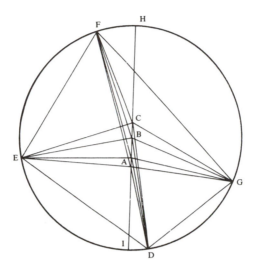

FIGURE 1. *The geometry of Kepler's vicarious orbit.*

was required. (This is equivalent to a small rotation of the angles around the equant point.) This we call the outer iteration. "If you are wearied by this tedious method," Kepler implores us, "take pity on me, who carried out at least 70 trials of it, with much loss of time."[4]

The question immediately arises as to why it took Kepler so many iterations. Is this simply an exaggerated claim to gain our sympathy? Several years ago, when I programmed Kepler's iterative procedure for an IBM 7094, I noticed that the computer required instead of 70 trials only nine iterations, the minimum possible. At the time I concluded that "Kepler was horribly plagued by numerical errors, that his trials accidentally diverged nearly as often as they converged."[5] In retrospect, this conclusion appears unreasonable, all the more so when my recomputation of the *Tabulae Rudolphinae* showed that these tables were virtually error-free.

As an alternative hypothesis, Kepler's 70 iterations could have resulted from the redundancy of data available to him. Since he eventually had observations for 12 Martian oppositions, possibly Kepler carried out the procedure repeatedly for different sets of four oppositions. If so, this would have been an early attempt to control the errors of observational data and, hence, exceedingly interesting to the history of astronomy.

The fact that Kepler does not state in the *Astronomia nova* why he required 70 iterations is in itself interesting. Most commentators have assumed, because of Kepler's sequential and at times autobiographical style, that Kepler has spared no detail in the chronicle of his researches. Examination of the manuscript material (Volume XIV of the Leningrad Kepler manuscripts[6]) shows, on the contrary, that the book evolved through several stages and represents a much more coherent plan of organization than a mere serial recital of his investigations would allow. Indeed, Kepler has spared us most of his agony with the vicarious orbit. The manuscripts show quite clearly why Kepler carried out at least 70 trials. They show, in fact, that he iterated almost as efficiently as the computer, and that the source of his difficulty lay elsewhere, in a direction unimagined in the absence of his working notes.

Of 894 pages in the Leningrad Kepler manuscripts Volume XIV, 754 are more or less directly associated with the *Astronomia nova*. The principal document is a notebook of 586 pages that apparently covers rather completely Kepler's Mars work in 1600 and 1601—that is, while Tycho was still alive. Kepler has numbered the pages to the end (to 538), but he has inserted occasional unnumbered supplementary quires. There are numerous cross references, so we know that the numbering is contemporary with the notes and also that the pages show sequential work. The many longhand calculations as well as crossed-out sections indicate that the material is essentially complete, and that Kepler generally did not resort to trial calculations on separate scratch paper.

I have found three places where the notes are dated: 6 July 1600 on page 199, at which time Kepler had returned to Graz after a three-month visit to Tycho in Prague; April 1601 on page 325, which was added later; and 29 March 1602 on page 474, apparently dating in the time an error was caught rather than when the page was originally written.[7] The notebook breaks off in mid-sentence at the bottom of page 538 without having mentioned an oval or noncircular orbit of Mars, which Kepler accepted sometime in 1602.[8]

(I should note in passing that the remaining 168 manuscript pages contain fragments from the calculation of the ephemeris for 1604, written in 1603; draft material prepared perhaps as early as 1602 for Chapters 27, 28, 32, and 35, which became 28, 27, 42, and roughly 46 in the finished *Astronomia nova*; and several miscellaneous but interesting pieces possibly from 1601 or 1602. This material is very incomplete, and there is apparently only a single folded sheet from 1605, the year of the ellipse.)

In the *Astronomia nova* itself, the various sets of observations are treated quite distinctly in Sections 2, 3, and 4. Section 2 centers on the use of the 12 acronychal observations to establish the vicarious hypothesis. In Chapters 26, 27, and 28 of Section 3, three groups of observations are introduced for triangulating to the orbit of Mars; these are used to prove that the Earth's accepted eccentricity should be bisected. At the beginning of Section 4, where the assault on Mars is resumed, two more groups of triangulation observations, for the Martian aphelion and perihelion, are introduced. In the manuscript material, no such neat separation is made, and from the very beginning Kepler alternates between the triangulation and the acronychal procedures, using observations on the same dates that ultimately appear in the book.

The beginnings of the vicarious orbit procedure are found around page 154 of the manuscript notebook,[9] *after* many pages devoted to the triangulation configuration that appears in Chapter 26 of the *Astronomia nova*. This work was done during Kepler's initial visit to Tycho, for it precedes the page 199 dated in July of 1600. Kepler intended to base his solution on the oppositions of 1587, 1591, 1593, and 1597, but within a few pages he exchanged 1597 for 1595 on the grounds that the latter observation was somewhat uncertain.[10] He started with a rough and rapid investigation of the allowed range of aphelion angles, which he narrowed between 29° and 27° 50' of Leo , and thereafter examined the two limits in some detail. Because the sum of the opposing quadrilateral angles was more than 180° in *both* cases, Kepler did not realize that the solution he sought lay between the two initial aphelion angles. This cost him an iteration, but afterwards his solution steadily if slowly converged.[11]

Never again would a set of inner iterations require as many as eight trials. With three and four more trials, respectively, Kepler completed the minimum number of outer iterations and found the desired fit, thus using 15 trials in all.

But in such a subtle matter, notes Kepler, it is desirable to compare the results with other observations, and therefore he calculated the predicted acronychal positions for 1582, 1585, 1600, and 1589. Alas the discrepancies for 1585 and 1589 amounted to 6'. Where had he gone astray? A reexamination of the fundamental observations then revealed an error of 8' in the position for 1587. Kepler started in again, but still he suspected something was wrong. "Nescio ubi sit error," he laments on page 198, apparently just before he returned to Styria in June of 1600.

Soon after his return to Graz, Kepler wrote to Herwart (12 July 1600)[12] about his first attempts, and he gave a geometrical proof of how to find the apsidal line and eccentricity from three observations. Since no similar diagrams appear in the manuscript notebook, we may surmise that part of his work at this time was recorded elsewhere, perhaps because he was traveling from Prague to Graz. In any event, the next pages of his notebook show this theorem applied, three observations at a time and in each combination, to ferret out the bad datum, apparently the angle *GAD* used instead of *GCD* in the 1595 observation.[13] With the mistake corrected, Kepler completed the outer iterations, making in all 13 more trials. Once again Kepler tested his solution against the other acronychal observations; the results he considered satisfactory:[14]

1582	Dec	28	3' —
1585	Jan	30	26"
1600	Jan	19	2' 44"
1589	Apr	15	41"

Then it was thus:

ln 28.57.20 Leo 179.33.26 fol 172
ln 29. 0. 0 Leo 179.55.53 fol 173 Diff 3° 11' 54" fol 174
ln 29. 0.25 Leo 180. 0. 0 fol 173 *CAE* 101.36.17 *BAE* 98.24.23
Fol: 174. 1' 16" to be added to Tycho's mean longitude.

Then it was thus:

ln 28°.32'.30" Leo 179.54.42 fol 175
ln 28.33.40 Leo 179.57.39 fol 176 Diff 27.59 f: 176
ln 28.33. 0 Leo 180. 0. 0 fol 176 *CAE* 101.7.53 *BAE* 100.39.54
ln 28.33.10 Leo 180. 4.23 fol 179 *CAE* 101.9. 3 *BAE* 100.52.45
                                                        16.18 diff

The parameters $AB = 11{,}058$ and $AC = 18{,}562$ are closer to the 3/5 ratio than are the final values of 11,332 and 18,564 recorded in the *Astronomia nova*.

Although the foregoing scheme works well enough for the longitudes, we know from Kepler's own later work that it fails for the distances. If we wish to approximate a Keplerian ellipse with a circle, the circle must be centered between the two foci of the ellipse. In the scheme above, the circle is intentionally not centered between the Sun and equant (which in a fashion correspond to the two foci of the ellipse), and hence the distances from the Sun are wrong. As Kepler was about to notice, predicted Martian latitudes are particularly sensitive to these distances. In fact, immediately after his interim result from the acronychal longitudes, Kepler turned to the latitudes and quickly recognized the discrepancy. "Magnum momentum est in hac ultima operatione," he writes on page 229.[15] This is one of those few exciting places in the notes where we see Kepler gaining an important insight into the Mars problem. Here, perhaps for the first time, the harsh conflict between what he was later to call the "vicarious hypothesis" and physical reality impressed itself into Kepler's consciousness. (It is interesting to note that Kepler became aware of the celebrated 8' discrepancy only at a much more sophisticated point in his researches.)

In spite of the evident conflict with the true orbit, Kepler had now discovered a more accurate scheme for predicting Martian longitudes than anyone before him. Why, then, did 40 more trials still lie ahead?

To understand this, we must recognize that Tycho Brahe had no operational way of knowing precisely when Mars was at opposition. Tycho's raw observations only approximated the opposition places, and Kepler was obliged to correct and interpolate them to obtain the acronychal positions he required. The time of opposition depended in part on the details of the solar theory, especially in Kepler's case since he had made the crucial decision to use the true Sun rather than the mean Sun.

Therefore, at this point, Kepler turned back to the triangulation, to the solar theory, and to a still more accurate reduction of the original observations. For example, to obtain the opposition position for 8 June 1591, he uses a dozen observations between 13 May and 16 July, a laborious exercise never even hinted at in the *Astronomia nova*.[16] The third and ultimate solution recorded in the extant portion of the notebook requires five outer iterations, two wasted because of a mistake in copying the 1593 position.[17]

Still unknown to Kepler was a further error in the 1587 position, which vitiated his result. Thus, after a cumulative total of 47 trials, Kepler gave up in disgust, not even checking the fit for the other opposition dates. "*N.B.*," he writes, "This sublety was completely in vain because the hypothesis itself can deviate more than we err if we would place the aphelion at 29°."[18]

At this point, April 1601, his work was interrupted by another trip to Styria. When Kepler came back to Prague in September, Tycho was interested in the solar eccentricity, and the notes show that Kepler turned his attention back to the triangulation for the Earth's orbit. He did not return to a direct assault on the vicarious hypothesis in the remaining 200 pages of his notebook, which breaks off probably late in 1601.[19] Nevertheless, he eventually came back to the vicarious orbit, perhaps sometime early in 1602,[20] and he added a 14-page supplement after page 324 of the notebook;[21] he has also added a marginal "Minime" ("By no means") to his previous disparaging remark. In the supplement, he carries out the comparison with the eight acronychal positions, noting at the same time the correction to the 1587 opposition.

In the supplement also appears a remarkable but typically Keplerian statement:[22]

> I believe indeed that fate ruled me here when in the year 1601 in the month of April I reworked everything with great care and repeated it a hundred times. However, I could not see the simplest error in subtraction until the places around the circle were sought by the new technique and were managed by the three unanimously against one; thus greatest necessity forced me to search for the fault.

If I understand the context correctly, Kepler is saying that, if the result had come out without numerical error, he might not have pressed on with his investigation. Because of the error, he searched on in frustration and eventually found the noncircularity of the orbit.

Unfortunately, the manuscript material does not cover the remaining 23 or more trials for the vicarious hypothesis. We know from a letter to Maestlin[23] that, by March of 1605, before he had found the ellipse, he had written the major part of the *Astronomia nova* and had the final parameters for the vicarious orbit. The Leningrad Volume XIV contains seven leaves relating to Kepler's Martian ephemeris of 1604, calculated in 1603. On Volume XIV, folio 375 the parameter $3'39''$ appears, which is the constant to be added to Tycho's longitudes that was determined in the aforementioned third solution of the vicarious orbit. (In the *Astronomia nova*, this constant[24] is $3'55''$). Apparently, Kepler had not yet recalculated the parameters late in 1603.

Since it should have taken Kepler about a dozen trials to get a complete solution of the vicarious orbit, and since about two dozen remained to reach the total of 70, we can suppose that he made two additional solutions in 1604.[25] Unless further related manuscripts are found, we can only guess that the two solutions resulted from further improvement in processing the initial observations.

Thus we see that in the vicarious orbit solution Kepler worked always with the same four oppositions, but the results were repeatedly tested against additional oppositions. In the course of five years' work, the reduction of the basic data was continually improved. His 70 iterations were spent (probably) in five separate solutions differing only in the values chosen for the initial times and angles. Kepler's first attempt came in the spring of 1600, during his inaugural visit to Tycho; the second presumably occurred in the fall of 1600; the third rather extended attempt began in the spring of 1601 but was interrupted and not actually completed until the winter of 1602, after Tycho's death. These three calculations are documented by the manuscripts, two additional solutions in 1604 are conjectural, but of course the final stages are printed in the *Astronomia nova* itself.

In conclusion, we can say that Kepler's "method" of treating redundant observations was largely persistence, coupled with a growing appreciation of the hazards inherent in observational data. In the manuscripts, we see the evidence of an extremely perceptive but ambitious young man who was already drafting his *Commentaries on Mars* in 1602; fortunately, divine fate prevented him from finishing his book until it was truly the *New Astronomy*.

Notes and References

[1] Undoubtedly Ptolemy had a great number of observations to choose from; the remarkable geometrical configurations that he displays make it unlikely that the observations recorded in the *Almagest* were the only ones he had available, but the rest have now been lost. See, for example, Owen Gingerich, "The Mercury Theory from Antiquity to Kepler." *Actes du XIIᵉ Congrès International d'Histoire des Sciences*, vol. 3A (Paris, 1971), pp. 57–64. [reprinted as selection 23 in this anthology]. For Copernicus there is documentary evidence of additional observations available to him, but whether he actually used them to test the accuracy of his theory or only selected the most suitable minimum number is unknown.

[2] A discussion of the role of observational error in the later chapters has been given by Curtis A. Wilson, "Kepler's Derivation of the Elliptical Path," *Isis*, vol. 59 (1968), pp. 5–25.

[3] The figure is reproduced from *Johannes Kepler Gesammelte Werke (JKGW)*, ed. by Max Caspar, vol. 3 (Munich, 1937), p. 153; see also R. Small, *An Account of the Astronomical Discoveries of Kepler* (1804) (reprinted Madison, Wisconsin, 1963).

[4] *JKGW*, vol. 3, p. 156.

[5] Owen Gingerich, "The Computer Versus Kepler," *American Scientist*, vol. 52 (1964), pp. 218–26 [reprinted as selection 21 in this anthology].

[6] I am indebted to Dr. P. G. Kulikovsky and Dr. V. L. Chenakal for providing me with an excellent microfilm of Volume XIV of the Leningrad Kepler material. In the *JKGW* this volume is always called "Pulkovo XIV" after its former location. I have partly adopted this now-anachronistic style in these notes, referring to the folio numbers within the volume by the prefix P because these numbers were assigned when the volume was at the Pulkovo Observatory. Assistance in these Keplerian studies and translations has been generously provided in grants 4074 and 4286 from the Penrose Fund of the American Philosophical Society.

[7] The correspondence between Kepler's own page numbers and the folio numbers of Leningrad XIV are: K199 = P172; K474 = P326v. [The date April 1601 is on P243v. Kepler's notebook begins on K3 = P68 and ends on K538 = P360v. Miss Martha List has pointed out an additional date to me, 23 Nov. 1601, on K381 = P276.]

[8] On page 63 (= P98) of the manuscript, Kepler does refer to " ♂ ovalem figuram," but this is a figure that bulges out rather than in and is not motivated by observational considerations. Perhaps it is related to the Copernican orbit, which Kepler knew bulged out; in any event, it may have provided a seminal idea, just like the Mercury figures in Peurbach's *Theoricae novae planetarum*, which Kepler cites in *JKGW*, vol. 3, p. 296.

[9] K154 = P149v.

[10] Indeed, for 1597, two observations from the same time differ by 31'; the reason, Kepler says in the *Astronomia nova*, is that Tycho had already left Hven, abandoning all the instruments except a radius. "Oh that he had remained there a little longer!" exclaims Kepler in *JKGW*, vol. 3, p. 119.

[11] The iterations are summarized on page 329 (P250) of the manuscript notebook as follows:

ln	29. 0.	Leo	184.47. 9	fol 163				
ln	27.50.	Leo	210.26.53	fol 164				
ln	30. 0.	Leo	219.58.34	fol 165				
ln	28.50.	Leo	183.31.10	fol 165				
ln	28.47.	Leo	182.32.34	fol 165				
ln	28.30.	Leo	180.12.27	fol 165				
ln	28.28.40	Leo	180.12. 5	fol 167	Diff 50'52"	fol 169		
ln	28.27.20	Leo	180. 0.10	fol 168	*CAE*	101.3.23	*BAE*	102.54.52

Folio 171. To be added to Tycho's longitude: 7.44".

[12] *JKGW*, vol. 14, pp. 128–36 (letter no. 168).

[13] The results of using three observations at a time are summarized on notebook page 210 (P177v):

Oppositions			Apogee	Eccentricity
1587, 1591, 1593			27° 8 Leo	17500
1587, 1591,		1595	20 52 Leo	28321
1587,	1593,	1595	4 14 Virgo	25541
	1591, 1593,	1595	7 2 Libra	86352

Clearly the 1595 observation is at fault. When this is corrected, the results, shown on page 214 (P179v), are:

1587, 1591, 1593		27° 8 Leo	17500
1587, 1591,	1595	26 36 Leo	18092
1587,	1593, 1595	27 52 Leo	18884
1591, 1593, 1595		25 26 Leo	18949

[14] Notebook pages 221–23 = P183–P184.

[15] K229 = P187.

[16] Notebook pages 270–72 = P213v–P214v.

[17] In the summary on pages 329–31 (P250–P251) the extent of these revisions in the times and angles may be seen and compared with the final values adopted in the book itself:

	1587 Mar 6			1591 Jun 8	
Solution 1	2^h25^m	25°30'10"	Here I erred	7^h35^m	26°44'11"
Solution 2	8 4	25 44 31	Here the book erred	7 31	26 43 23
Solution 3	9 16	25 52	*	7 43	26 43
Astr. nova	7 23	25 43		7 43	26 43

*Added later: "Here I erred. . ."

	1593 Aug 25			1595 Oct 30/31	
Solution 1	16^h52^m	12°15'55"		23^h17^m	17°27'24" And here I erred
Solution 2	16 52	12 15 57		1 0	17 31 46
Solution 3	16 52	12 12	Here I erred by 4'	0 34	17 31 30
Solution 3a	16 52	12 16			
Astr. nova	17 27	12 16		0 39	17 31 40

[18] "Haec subtilitas tota fuit frustranea cum hypotheses ipsa plus aberrare possit quam erramus si Aphelium ponamus in 29°. . ." The quotation follows the summary on page 331 (P251).

[19] In a letter to Magini from Graz dated 1 June 1601 (*JKGW*, vol. 14, pp. 172–84, no. 190) Kepler says he cannot undertake certain calculations because he doesn't have his books with him; also he expects to stay at most three more weeks. The letter is primarily concerned with the eccentricity of Mars, including the use of the four acronychal observations. His visit to Graz dragged on until the end of August, and probably he added nothing to his notebook during these months. In a letter to Longomontanus written at the beginning of 1605 (*JKGW*, vol. 14, no. 323. pp. 134–43) Kepler reviews the chronology of his Mars work, and says that from September on he worked on the solar theory, which agrees with the notebook. According to this letter, he came upon the oval orbit early in 1602, but nothing of this is found in the notebook, nor is the term "vicarious hypothesis" used. A letter to Maestlin on 10/20 December 1601 (*JKGW*, vol. 14, pp. 202–8, no. 203) mentions bisecting the solar eccentricity, but not the vicarious hypothesis. Hence the end of the notebook seems to come late in 1601 or at the beginning of 1602. The date 29 March 1602 written in the margin of page 474 (P326v) seems to have been added later.

[20] The supplement pages inserted in the notebook explicitly mention the vicarious and physical hypotheses (P246), so they must come later than the December 1601 letter to Maestlin (cf. supra), but before 1 October 1602 when Kepler sent Fabricius (*JKGW*, vol. 14, 263–280, no. 226) the same eccentricities, $AC = 18557$, $AB = 11613$, that appear on page 10 of the supplement (P245v). On the following page is an explicit reference to "Caput 27," so at this early date Kepler must have been well advanced in planning his *Commentaries on Mars*; the early arrangement of chapters is found at the beginning of this Volume XIV and has been transcribed by Caspar in *JKGW*, vol. 3, pp. 457–60. "Ex vicaria" and "ex physica" are explicitly mentioned only from marginal notes added to the table of contents later, so the table of contents itself must date from 1601 and the additions from 1602.

[21] K323 = P240.

[22] Notebook supplement page 6 (P243v): "Credo equidem fatum dominari in hisce. Cum anno 1601 mense Aprili summa diligentia omnia reviderem et ab eo tempore centies repeterem, errorem tamen subtractionis levissimum videre non potui. Donec novo artificio circumstantia loca conquisita et tria unanimiter contra unam testata sunt; itaque summa necessitas me impulit ad quaerendum vitium." The passage is somewhat puzzling because the scheme of three against one was introduced much earlier in the notes and is not used in this particular section.

[23] *JKGW*, vol. 15, pp. 170–6 (letter no. 335, 5 March 1605).

[24] *JKGW*, vol. 3, p. 166.

[25] In Codex 10686[2] of the Osterreichische Nationalbibliothek in Vienna there appears a manuscript ephemeris of Mars for 1605, presumably calculated in 1604. On folio 2 there appears the remark "medio motui adde 3'55" ex sententia Kepleri," presumably referring to the ephemeris that follows. If the statement is contemporary with the ephemeris, then the final vicarious orbit solutions (which adopt the correction 3'55") were made in 1604; this seems entirely likely.

Mercury Theory from
Antiquity to Kepler

T he great achievement of Copernicus was the recognition that the Earth was but one of the several planets revolving about the Sun. Actually, because he was still so enmeshed in the technical structure of ancient astronomy, his planets revolved about the center of the Earth's orbit rather than about the Sun. For this reason we can justly call the Copernican system heliostatic rather than heliocentric. Nowhere is this shown more vividly than in his theory of Mercury, which retains such Ptolemaic vestiges as an epicyclic structure operating with the Earth-related period of one year. In order to explain the reasons for this peculiarity, I shall briefly survey the theory of Mercury from antiquity until the time of Kepler.

Figure 1 shows the heliocentric orbit of the Earth and a circular orbit of Mercury appropriately displaced along its line of apsides. The figure indicates six of the basic observations that Ptolemy used to establish the orbit of Mercury; the angles of elongation are labeled according to the *Almagest*. Apparently, Ptolemy had many observations to choose from, since he was able to find both a morning and an evening elongation for the same date in February (although of course on two different years) and also for another date in June. A very curious situation is immediately apparent from the figure—namely, that the arrangement is completely symmetrical between the February

Selection 23 reprinted from *Actes du XIIe Congrès International d'Histoire des Sciences, Paris, 1968*, vol. 3A (1971), pp. 57–64.

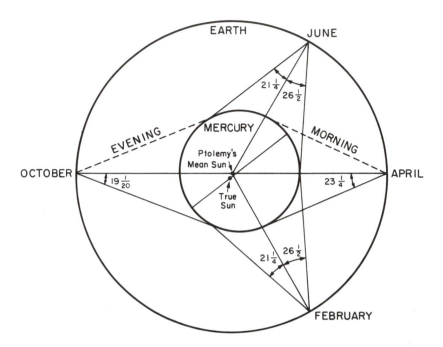

FIGURE 1. *Six of Ptolemy's Mercury observations are labeled according to the* Almagest *on modern heliocentric orbits; some of the actual angles differ by a degree from Ptolemy's values. Geometric circumstances prevented Ptolemy from obtaining evening observations in October and morning observations in April.*

and June dates. If the data are correct, then the line of apsides falls exactly along the horizontal line. A few of these difficult Mercury observations were quite precise, but others were in error by several degrees. The average error for Ptolemy's Mercury observations is a little over 1°; as luck would have it, he erred by about 30° in establishing the line of apsides.

To check this orientation, Ptolemy chose a pair of observations corresponding to these supposed perigee and apogee directions. At this point, nature conspired against him. At sunset in April, the ecliptic rose almost perpendicularly from Ptolemy's western horizon. At elongation, Mercury could easily be seen in the darkening sky before it sank too low toward the horizon. But at sunrise the ecliptic cuts an oblique line 23°.5 *below* the equator. Although its elongation is the same angular distance from the Sun, the sky invariably brightened before Mercury rose far enough above the horizon haze to be sighted. A similar situation occurs in the evening twilight six months later.

Ptolemy had to assume that the angles he could not measure were symmetrical and equal to those observations he could make. Hence he concluded that, in April, the orbit of Mercury subtended at an angle of $46°.5$, whereas both in February and in June the angle was larger, $47°.5$. Thus the orbit of Mercury apparently had one apogee point in October, but two perigee points in February and June. The fact that Mercury's orbit is actually elliptical, rather than circular, is insufficient to make Ptolemy's conclusion correct. Yet, strangely enough, Delambre is practically the only commentator to mention that Ptolemy's conclusion was wrong.[1] Notice also, by the way, that an error of 5° in Mercury's heliocentric longitude will make rather little difference in the observed elongation, but it would have considerable effect at conjunction. Since the model is necessarily based on elongation observations, it will have large errors at inferior conjunction, but these would in general be unobservable.

The mechanism adopted by Ptolemy for Mercury is shown in Figure 2. In keeping with the other planets, Mercury rides on a circular epicycle centered on a larger deferent circle. Mercury travels around the epicycle in 116 days, its synodic period; and the epicycle goes around the deferent once a year, thus keeping up with the Sun. In order for Mercury's epicycle to subtend a large angle in June and February and the smallest angle in October, Ptolemy adopted an epicyclet mechanism in the center of the orbit—here doubled in size for clarity; this small crank simply pushes the epicycle in and out during the course of a year. Various commentators[2] have graphed the oval-shaped effective deferent produced in this manner, and Hartner[3] has shown that it is almost indistinguishable from an ellipse.

How successful was the Ptolemaic theory for Mercury? One way to find out is to compare modern values of planetary longitudes with the predictions of the *Alfonsine Tables*. Although the *Alfonsine Tables* were compiled late in the thirteenth century, both their parameters and theory for Mercury are purely Ptolemaic. The Alfonsine precession was sufficiently accurate that Ptolemy's error in placing the line of apsides did not get much worse by the year 1300. My comparison between the modern positions and Ptolemaic predictions was rendered particularly easy by the existence of the modern recomputation of planetary positions carried out by Bryant Tuckerman of IBM[4] and by the fact that I had computed, in association with Emmanuel Poulle,[5] a daily Alfonsine ephemeris from 1300 to 1600.

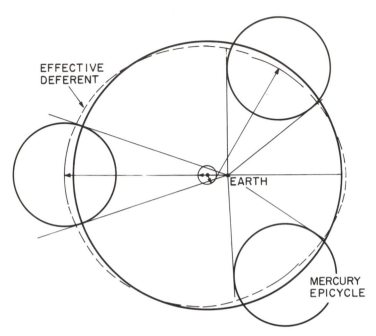

FIGURE 2. *The Ptolemaic mechanism for Mercury: the central epicyclet has been enlarged for clarity.*

The graph in Figure 3 shows the errors in longitude for Mercury for a time span a little longer than a year. Near the bottom, intervals corresponding to the synodic period of Mercury have been indicated; each vertical line represents the moment when Mercury is at inferior conjunction with the Sun, where the faults of the theory are greatly magnified. At the top of the graph, intervals corresponding to the 88-day sidereal period of Mercury have been marked—that is, intervals between successive perigee points. Since the time for three inferior conjunctions nearly equals the time for four perigee passages, the pattern of errors repeats over and over again with the period of 348 days. Notice the great stability of the error pattern over a 255-year time span.

We are now ready to ask how Copernicus fared with his theory of Mercury. In the transformation of the geocentric Ptolemaic system into a heliostatic one, it is patently absurd to retain a small epicylet that expands and contracts the orbit of Mercury with a period of one earth-year. Such an anomaly, crying out for correction, could scarcely

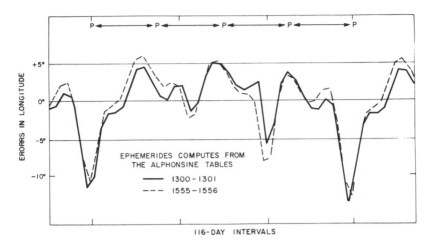

FIGURE 3. *Errors in the longitude of Mercury as predicted by the Ptolemaic* Alfonsine Tables *during the Middle Ages.*

have gone unnoticed by Copernicus. Yet he retains all this Ptolemaic awkwardness. Clearly, Copernicus knew what observations he would have liked to have:[6]

> The ancients had the advantage of a clearer sky; the Nile—so they say—does not exhale such misty vapors as those we get from the Vistula. Nature has denied us that advantage since we inhabit a more severe region where the rarity of fair weather coupled with the great obliquity of the sphere permits us to observe Mercury less frequently, so that it cannot be seen at the maximum elongation from the Sun when rising in Aries and Pisces or when setting in Virgo and Libra; nor can it be seen in Cancer or Gemini (on account of the summer twilight) . . . . The planet has tortured us with its many riddles and with the painstaking labor involved as we explored its wanderings.

What observations of Mercury did Copernicus actually have? Apparently, Copernicus had available an as yet unpublished manuscript by Johann Schöner,[7] giving scattered observations by Bernhard Walther and himself. Most of Walther's thirty-some sightings of Mercury were so vague as to be entirely useless to Copernicus. Only a cluster of six observations, made at the end of August and in early September 1491, contained adequate numerical precision, and Coper-

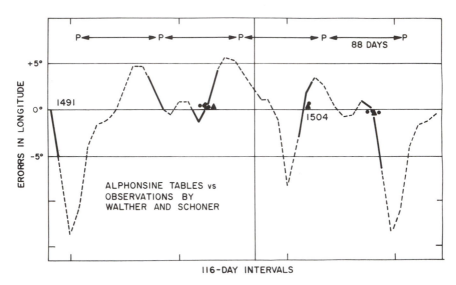

FIGURE 4. *Copernicus used the three Mercury observations designated by triangles, one from Bernhard Walther (left) and two from Johann Schöner (right).*

nicus selected one of these. Another six observations were obtained in 1504 by Schöner himself, and from these Copernicus was able to extract both a morning and an evening observation. Figure 4 shows the errors in the predicted positions from *Alfonsine Tables* in these same years; a solid line indicates when Mercury was actually observable. Note that the largest errors, approaching 15° in longitude, occur at inferior conjunction when Mercury is invisible. The observations of Walther and Schöner are rather accurate, but with few exceptions they fall at times when the Ptolemaic theory is also rather accurate. However, in the observation selected by Copernicus from Walther, the discrepancy amounts to 3°, which had led Walther to exclaim "Behold the great difference from the table."

This handful of data was an insufficient observational base to allow a radical geometric departure for a new Copernican Mercury theory. It must have been with considerable reluctance that Copernicus settled instead for a timid geometrical transformation of the Ptolemaic Mercury model. Thus Mercury continued as an anomaly among the planets, with its own peculiar and complex mechanism.

Nevertheless, the *Prutenic Tables*, which are Copernican in their parameters and mechanism,[8] allowed Stadius to compute an ephemeris

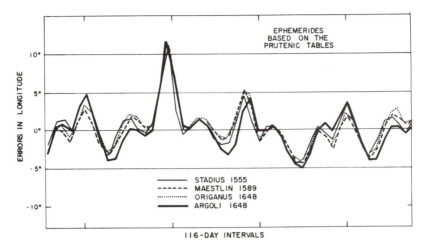

FIGURE 5. *Errors in the longitude of Mercury as predicted by four astronomers who based their ephemerides on the Copernican* Prutenic Tables.

whose errors were noticeably diminished from those of the *Alfonsine Tables*. Not only is the maximum error at the conjunctions reduced, but the predictions are especially improved at the elongations during the weeks when Mercury was actually observable. The improvement results from the fact that the three new observations of Mercury provided Copernicus with a greatly extended time base and allowed him to determine with considerably better accuracy the mean motion of this planet. Copernicus reset the clock, so to say, without repairing its mechanism. Figure 5 graphs the errors in four ephemerides based on *Prutenic Tables*. The patterns of error in the ephemerides of Stadius,[9] Maestlin,[10] Origanus,[11] and Argoli[12] are remarkably similar over a century.[13]

The real breakthrough in establishing a successful Mercury model came only partly through superior observations. Tycho observed this planet about 85 times, and from these precious observations Kepler was able to establish the line of apsides to within 2° (see Figure 6). Nevertheless, it was his belief in the theory of elliptical orbits that carried the predictions accurately through the periods of invisibility.[14] Furthermore, Kepler was able to predict a transit of Mercury over the disk of the Sun, during the hitherto unobservable inferior conjunction. Kepler did not live to see his prediction fulfilled, but the transit of

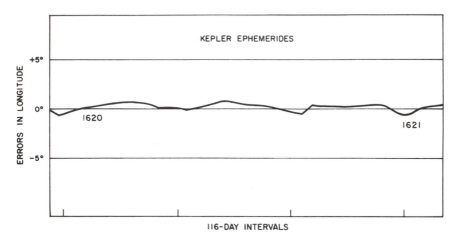

FIGURE 6. *Errors in the longitude of Mercury were greatly reduced by Kepler, who based the predictions on his* Rudolphine Tables.

Mercury was actually witnessed in Paris on 7 November 1631, by Pierre Gassendi. In a letter to Schickhard, Gassendi wrote:[15]

> But Apollo, acquainted with [Mercury's] knavish tricks from his infancy, would not allow him to pass altogether unnoticed. To be brief, I have been more fortunate than those hunters after Mercury who have sought the cunning god in the sun. I found him out, and saw him, where no one else had hitherto seen him.

Riccioli[16] relates that Ptolemy, Copernicus, and Longomontanus each erred by about 5° in predicting this event, whereas Kepler missed by less than 10 minutes of arc. The evidence of this *experimentum crucis* was overwhelming. Both geocentric and makeshift heliostatic models were doomed. In this almost forgotten way, the heliocentric solar system was convincingly established.

Notes and References

[1] J. B. J. Delambre, *Histoire de l'astronomie ancienne*, vol. 2 (Paris, 1817), p. 321.

[2] E.g., G. Peurbach, *Novae theoricae planetarum* (Nuremberg, 1472); D. J. Price, *The Equatorie of the Planetis* (Cambridge, 1954), p. 102. A sixteenth-century Castilian manuscript, Escorial V. II. 9, shows the effective deferent as a rounded lozenge.

[3] W. Hartner, "The Mercury Horoscope of Marcantonio Michiel of Venice," in *Vistas in Astronomy*, vol. 1 (London, 1955), pp. 84–138.

[4] B. Tuckerman, *Planetary, Lunar and Solar Positions. A.D. 2 to A.D. 1649 at Five-Day and Ten-Day Intervals*, vol. 59 of *Memoirs of the American Philosophical Society* (Philadelphia, 1964).

[5] E. Poulle and O. Gingerich, "Les positions des planètes au moyen âge: application du calcul electronique aux tables alphonsines," *Académie des inscriptions et belles-lettres comptes rendus des séances* (1967), pp. 531–48.

[6] N. Copernicus, *De revolutionibus orbium coelestium* (Nuremberg, 1543), Lib. V, Chap. 25, fol. 164. I would like to thank Miss Ann Wegner for her assistance with the translations used for this paper, and Miss Barbara Welther and Mr. Kenneth Manning for plotting the various error graphs.

[7] J. Schöner, *Scripta clarissimi mathematici M. Joannis Regiomontani . . .* (Nuremberg, 1544).

[8] E. Reinhold, *Prutenicae tabulae coelestium motuum* (Tübingen, 1551).

[9] J. Stadius, *Ephemerides Novae . . . ab Anno 1554. usque ad Annum 1600* (Cologne, 1570).

[10] M. Maestlinus, *Ephemerides novae ex tabulis Prutenicis ab anno 1577 ad annum 1590 supputatae* (Tübingen, 1580).

[11] D. Origanus, *Ephemerides annorum posteriorum 30 incipientum ab anno Christi 1625 et definintium in annum 1654* (Frankfurt an der Oder, 1609).

[12] A. Argoli, *Ephemerides exactissimae coelestium motuum ad longitudinem Almae urbis, et Tychonis Brahe hypotheses, ac deductas è Coelo accuratè observationes* (Lyon, 1659). For Mercury, at least, Argoli was still using the *Prutenic Tables*, as the error pattern reveals.

[13] Surprisingly, in the same Stadius ephemeris the pattern of errors beginning in 1596 is suspiciously like those in the *Alfonsine Tables*. Astronomers like Tycho Brahe and Johannes Antonius Magini assumed that Stadius was using the *Prutenic Tables*, although they complained that he did not know how to compute correctly from them. Apparently in 1596 Stadius simply switched tables.

[14] In addition, Kepler's use of the true Sun instead of the mean Sun as the center of coordinates (that is, a heliocentric rather than heliostatic model) was indispensable for a correct latitude theory. This played a key role, although of lesser importance at the nodes where any transit must take place.

[15] R. Grant, *History of Physical Astronomy* (London, 1852), p. 415.

[16] J. B. Riccioli, *Astronomicae reformatae* (Bologna, 1665), p. 348.

Kepler, Galilei, and the Harmony of the World

I n William Shakespeare's *Merchant of Venice*, Lorenzo, gazing on the star-studded, moonlit sky, exclaims:[1]

> Sit, Jessica. Look how the floor of heaven
> Is thick inlaid with patines of bright gold;
> There's not the smallest orb which thou behold'st
> But in his motion like an angel sings.

The notion of heavenly harmonies is an ancient one, but nowhere do we find it more deeply and continually expressed than in the work of Shakespeare's contemporary, Johannes Kepler.

In the *Epitome of Copernican Astronomy* of 1621, which is Kepler's longest and most mature explanation of his own formulation of the heliocentric cosmology, the astronomer poses and answers a most astonishing series of questions. "What is the reason for the size of the Sun?"[2] The Sun is first in archetypal order, if not in temporal, responds Kepler; Moses makes light the work of the first day of creation, by which we can understand the solar body. But, if it was created first, then the solar body has no ratio to other bodies; had it been created twice as great, then the whole world and man in it would have had to

---

Selection 24 reprinted from *Music and Science in the Age of Galileo*, ed. by V. Coelho (Dordrecht, 1992), pp. 45–63.

be twice as great. In other words, Kepler has grasped the subtle rela-tivistic argument that, for a single isolated body, size has no meaning.

But his catechism continues as he asks about the size of the Earth in ratio to the Sun; this, he declares, depends on our vision, the fact that the Sun appears as half a degree in the sky, or 1/720 part of a circle. "What do you think is the reason for this number?"[3] Kepler replies that we must first seek an archetypal cause; he would prefer this to be geometrical, for example, based on a 720-sided polygon. If a geomet-rical method existed for inscribing a 45-sided figure, then its sides could be repeatedly bisected to form a 90-, 180-, 360-, and finally a 720-sided polygon, but, alas, there is no geometrical method for in-scribing a 45-sided polygon within a circle. Hence we must look else-where, to musical harmonies. And 720 turns out to be, in Kepler's opinion, the smallest number to encompass the *systema diapason duplex*—that is, the ratios of the harmonies in both the major and minor scales.

To see what Kepler means by this very odd and yet quite charac-teristic question and its extraordinary answer, let us turn to an account of his life story and the remarkable intertwining of astronomy and celestial harmony.

Kepler was born in 1571 in Weil der Stadt, a small village west of Stuttgart.[4] (He was therefore seven years younger than Galileo, and, since he died earlier, the Italian astronomer's life completely encom-passed his own years.) Sent to the local German school, he proved bright enough to be transferred to the Latin school, and subsequently he won a scholarship to the nearby Tübingen University. Since the university housing was overcrowded, he spent his undergraduate days at a nearby preparatory school and took his baccalaureate by exami-nation. Having moved to Tübingen for a Master's degree, Kepler met the astronomer Michael Maestlin, who openly taught about the helio-centric Copernican system. Now since the publication of Copernicus's *De revolutionibus* five decades earlier, its radical cosmology had been considered an interesting, but fictitious, mathematical scheme that did not speak to the actual physical, Earth-centered world.[5] Kepler felt otherwise, for he believed in a real heliocentric universe. A few years later, as a 24-year-old, he wrote biographically:[6]

> When I was studying under the distinguished Master
> Michael Maestlin at Tübingen six years ago, seeing the many
> disadvantages of the commonly accepted theory of the uni-
> verse, I became so delighted with Copernicus, whom Mae-

stlin often mentioned in his lectures, that I not only defended his opinions in the debates of the physics candidates,
but even wrote a thorough disputation about the first motion, maintaining that it happens because of the earth's rotation. Thus I have gradually collected, partly through hearing Maestlin and partly by my own efforts, the advantages
that Copernicus has mathematically over Ptolemy. At last in
the year 1595 I pondered this subject with the whole force of
my mind. And there were three things above all for which I
stubbornly sought the causes as to why it was this way and
not another: the number, the dimensions, and the motions of
the orbs.

After taking his M.A., Kepler had continued in the theology program at Tübingen, but midway through his third and final year he had
been shipped out to the provinces (under some protest) as a high-
school mathematics teacher. There, in Graz in southern Austria, he
pondered the question of the number and spacing of the planets. In the
ancient system of Aristotle and Ptolemy, this had not been seen as a
pressing problem. Astronomers envisioned a system of aethereal
spheres nested together as tightly as their epicyclic mechanisms would
allow, and hence they saw no further need to fret about the spacing of
the planets. They also assumed that God caused the outermost sphere
to rotate every 24 hours, with the motions transmitted down through
the spheres to the planets, including the Sun and Moon.

Because in the Copernican system the outermost stars were fixed
and the planets went faster the closer they were to the center, any
driving force relevant to the Copernican system presumably had to
come from the Sun itself. Furthermore, the spacing of the planets no
longer depended on a tightly nested arrangement, but, on the contrary,
there seemed to be a vast amount of unnecessary, empty space; consequently, the planetary distances appeared to be completely arbitrary.

In order to remedy these lacunae in the heliocentric theory, Kepler
first sought some kind of plane geometry that could account for the
planetary spacings, essentially a divine blueprint. "And then it struck
me," he wrote. "Why have plane figures among three-dimensional
orbits? Behold, reader, the invention and the whole substance of this
little book!"[7]

What Kepler had noticed was that there are five, and only five,
regular solids—that is, polyhedra having identical regular polygons for
each face. A little reflection shows that there can be three solids with
faces made of equilateral triangles (depending on whether three, four,

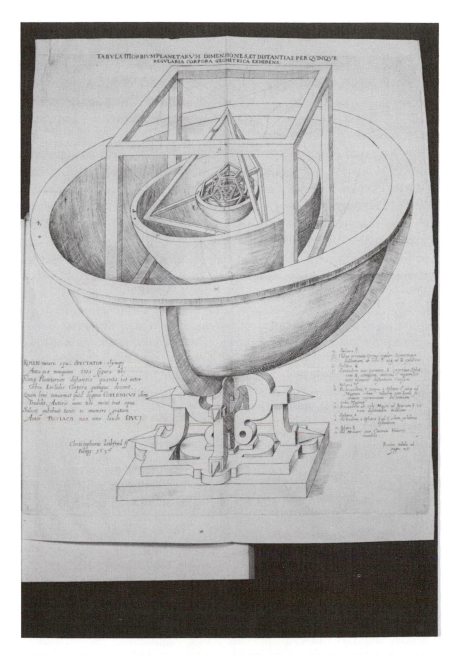

FIGURE 1. *The six planetary spheres with the five nested polyhedral spacers, shown in Kepler's* Mysterium cosmographicum *(1621). The cube separates Saturn and Jupiter, the tetrahedron Jupiter and Mars, and so on. In the first edition (1596), the same figure appears but in a mirror image.*

or five triangles are joined at each vertex), plus a solid with square faces and one with pentagonal faces. What he subsequently discovered was that these could be inscribed and circumscribed as five spacers between the spheres holding the six planetary orbits (see Figure 1). At one stroke, Kepler had answered why there were only six planets and why they were so spaced. With the help of his former teacher, Maestlin, Kepler soon published these ideas in his *Mysterium cosmographicum*, the *Secret of the Cosmos*.[8]

Kepler asked other questions as well, such as why the zodiac divides the sky into twelve parts. And in this connection musical theory reared its head. Kepler began by arguing that harmony, like the archetypal celestial arrangements, is grounded in geometry, not in arithmetic or the mere number juggling of his predecessors.[9] They had discovered the harmonies, but not the underlying cause of the harmonies—they had found the fact in itself or the *di oti* of Aristotle's *Posterior Analytics*, but not the reason why, the *to oti*.[10] Kepler declared:[11]

> I followed the evidence of my ears at a time when, in establishing the number of the divisions, I was still struggling over their causes, and did not do the same as the ancients did. They advanced to a certain point by the judgement of their ears, but soon abandoning their leadership completed the rest of the journey by following erroneous Reason, so to speak dragging their ears astray by force and ordering them to turn deaf. Indeed, I have taken extra pains to ensure that anybody may have a ready opportunity to consult his hearing . . . so that he can be sure that we are struggling over the causes of what rests on the dependable test of the senses, and are not improvised fictions of my own (a charge of which the Pythagoreans stand accused) and intruded in the place of truth.

Kepler considered the monochord, the single string beloved of the musical theorists. The string as a whole represents the fundamental, tuned at G as on the lower course of an early Renaissance lute. Kepler then divided it into successive parts that were harmonious with the whole and with each other, using the integer segments 1:5, 1:4, 1:3, 1:2, 3:5, 2:3, and 1:1. For example, the longer segment of the division 1:5 then gave a length 5/6 of the whole, and so on. (He tacitly rejected 1:6 and 2:5 because a string length of $n/7$ had been considered discordant since antiquity, but later he labored long and hard attempting to rationalize this somewhat arbitrary decision.[12]) The resulting notes are shown in the *Mysterium cosmographicum*:[13]

B♭	B	C	D	E♭	E	G'	G
5/6	4/5	3/4	2/3	5/8	3/5	1/2	1
100	96	90	80	75	72	60	120

Given these fractions, the lowest common multiple is 120. Below the diagram I have indicated both the fraction of the fundamental and integer length of string required for the lower note in each pair, assuming the fundamental length to be 120. The arrangement does not yield a full scale, since it lacks A and F, but it gives an array of consonances including the fifth, D (80), and the octave, G' (60); if the fourth, C (90), is taken as a higher fundamental, then the octave G' is its fifth. These three—G, D, and C—represent the perfect tones. The procedure also yields a minor third, B♭ (100), and a major third, B (96), which are too close to be harmonious taken together, so Kepler considers them as a single imperfect tone. Similarly, the minor sixth, E♭ (75), and the major sixth, E (72), represent a single imperfect tone; they lie a major and a minor third below the octave, respectively. Thus, the three perfect and two imperfect tones come to five, which Kepler saw as exactly the number of Platonic polyhedra. Furthermore, when only the perfect tones are considered, they can be represented with an integer fundamental of 12 (that is, 12, 9, 8, 6), and twelve is the division of the zodiac. Hence (for Kepler) this was the answer as to why the zodiac divides the sky into twelve parts.

If this seems a bit farfetched, we have only to look at the notes Kepler added to the second edition of the *Mysterium*, published in 1618: "It is pleasant to contemplate my first efforts at my discoveries, even though they were wrong," he says; "Behold how I anxiously sought the genuine and archetypal causes of the concordances (which I was studying) like a blind man, as if they were absent. The plane figures are themselves the causes of the concordances, not because they are the surfaces of solid figures. . . . It is not surprising that the fitting of harmonies to the solids is not obvious; for what is not in the bosom of Nature cannot be drawn out."[14] In any event, because Kepler in the

1590s had not yet come to terms with the entire musical scale, he was still far from his special number 720.

When Kepler's book was ready, early in 1597, he sent two complimentary copies to Italy along with a friend, Paul Hamburger, with instructions to give them to anyone appropriate.[15] Hamburger was already on his way back when he realized that he had not carried out this obligation, but inquiry revealed that the books could appropriately be given to a young Pisan professor named Galileo Galilei. The Italian mathematician hastily penned a "thank you" note to send back with the emissary, saying that he had had time to read only the introduction, and that he, too, was a Copernican, albeit secretly.[16] The recipient was obviously unknown to Kepler, for when he received the letter, he communicated with some bemusement to Maestlin that he had just heard from an Italian whose last name was the same as his first.[17] To the Pisan, Kepler wrote concerning his covert Copernican sentiments, encouraging him to "Stand forth, O Galileo!"[18] But nothing more was heard from that quarter for more than a decade, until Galileo's telescopic discoveries burst upon the scene.

Meanwhile, Kepler had also sent a copy of his *Mysterium* to Tycho Brahe, the greatest observational astronomer of the century. Tycho, in Denmark, urged Kepler to come for a visit, but the distance was much too far for the young and newly married Kepler to entertain. "That is why I consider it an act of Divine Providence that Tycho came to Prague," Kepler later wrote.[19] He did not mention as an act of Providence the fact that he, along with the other Protestant teachers in Graz, suddenly became victims of the Counter Reformation and were given until sundown to leave town. Thus fate bound together this pair of astronomers—grand, self-possessed, already famous Tycho Brahe, and Johannes Kepler, the visionary young seeker for the causes of the cosmos, just on the threshold of his own fame.

Kepler came to the imperious Tycho with mixed feelings; he would rather have had a faculty position at Tübingen, but, on the other hand, Tycho had the best observations, though he lacked an architect for building a new astronomical structure.[20] Fortunately, Tycho and his chief assistant, Christian Longomontanus, were concentrating on Mars—the only planet with a sufficiently eccentric orbit and with approaches close enough to Earth to unlock the planetary secrets, as Kepler later appreciated. Kepler bet Longomontanus that he could solve the problem of Mars within a week,[21] but in reality it took most of six years, from 1600 into 1605. Although Kepler soon had an orbit considerably better than any that had gone before, it still yielded errors of 8' of arc (approximately a quarter of the Moon's diameter). Later he wrote, "Divine Providence granted us such a diligent observer in Tycho Brahe that his observations convicted this calculation of an error of 8'; it is only right that we should accept God's gift with a grateful mind. . . . Because these 8' could not be ignored, they alone have led to a total reformation of astronomy."[22]

Kepler did not let the problem go until he had found the elliptical path and the so-called law of areas. By the time he had finished teasing out an orbit half-concealed by a clutter of inevitable observational errors, not only the theory of Mars but Kepler himself had surely been through the refiner's fire. He could handle the tedious task of making the planetary tables so that Emperor Rudolf II could have more accurate horoscopes, but he also knew that his polyhedra, the regular Platonic solids, did not fit between the planetary orbs with as much precision as his art and science could now command.

In 1609 his greatest book was published, the *Astronomia nova* or *Commentary on the Motion of Mars*, one of the few with virtually no mention of musical harmonies. By then Tycho Brahe was long since dead, and Kepler had inherited his position as Imperial Mathematician (but with only a fraction of Tycho's salary).

The following few years marked a period of both excitement and great turmoil in Kepler's life. In the spring of 1610, Galileo published his *Sidereus nuncius*, which carried the news of his astonishing telescopic discoveries. Kepler was not one of the lucky recipients of the new optical tube—Galileo reserved them for men of political influence—although eventually Kepler borrowed one long enough to confirm the novel findings. In a public letter to Galileo, published as the *Conversation with the Sidereal Messenger*, Kepler paraphrased Galileo, saying, "I behold 'great and most marvelous sights proposed to philosophers and astronomers,' including myself, if I am not mistaken; I behold 'all lovers of true philosophy summoned to the commencement of great observations.' "[23] In spite of his lack of a telescope, Kepler announced that he accepted the new discoveries, and in gratitude Galileo responded that "You were the first one, and practically the only one, to have complete faith in my assertions."[24] It was one of the very rare times that the Italian deigned to write directly to his German contemporary.

But, within months of these cosmic events, Kepler's world suddenly collapsed: his wife and several children became seriously ill and died, Prague turned into a war zone as the Thirty Years' War heated up, and his patron Rudolf II was forced to abdicate. It was, Kepler said, "an

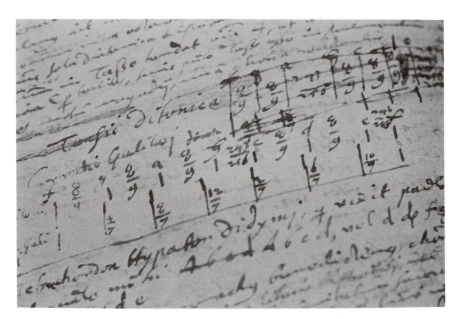

FIGURE 2. *Manuscript page of Kepler's notes on Galilei's* Dialogue on Ancient and Modern Music, *showing tetrachords. St. Petersburg Kepler Archive, Volume XXI, fol. 136v.*

altogether dismal and calamitous year."[25] The lonely astronomer sought a quieter home in Linz, where the authorities pressed him to get on with the *Tabulae Rudolphinae* for computing planetary positions. "Don't sentence me completely to the treadmill of mathematical calculations," he pleaded to a correspondent, "leave me time for philosophical speculations, my sole delight."[26]

In 1617, Kepler was obliged to travel back to Württemberg to look after his Mother's witchcraft trial, which was a showcase for human fears, greed, and stupidity; he took along a book by Galilei for reading on the journey[27]—not by the astronomer Galileo, but by his father, the musician Vincenzo. Kepler had remained interested in the relationship of musical harmony to geometry, and soon after writing the *Mysterium* he had outlined a more ambitious book on celestial harmonies. In his correspondence, Kepler continually returned to his notions about harmony in the cosmos. Vincenzo Galilei's *Dialogo della musica antica e della moderna* surely helped rekindle his interest in musical theory. Even though Galilei rejected the multi-voiced polyphony of his day, which was to play such a fundamental role in Kepler's harmony of the world, nevertheless Galilei's book became Kepler's most quoted musical source. By now Kepler realized that in his earlier works he had not addressed a critical question: the theoretical foundation of the musical scale itself. He had thus far only established certain harmonic intervals. Yet, he wrote, "since the harmonic proportions are infinite, our knowledge, as far as it goes, is still rough, unpolished, unnoticed, and unnamed, and heaped together or, rather, scattered like some mass of rough stones or timber; the next thing is for us to proceed to polish them, to attach names to them, and finally to construct from them the splendid edifice of the harmonic system, that is, the musical scale."[28]

Let us look at how Kepler proceeded to construct the musical scale mathematically, keeping in mind that the construction he defends is not new with him but, rather, is one of several mentioned by Ptolemy and the one discussed and favored by Kepler's Italian contemporary Giuseppe Zarlino. Let us consider the ratios between certain of the harmonies in Kepler's earlier table:

G	$B^\flat$	B	C	D	$E^\flat$	E	G'
120	100	96	90	80	75	72	60

The ratio of the interval D/C is 8/9, whereas C/$B^\flat$ is 9/10 and C/B is 15/16. These are, respectively, the major whole tone, the minor whole tone, and the semitone, and for Kepler they provide the basic units for building the scale.[29] Note that successive ratios of 8/9, 9/10, and 15/16 yield, by multiplication, 3/4, which is the interval of the fourth between G and C. This pattern suggests that an intermediate note (A) between G and $B^\flat$ or B ought to have a ratio of 8/9 with the fundamental G; the resulting diatonic sequence, GABC, is the so-called tetrachord from ancient Greek musical theory. Likewise, the high G can have a ratio of 8/9 with the other intermediate note, F, and the sequence DEFG is also a tetrachord. Kepler noted that the tetrachords could have different patterns of whole and half tones, just as in the major and minor scales,[30] and he prepared a large grid of possible combinations of the different tetrachords.

If the fundamental is held as 120, it is no longer possible to work out these ratios with integers, but we can do it easily with the string length renumbered as 720. Using the following sequence of ratios to generate the scale, we get the string lengths shown on the subsequent line:

	8/9	9/10	15/16	8/9	15/16	9/10	8/9	
720	640	576	540		480	432	405	360
G	A	B	C		D	$E^\flat$	F	G'

This combination of tetrachords does not create a very pleasant scale, as the two semitones or half steps (BC and DE$^\flat$) are too near each other. Hence Kepler makes the following interchange in the first tetrachord to produce a minor scale:

8/9	9/10	15/16	8/9	15/16	9/10	8/9
720	640	600	540 | 480	450	405	360
G	A	B$^\flat$	C | D	E$^\flat$	F	G'

Alternatively the semitone in the upper tetrachord can be rearranged to produce what he called a "major" scale (*cantus durum*)–a scale in the Mixolydian mode:

8/9	9/10	15/16	8/9	15/16	9/10	8/9
720	640	576	540 | 480	432	405	360
G	A	B	C | D	E	F	G'

These particular string lengths correspond to what is called *just intonation*, and Kepler found them both mathematically and musically satisfying. Whether singers unaccompanied by tuned instruments could actually keep these pitches was another matter. Kepler, having read both Zarlino and Galilei, was aware of their acrimonious debate as to whether contemporary *a capella* singing was in just intonation, as the former believed, or a tempered scale, as Vincenzo Galilei correctly argued. Kepler, for reasons of mathematical harmony, preferred the just scale, and he felt that Galilei's attempts at tempering the scale were, from a theoretical viewpoint, "ruinous." Kepler remarked:[31]

> See another very clever tempering of this sort by Vincenzo Galilei, made not in ignorance of the mathematical size of the notes, but with a particular intention. And I indeed recognize its mechanical function, so that in instruments we can enjoy almost the same freedom of tuning as can the human voice. However for theorizing, and even more for investigating the nature of melody, I consider it ruinous; and the effect of it is that the instrument never truly attains the nobility of the human voice.

We have, incidentally, just seen the origin of the magic number 720, the *systema diapason duplex*, that Kepler claimed was the reason for the Sun's apparent size in the sky being 1/720 of the entire ecliptic circle. But Kepler's love for the just intonation went much deeper than this. For years he had puzzled over the precise spacing of the planets, supposing that the five Platonic solids had provided a rough template for their arrangement. However, the fine tuning eluded him until he finally began to work out the harmonic details of their fastest and slowest motions. Kepler claimed that these planetary speeds, properly transposed, fell into a scale of just intonation. For example, the slowest motion in the solar system is Saturn at aphelion (farthest from the Sun), which Kepler placed in correspondence with a low G. The fastest planetary motion—Mercury at perihelion (closest to the Sun)—is $2^7 \times 720/432$ times swifter, which corresponds to an E seven octaves higher. Saturn at perihelion goes $720/576$ times faster than at aphelion, and Jupiter $2 \times 720/576$ times faster, so each would correspond to B, though Jupiter's B would be an octave higher than Saturn's.

It is revealing to trace through some of Kepler's calculations in detail, but it is rather more instructive, and even shocking, to watch his specific application of the planetary speeds to his scale of just intonation. Since he rejected the actual speeds of the planets in favor of the apparent angular speeds as seen from the Sun, all that mattered was the period of each planet and the eccentricity. These data are readily found in his *Epitome of Copernican Astronomy* and are displayed here in Table 1. The mean daily motion follows directly from the period,[32] and speeds at aphelion and perihelion are found by taking the eccentricity into account.[33] Each speed is then scaled, an octave at a time, by repeated halving until a further step would take it below the 1' 46" Kepler had found for Saturn's aphelion. The discrepancies between the exact calculations and Kepler's, shown in the alternate lines in italics,[34] arise primarily from a variety of approximations used by Kepler in calculating the eccentric velocities, except for Mercury, where he has quite unconvincingly argued that the extremes must reflect a slightly diminished mean motion.[35]

Kepler next converted his major and minor scales to speeds, taking the motion of Saturn at aphelion to correspond with low G (and then, similarly, the motion of Saturn at perihelion). It is more revealing, however, to work the other way and to convert the speeds to string lengths in the scale in which 720 is the fundamental, and to list them in descending order, as done in Table 2.

Kepler found more "hits" by scaling from the Earth's aphelion speed rather than Saturn's aphelion. "But who wants to quarrel about 1" in the motion?" he asks.[36] The first pattern gives notes in the major scale, but it has no place for the perihelion of Venus or of the Earth. The second pattern, however, uses both of these and comes out as a

TABLE 1. Planetary Speeds at Aphelion and Perihelion

Planet	Period				(in days)	Mean Daily Motion			Eccentricity
Saturn	29^y	174^d	4^h	58^m =	10759^{d}208	2'	0"	27'''	0.05700
Jupiter	11	317	14	50 =	4332.625	4	59	8	0.04822
Mars	1	321	23	32 =	686.979	31	26	31	0.09263
Earth		365	5	49 =	365.2424	59	8	20	0.01800
Venus		224	17	53 =	224.7451	96	6	32	0.00694
Mercury		87	23	15 =	87.9688	245	32	30	0.21000

Planet	Aphelion		Aphelion Scaled		Perihelion		Perihelion Scaled	
Saturn	1'	48"	1'	48"	2'	15"	2'	15"
	1	*46*	*1*	*46*	*2*	*15*	*2*	*15*
Jupiter	4	32	2	46	5	30	2	15
	4	*30*	*2*	*15*	*5*	*30*	*2*	*45*
Mars	26	20	3	18	38	11	2	23
	26	*14*	*3*	*17*	*38*	*1*	*2*	*23*
Earth	57	4	1	47	61	20	1	55
	57	*3*	*1*	*47*	*61*	*18*	*1*	*55*
Venus	94	47	2	58	97	27	3	3
	94	*50*	*2*	*58*	*97*	*37*	*3*	*3*
Mercury	167	42	2	37	393	26	3	4
	164	*0*	*2*	*34*	*384*	*0*	*3*	*0*

minor scale. There is the lack of the F in either pattern, but "indeed in music F$^\sharp$ often replaces F, as is seen everywhere."[37] (What Kepler was seeing and hearing was of course the modern G-major scale with the F$^\sharp$.)

Kepler illustrates his wonderful conception with the runs of notes shown in Figure 3. To the casual reader, it appears that Kepler has shown that the just intonation lies in the celestial harmonies. But a careful inspection of Table 2 reveals what a shambles the scheme really is. All too many correspondences are approximate, as even Kepler admits. (These are designated with "∼" in the table.) In a section that follows, Kepler gives no fewer than 50 propositions to justify every deviation, and to argue for an intricate set of interlocking harmonies

TABLE 2. String Lengths Corresponding to Planetary Speeds at Aphelion (A) and Perihelion (P).

Just Scale		Saturn Aphelion ("Major Scale")			Saturn Perihelion ("Minor Scale")		
G	360	Earth A	360	G	Jupiter A	360	G
		Mars A	391	F#			
F	405						
		Venus P	421				
		Mercury P	428	~ E	Earth P	423	~ E
E	432	Venus A	433	E			
E♭	450				Earth A	454	~ E♭
					Saturn A	459	
		Jupiter P	468	~ D			
D	480				Mars A	493	~ D
		Mercury A	501	~ C#			
					Venus P	531	~ C
C	540	Mars P	539	C	Mercury P	540	C
		Saturn P			Venus A	547	
B	576	Jupiter A	571	B			
B♭	600				Jupiter P	590	~ B♭
A	640				Mercury A	631	A
		Earth P	671		Mars P	680	
G	720	Saturn A	727	G	Saturn P	720	G

and tonal intervals. But surely *any* intonation could be hammered into such a frame. From anyone else, the carefully crafted excuses and scales would be considered the edifice of a madman.

But to Kepler, this was truly Divine Harmony, a geometrical vision into the mind of God and into the hidden workings of the universe. "Geometry is coeternal with the Mind of God before the creation of things; it is God himself, has supplied God with the models for the creating of his world, and has been directly transferred to man with the image of God," Kepler wrote in Book IV of the *Harmonice*.[38]

How did Kepler's system of geometrical planetary harmonies work in the real cosmos? Naturally, a planet could resonate with only one note at a time, and only rarely one of the potentially harmonious notes at the extremes of its motion. The notes were, of course, silent: "There are no sounds in the heavens, nor is the movement so turbulent that any noise is made by rubbing against the aether."[39] Still, the most wise

HARMONICIS LIB. V. 207

mnia (infinita in potentiâ) permeantes actu : id quod aliter à me non CAP. VI
potuit exprimi, quam per continuam seriem Notarum intermedia-

rum.   Venus ferè manet in unisono non æquans tensionis amplitu-
dine vel minimum ex concinnis intervallis.

FIGURE 3. *According to Kepler, each planet "sings" a range of notes depending on its varying speeds. Mercury, with the most eccentric orbit and highest speed, has the highest and largest range. It does not descend faster than it ascends— the depiction is simply the result of the crowding to allow place for the moon at the right. From Kepler's* Harmonice mundi *(1619).*

Creator could appreciate these majestic concordances. And with seven planets simultaneously singing their silent tones, sometimes in consonance, mostly in dissonance, the celestial harmonies resembled a grand cosmic polyphony. Swept on by the grandeur of his vision, Kepler exclaimed:[40]

> It should no longer seem strange that man, the ape of his Creator, has finally discovered how to sing polyphonically, an art unknown to the ancients. With this symphony of voices man can play through the eternity of time in less than an hour and can taste in small measure the delight of God the Supreme Artist by calling forth that very sweet pleasure of the music that imitates God.

Max Caspar, in his biography *Kepler*, gave an extended and perceptive summary of the *Harmonice*, concluding:[41]

> Certainly for Kepler this book was his mind's favorite child. Those were the thoughts to which he clung during the trials

of his life and which brought light to the darkness that surrounded him. . . . With the accuracy of the researcher, who arranges and calculates observations, is united the power of shaping of an artist, who knows about the image, and the ardor of the seeker for God, who struggles with the angel. So his *Harmonice* appears as a great cosmic vision, woven out of science, poetry, philosophy, theology, mysticism. . . .

For Kepler, the *Harmonice* was not yet finished—there was still more to come. In the course of these harmonic investigations, he discovered that the square of the time (in years) required for a planet to orbit the Sun equaled the cube of its average distance from the Sun (in astronomical units).[42] "If you want the exact time," Kepler candidly remarked, "it was conceived on March 8 of this year, 1618, but unfelicitously submitted to calculation and rejected as false, and recalled only on May 15, when by a new onset it overcame by storm the darkness of my mind with such full agreement between this idea and my labor of 17 years on Brahe's observations that at first I believed I was dreaming and had presupposed my result in the initial assumptions."[43]

This "harmonic law," one of the permanent achievements of Kepler's astronomy, gave him great pleasure, for it neatly linked the planetary distances and their periods—the distances that played such a central role in the nested Platonic solids of the *Mysterium*, and the velocities or periods that figured so prominently in the celestial harmonies of the *Harmonice mundi*. The discovery made Kepler so ecstatic that he immediately added these rhapsodic lines to the Introduction of Book V:[44]

> Now, since the dawn eight months ago, since the broad daylight three months ago, and since a few days ago, when the full Sun illuminated my wonderful speculations, nothing holds me back. I yield freely to the sacred frenzy; I dare frankly to confess that I have stolen the golden vessels of the Egyptians to build a tabernacle for my God far from the bounds of Egypt. If you pardon me, I shall rejoice; if you reproach me, I shall endure. The die is cast, and I am writing the book—to be read either now or by posterity, it matters not. It can wait a century for a reader, as God himself has waited six thousand years for a witness.

Kepler himself gave the harmonic law relatively little emphasis, and it remained for later scientists to single out its importance. Nevertheless, it represents the culmination of a lifelong search and illustrates his imaginative approach to the mysteries of the universe. The harmonic law would prove to be a foundation stone for Isaac Newton's grand gravitational synthesis. Thus Kepler's great cosmic vision of celestial harmony—part fantasy and chimera—had indeed ultimately brought him closer to the eternal architecture of his Creator.

Notes and References

[1] *Merchant of Venice Act,* V, 58–61.

[2] The standard modern edition for the Kepler works and letters is *Johannes Kepler Gesammelte Werke* (Munich, 1937–), a multivolume series not yet fully completed; I shall cite it as *JKGW,* where numbers after the colon refer to the numbered lines in the edition. *Epitome of Copernican Astronomy, JKGW,* vol. 7, p. 277:7ff. This part of the *Epitome* has been translated in *Great Books of the Western World,* vol. 16 (Chicago, 1938), p. 873. This volume will be cited as *GBWW.* Another important source is Michael Dickreiter, *Der Musiktheoretiker Johannes Kepler* (Bern and Munich, 1973).

[3] *Epitome of Copernican Astronomy, JKGW,* vol. 7, 277:36ff; *GBWW,* p. 874.

[4] The facts of Kepler's life are found in the standard biography, Max Caspar's *Kepler,* trans. by Doris Hellman (New York, 1959); a reprint with extensive new annotations by Owen Gingerich and Alain Segonds is forthcoming (1993). See also Owen Gingerich, "Kepler," in *Dictionary of Scientific Biography,* vol. 7 (1973), pp. 289–312.

[5] See, for example, Owen Gingerich, "From Copernicus to Kepler: Heliocentrism as Model and as Reality," *Proceedings of the American Philosophical Society,* vol. 117 (1973), pp. 513–22 [reprinted as selection 16 in this anthology].

[6] Abridged from *Mysterium cosmographicum, JKGW,* vol. 1, p. 9:11ff.; my translation.

[7] *Mysterium cosmographicum, JKGW,* vol. 1, p. 13:5–7.

[8] *Mysterium cosmographicum* (Tübingen, 1596). For an English translation, including the notes added by Kepler in the revised 1618 edition, see A. M. Duncan (trans.), *Mysterium Cosmographicum: Secret of the Universe* (New York, 1981); this English version will be cited as *Secret.*

[9] See D. P. Walker, "Kepler's Celestial Music," *Journal of the Warburg and Courtald Institutes,* vol. 30 (1967), pp. 228–50, reprinted in his *Studies in Musical Science in the Late Renaissance* (London, 1978), pp. 34–62.

[10] For a brief discussion of this point in a related context, see Owen Gingerich and Robert S. Westman, *The Wittich Connection: Priority and Conflict in Late Sixteenth-Century Cosmology,* vol. 78, part 7 of *Transactions of the American Philosophical Society* (1988), pp. 42–43. Kepler himself uses the Greek *di oti* and *to oti* in his *Harmonice mundi, JKGW,* vol. 6, p. 94.

[11] *Harmonice mundi, JKGW*, vol. 6, pp. 119:39–120:9, Book 3, Chapter 2. English translation by E. J. Aiton, A. M. Duncan, and J. V. Field.

[12] Walker, "Kepler's Celestial Music" (ref. 9), describes in considerable detail Kepler's rejection of the division into sevenths.

[13] *Mysterium cosmographicum, JKGW*, vol. 1, p. 40; *Secret*, p. 132.

[14] *Mysterium cosmographicum* (1618), p. 46, note 13 and p. 47, note 18; revised from the English translation of A. M. Duncan (1981), pp. 141, 143.

[15] For further details, see the chapter "Galileo, Kepler, and Their Intermediaries," esp. pp. 123–27 in Stillman Drake, *Galileo Studies* (Ann Arbor, 1970).

[16] Galileo to Kepler, 4 August 1597, *JKGW*, vol. 13, p. 130:15–16.

[17] Kepler to Maestlin, early October 1597, *JKGW*, vol. 13, p. 143:121.

[18] Kepler to Galileo, 13 October 1597, *JKGW*, vol. 13, p. 145:51.

[19] *Astronomia nova* (Prague, 1609), *JKGW*, vol. 3, p. 109:7–8.

[20] Kepler tells himself this in a private memo entitled "Reflections on a stay in Bohemia," now published in the documents volume of the collected works, *JKGW*, vol. 19, p. 37.

[21] Kepler to Longomontanus, early 1605, *JKGW*, vol. 15, no. 323:188–89.

[22] *Astronomia nova* (Prague, 1609), *JKGW*, vol. 3, p. 178:1ff.

[23] Edward Rosen (translator), *Kepler's Conversation with Galileo's Sidereal Messenger* (New York, 1965), p. 11.

[24] Galileo to Kepler, 19 August 1610, *JKGW*, vol. 16, no. 587:1–2.

[25] Kepler to Peter Crüger, 1 March 1615, *JKGW*, vol. 17, no. 710:13.

[26] Kepler to Vinzenz Bianchi, 17 February 1619, *JKGW*, vol. 17, no. 827:249–51

[27] Kepler to Wacker von Wackenfels, early 1618, *JKGW*, vol. 17, no. 783:21ff. Vincenzo Galilei's *Dialogo della musica antica e della moderna* was published in Florence in 1581 (facsimile, New York, 1967) and reprinted there in 1602. On Galileo's father, see D. P. Walker, "Vincenzo Galilei and Zarlino," in his *Studies in Musical Science in the Late Renaissance* (London, 1978), pp. 14–26.

[28] *Harmonice mundi, JKGW*, vol. 6, p. 114:5–11, Book 3, Chapter 2. English translation by E. J. Aiton, A. M. Duncan, and J. V. Field.

[29] The musical notation is nicely discussed by Elliott Carter, Jr. in a long technical footnote in *GBWW*, pp. 1026–28.

[30] *Harmonice mundi, JKGW*, vol. 6, p. 152. The notion of major and minor scales was just in the process of formation, and Kepler's use of "durum" (or hard) and "molle" (or soft) does not fully match the modern use; D. P. Walker (ref. 9) points out that Kepler generally uses "durum" when a scale or chord contains a B and "molle" when it contains a B^b.

[31] *Harmonice mundi, JKGW*, vol. 6, p. 145:6–12, Book 3, Chapter 8. English translation by E. J. Aiton, A. M. Duncan, and J. V. Field.

[32] Kepler gives the periods in terms of Egyptian years of exactly 365 days.

[33] The aphelion and perihelion columns are correctly found by dividing the mean speeds by $(1 + e)^2$ and $(1 - e)^2$, respectively.

[34] *Harmonice mundi, JKGW*, vol. 6, p. 321; *GBWW*, p. 1031.

[35] Kepler diminishes the Mercury eccentricity to 0.1736.

[36] *Harmonice mundi, JKGW,* vol. 6, p. 318:22–23; *GBWW,* p. 1036.

[37] *Harmonice mundi, JKGW,* vol. 6, p. 319:4; *GBWW,* p. 1036.

[38] *Harmonice mundi,* abridged from *JKGW,* vol. 6, p. 323:32–35, Book 4, Chapter 1. English translation by E. J. Aiton, A. M. Duncan, and J. V. Field. This echoes a theme already sounded in his *Conversation with Galileo's Sidereal Messenger,* p. 43, where he wrote, "Geometry is unique and eternal, and it shines in the mind of God. The share of it which has been granted to man is one of the reasons why he is the image of God."

[39] *Harmonice mundi, JKGW,* vol. 6, p. 311:33–34; translation from *GBWW* p. 1030.

[40] *Harmonice mundi, JKGW,* vol. 6, p. 328; *GBWW,* p. 1048; my translation.

[41] Max Caspar, *Kepler,* pp. 288–90, slightly retranslated.

[42] More precisely, he found that the ratio between the periodic times for any two planets is precisely the ratio of the 3/2 power of their mean distances. *Harmonice mundi, JKGW,* vol. 6, p. 302:21–23; *GBWW,* p. 1020.

[43] *Harmonice mundi, JKGW,* vol. 6, p. 302:14–15; *GBWW,* p. 1020.

[44] *Harmonice mundi, JKGW,* vol. 6, 290; *GBWW,* p. 1010; my translation.

Circumventing Newton

L earned heads, therefore, are apt to nod in agreement when the statement, attributed to Einstein, is quoted to the effect that even had Newton or Leibnitz never lived, the world would have had the calculus, but that if Beethoven had not lived, we would never have had the C-Minor Symphony."[1]

Despite the ancient affinity between music and physics, running from Pythagoras and Ptolemy to Kepler and Galileo, the *differences* between the arts and sciences generally strike the casual observer as far more apparent than their similarities. Probably most scientists would agree with the opening quotation and would disagree that there is an accurate parallel between the statements "If Newton had never lived, his theory of universal gravitation would never have been formulated," and "If Beethoven had never lived, his *Ninth Symphony* would never have been written."

How close is the analogy between these two important domains of human creation? Let us further examine the juxtaposition of these statements. The particular phrasing of the inquiry invokes certain subtleties: for example, to say "If Newton had never lived, the *Principia* would never have been written" would no doubt provoke the learned heads to declare we are treading close to tautology. The ideas of the *Principia* might have been expounded by someone, though of course not in Newton's exact words. But similarly we could argue that the general problems of marshaling a large choral force within the framework of an orchestral symphony, say, would have been recognized and

Selection 25 reprinted from *American Journal of Physics,* vol. 46 (1978), pp. 202–6.

"solved" by another composer had Beethoven not lived so long.

Indeed, Newton's ideas are rather easily rephrased; textbook writers have done it so successfully that the *Principia* itself is little read nowadays. But where is the substitute for Beethoven's *Ninth*? Gunther Stent,[2] in arguing from a similar viewpoint, has noted that scientific ideas are more susceptible to paraphrase than a great work of art. To rewrite the ideas of Shakespeare's *Timon* would require a genius equal to the genius of the original creator; it took the writing of *King Lear* to paraphrase successfully the ideas of *Timon* and to supersede *Timon* in the Shakespearean dramatic repertoire.

Such genuine differences between art and science need not obscure our fundamental theme: the nature of imagination and creativity in the scientific enterprise. Norman Campbell's penetrating remarks are well worth remembering: "Beethoven's music did not exist before Beethoven wrote it, and Newton's theory did not exist before he thought of it. Neither resulted from a mere discovery of something that was already there; both were brought into being by the imaginative creation of a great artist."[3]

Most scientists subscribe to the view that the external world and its laws are simply there, and it is the job of the scientist to find them. If Newton had not formulated his laws of motion and theory of gravitation, others would have approached the same problems and would have arrived at equivalent solutions. Of this I have no doubt. Yet this approach may have been much later and in a considerably different form. What is great about Newton's *Principia* is not only the brilliance of its central concept, truly *universal* gravitation, but the astonishing scope and thoroughness with which the idea is systematically explored. The contrast between what others might have done and what Newton did is the contrast between the symphonic themes and the symphony itself.

To show that Newton's achievement represented an individual choice of form and concepts, and that it was not merely the discovery of a set of equations preexisting in nature, requires some demonstration of possible alternative routes. The thrust of this article is to present the outline of such a demonstration as I have used it (in the framework of the Beethoven-Newton question) in my Natural Sciences course at Harvard University. What I propose is not an episode in the history of science, because in fact history had its Newton and not my mythical alternative. Rather, I shall show a conceivable alternative for explaining basic celestial phenomena (taken here as Kepler's laws). This alternative is based on conservation laws and lacks such fundamental Newtonian concepts as force and inertia. By showing Newton's own route as but one of the possible paths, his own formulation becomes more personal, more uniquely his. First, however, I shall outline the standard Newtonian path as it proceeds in my course.

From Newton's Laws to Kepler's Laws

For purposes of the demonstration we take Kepler's three laws (the law of the ellipse, the law of areas, and the harmonic law, that is, $a^3/P^2 = \text{const}$) as a reasonable approximation of motions in the solar system, and we first attempt to show how these follow from Newton's laws. This poses something of a challenge because, although most of our students have had high-school physics, they are not prepared to go beyond rather simple algebra.

Getting Kepler's law of areas is the easiest of the three. This can be done geometrically, using successive impulses directed along a central force, as Newton himself did.[4]

Kepler's harmonic law is readily found for the case of circular orbits, providing that the equal and opposite forces of Newton's third law are invoked to establish the acceleration of the Sun caused by the planet as well as the acceleration of the planet. (It is interesting to note that, although the idea of force is introduced conceptually, in this problem it is a construct that promptly disappears.) We add the acceleration of the Sun and of the planet to get the total acceleration, which then equals the central acceleration, that is,

$$Gm_1/a^2 + Gm_2/a^2 = \text{acceleration} = v^2/a.$$

Here a is the semimajor axis of the planetary ellipse or the radius in the circular case, v the circular velocity, m_1 the planet's mass, m_2 the solar mass, and G the constant of universal gravitation. In a circular case $v = 2\pi a/P$ where P is the period. This leads at once to Newton's revision of Kepler's third law, an equality not limited to circular orbits despite its restricted derivation:

$$a^3/P^2 = (G/4\pi^2)(m_1 + m_2).$$

This equation reveals the approximate nature of Kepler's result and the greater power of the Newtonian formulation. In fact, what is most impressive about Newton's approach to celestial physics is the way he went beyond the previous results and the remarkable range with which he forged ahead in seeking out the ramifications of his laws, for example, in lunar theory.

Obtaining an elliptical orbit from Newton's laws without calculus presents a thornier problem. Fortunately, we have access to computer terminals with graphic displays, and few of our students have difficulty in following the computer code.[5] (The understanding of this numerical technique is greatly facilitated by first calculating by hand, step-by-step, a parabolic trajectory with constant g.) The computer program gives sufficient evidence that Newton's laws do generate elliptical orbits.

From Conservation Laws to Kepler's Laws

As an alternative route to Kepler's three laws, from an entirely different physical structure, we can take the conservation laws as our set of axioms. Although the conservation laws are usually studied as derivatives of Newton's laws, they can be taken as fundamental postulates in their own right. We shall need the conservation of momentum and of angular momentum, and the conservation of energy in the form

$$W = T + V$$

where W is the total energy, T the kinetic energy, and V the potential energy. In the absence of Newton, we must suppose that some other genius has found the form and universal nature of the gravitational potential energy.

The most fundamental difference in outlook between the Newtonian formulation and the conservation laws as axioms is the absence here of the concept of force. The planet, instead of interacting with the distant Sun, somehow gauges its behavior from the field properties of the space in which it moves. (The introduction of this idea provides a stepping-stone to the curved space of general relativity.) Nor does inertia or the naturalness of straight-line motion play any role in this formulation. The nearest substitute for the concept of inertia is the conservation of angular momentum. The infinitesimal case is shown in Figure 1. If the planet moves from a'' to a' in one instant, conservation of angular momentum requires that the planet be found somewhere along the line $abcd$ at the end of the next equal instant. (That this satisfies the law of areas and hence the conservation of angular momentum is seen by completing the triangles $a'b'a$, $a'b'b$, $a'b'c$, etc.; they have a common base $a'b'$ and a common altitude and consequently have equal areas.)

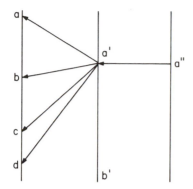

FIGURE 1. *Conservation of angular momentum.*

How can we show that the conservation laws lead back to Kepler's laws? Again, the area law is the easiest, since it is just an alternative way of looking at the conservation of angular momentum.

To find Kepler's third law, we begin by writing out more explicitly the conservation of energy, but ignoring for the moment the kinetic energy of the Sun:

$$W = \tfrac{1}{2}m_1 v^2 - G m_1 m_2 / r. \tag{1}$$

The total energy is generally evaluated in energy conservation problems by taking an extremum where one term or the other vanishes. We can do this here too by imagining a limiting orbit so elongated that at aphelion the kinetic energy becomes arbitrarily small, and so

$$W = - G m_1 m_2 / 2a.$$

We then have the desired energy equation of the orbit for the case where the planet's mass m_1 is negligible compared to the solar mass m_2:

$$v^2 = G m_2 (2/r - 1/a). \tag{2}$$

For circular orbits where $r = a$ we can quickly find

$$a^3 / P^2 = \text{const}$$

but the constant does not contain m_1 unless we include a term $\tfrac{1}{2}m_2 v_2^2$ in the original energy balance (1) and then use conservation of momentum to combine the individual velocity components to get the relative velocity.

Note that by setting $a = \infty$ we obtain the expression for escape velocity. In an astronomically oriented natural sciences class, this leads to such wonderful things as the Schwarzschild radius as well as the density required to close the universe!

Obtaining an elliptical orbit from the conservation laws without calculus presents a serious obstacle, and our solution is schematic only, although once again, as in the Newtonian case, we could in principle use the computer to demonstrate that it works out. In the Newtonian case, we used the gravitational accelerations at each step to calculate the incremental deviations from straight-line motions. Here we require an incremental path that preserves the total energy and angular momentum. The angular momentum requirement tells us the line we must reach (*abcd* in the diagram): each point corresponds to a unique potential energy, and the vectors $a'a$, $a'b$, $a'c$, $a'd$ each correspond to a particular kinetic energy. The computer could be programmed to search for the point with the required energy sum. Hence the trajectory is seen as a series of points satisfying the energy and angular momentum invariants rather than as a continuing deviation from an inertial path. Conceptually, the physics of the orbit is entirely different in the two schemes. In particular, the concept of force does not enter the conservation-law formulation.

That this procedure leads to an elliptical orbit is left as a matter of statement in our natural sciences class. The foregoing presentation determines the entire mathematical level of the course, and this sequence is more mathematical than any other part. Many students who are rather put off by mathematics nevertheless feel that the effort to understand this intellectual climax of the course is very rewarding, but they are prepared to accept on faith that the conservation laws will lead to Kepler's law of elliptical orbits.

Epilogue

Though the foregoing exercise may prove little about the Newton-Beethoven analogy, it at least illuminates one facet. By showing an alternative physical framework to explain certain basic celestial phenomena, Newton's accomplishment becomes more personally his own. Newton's concept of a *universal* gravitation, systematically built on a corpuscular view of matter, is a creative intellectual achievement of the first order. In the absence of a Newton, many of his ideas might have been proposed by others, probably in some piecemeal fashion. But the path could be entirely different, as the existence of our alternative scheme shows. There is no *necessity* in nature that the Newtonian world system had to be found. The glory and impact of Newton's work is that he accomplished such an integrated science so single-handedly. Newton's *Principia* is surely a personal achievement that places Newton in the same creative class as Beethoven or Shakespeare.

Our opening quotation, which compared the calculus and Beethoven's *Fifth*, is from I. B. Cohen's *Franklin and Newton*; in its original setting it contained a footnote to the effect that "Albert Einstein was willing to admit this hypothesis in respect to only the early special theory of relativity, but not the general theory of relativity. Einstein actually wrote that if he had not discovered the special theory of relativity Paul Langevin would have."[6]

A similar opinion was related by John U. Nef in a discussion on science and art[7]:

> I am told that Heisenberg is a very good player on the piano, by the way. He was in residence at Cambridge not too long ago and they asked him if he would play.
>
> He sat down at the piano and played from beginning to end Opus 111, the last sonata of Beethoven, which is an absolutely unique work. All the dons were more and more overwhelmed by this music, and there wasn't a sound when he finished.
>
> Heisenberg is reported in this connection to have discussed the difference between science and art. "If I had never lived, someone else would probably have formulated the principle of determinacy. If Beethoven had never lived, no one would have written Opus 111."

I queried Professor Heisenberg about this incident, and he replied that he couldn't remember if it was actually Opus 111 he had played on that occasion, and he modestly added that he probably would have referred to relativity rather than his own work. But whether Langevin or Poincaré or Minkowski would have found the special theory of relativity is a moot point, history having gone the way it did. It is hard to imagine that the designers of accelerators would not by now have found an empirical relation between inertial mass and velocity, for example, but the conceptual formulation might well differ from Einstein's. Certainly the grand fabric into which relativity was woven—I mean both Einstein's peculiar mixing of operationalism with *a priori* postulates and the related strands running through all his 1905 papers—was a personal expression of style and imagination, with few peers in this century. Without Beethoven, we would have had piano sonatas in the classical style, but not Opus 111; without Einstein, we would have had relativity, but not with the impact of his unique presentation.

As Norman Campbell pointed out, "Science would not be what it is if there had not been a Galileo, a Newton or a Lavoisier, any more than music would be what it is if Bach, Beethoven, and Wagner had never lived. The world as we know it is the product of its geniuses and to deny that fact, is to stultify all history, whether it be that of the intellectual or the economic world."[8]

Nevertheless, to draw the analogy between art and science too closely would, I believe, be a mistake. The synthesis of knowledge achieved in a major scientific theory is not fully the same as the ordering of components in an artistic composition. Both are restricted by the materials and tools available and by the modes and styles of the era. But the scientific theory has a referent in the world of nature. It is subject to experimentation, extension, falsification.

Scientific change is not merely brought about through alterations of style or convention. There is an ongoing interaction between theory and observation that is either lacking or present in a far different form in the arts. In another passage, Campbell declares that the Newton-Beethoven analogy seems to him exact. "Beethoven's work *had* to be tested: the test of artistic greatness is appeal to succeeding generations free from the circumstances in which the work was conceived; it is very nearly the test of universal agreement."[9] I fear Campbell has overdrawn his analogy. As a corpus of understanding, science exhibits significant qualities of progress not shared by the arts. Much as Aristotle's cosmology can be admired for its synthesis of Greek knowledge, it is obsolete by today's standards. In contrast, we would never dream of scrapping the Venus de Milo, the *Iliad*, or Bach's *Mass in B Minor* just because we now have Rodin's *Thinker*, Dostoevsky, or Bruckner.

The conclusion, then, is tinged with ambiguity. Surely there is a closer parallel between the mighty creativity of Newton's *Principia* and Beethoven's *Symphony No. 5 in C Minor* than strikes the casual observer. Yet the analogy reaches a limit when we examine the relationships of science and art to the world in which they are embedded. To recognize both the analogy and its boundaries is to gain a more sensitive view of the nature of scientific creativity.

Appendix: The Rigorous Solution

Physicists, I am sure, will believe that the conservation laws will lead to elliptical orbits, but they may be worried about the completeness of the set of axioms. Because this development is not commonly found in texts on mechanics, it is given below.

Besides the conservation of momentum, of angular momentum, and of kinetic and potential energy, we need to know the form of the gravitational potential, namely,

$$V = - Gm_1 m_2 / r \equiv - k/r,$$

where r is the separation of the masses m_1 and m_2. (This is, of course, equivalent to having the law of gravitation in the standard Newtonian formulation.)

The conservation of linear momentum will help get the reduced mass μ, which essentially converts this from the two-body problem to a one-body problem. First we simply integrate over time to get

$$m_1 \mathbf{r}_1 = m_2 \mathbf{r}_2,$$

where the vectors are taken from the center of mass. Then

$$\mathbf{r} = \mathbf{r}_1 + \mathbf{r}_2 = \mathbf{r}_1 + \frac{m_1}{m_2} \mathbf{r}_1 = \frac{m_1 + m_2}{m_2} \mathbf{r}_1.$$

Hence

$$\mathbf{r}_1 = \frac{m_2}{m_1 + m_2} \mathbf{r} \quad \text{and} \quad \mathbf{r}_2 = \frac{m_1}{m_1 + m_2} \mathbf{r}. \tag{3}$$

Both angular momentum and kinetic energy will involve terms of the sort $(m_1 r_1^2 + m_2 r_2^2)$, and with (3) this becomes

$$\frac{m_1 m_2^2}{(m_1 + m_2)^2} r^2 + \frac{m_2 m_1^2}{(m_1 + m_2)^2} r^2 = \frac{m_1 m_2}{(m_1 + m_2)} r^2 \equiv \mu r^2.$$

Thus the angular momentum is

$$J = m_1 r_1^2 \dot{\theta} + m_2 r_2^2 \dot{\theta} = \mu r^2 \dot{\theta}$$

and similarly the kinetic energy is

$$T = \tfrac{1}{2} \mu (\dot{r}^2 + r^2 \dot{\theta}^2).$$

The conservation of energy is then[10]

$$\tfrac{1}{2} \mu (\dot{r}^2 + r^2 \dot{\theta}^2) - k/r = W.$$

In order to reduce this to an expression in only two differentials we wish to introduce the angular momentum; we note that

$$\dot{r} = \frac{dr}{d\theta}\dot{\theta},$$

so that

$$\frac{\mu}{2}\left[\left(\frac{dr}{d\theta}\right)^2 + r^2\right]\dot{\theta}^2 - \frac{k}{r} = W.$$

We can now use angular momentum as a constant:

$$\left[\left(\frac{dr}{d\theta}\right)^2 + r^2\right]\frac{J^2}{2\mu r^4} - \frac{k}{r} = W.$$

Since the dependence on time has been eliminated, this is the differential equation for the orbit, and it remains to show that it corresponds to an ellipse. Algebraic rearrangement gives a rather clumsy quadratic expression that can nevertheless be found easily in standard tables (e.g., no. 259 in the *Handbook of Chemistry and Physics*):

$$\theta = \int \frac{J\,dr}{r(2\mu Wr^2 + 2\mu kr - J^2)^{1/2}} = \sin^{-1}\left(\frac{\mu kr - J^2}{r(\mu^2 k^2 + 2J^2\mu W)^{1/2}}\right) - \alpha.$$

We arbitrarily take $\alpha = \pi/2$ so that $\sin(\theta + \alpha) = \cos\theta$. This yields

$$r = \frac{J^2/\mu k}{1 + (2J^2 W/\mu k^2 + 1)^{1/2}\cos\theta}, \tag{4}$$

which is a form of the general polar equation of an ellipse from one focus where θ is measured from the perihelion,

$$r = \frac{a(1 - e^2)}{1 + e\cos\theta}. \tag{5}$$

where a is semimajor axis and e the eccentricity. Thus we have derived *Kepler's first law* from the conservation laws. Furthermore, identification of terms in (4) and (5) gives

$$J^2/\mu k = a(1 - e^2) \tag{6}$$

and

$$2J^2 W/\mu k^2 + 1 = e^2,$$

from which it follows at once that

$$W = -k/2a.$$

In words, the total energy of the system depends only on the semimajor axis of the elliptical orbit.

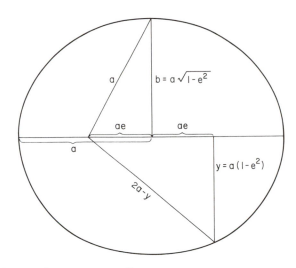

FIGURE 2. *Basic relations in an ellipse.*

As is well known, the law of conservation of angular momentum is equivalent to a constant areal velocity,

$$r^2 \dot{\theta} = J/2\mu, \tag{7}$$

so no further proof is necessary to show that *Kepler's second law* is immediately implied by the axioms. To get the third law, we need an expression containing the planetary period P, which is obtained by integrating (7) to get

$$\frac{JP}{2\mu} = \tfrac{1}{2} \int_0^{2\pi} r^2 \, d\theta = \pi a b,$$

$\pi a b$ being the area of an ellipse in terms of the semimajor and semiminor axes. Hence, using the definition of b from Figure 2 and the expression for J from (6), we have

$$P = \frac{2\mu\pi ab}{J} = \frac{2\mu\pi a^2 (1 - e^2)^{1/2}}{[\mu k a (1 - e^2)]^{1/2}} = 2\pi a^{3/2} \left(\frac{\mu}{k}\right)^{1/2}.$$

This rearranges into Newton's revised form of *Kepler's third or harmonic law*:

$$\frac{a^3}{P^2} = \frac{G}{4\pi^2} (m_1 + m_2). \tag{8}$$

Notes and References

[1] Bernard Cohen, *Franklin and Newton*, vol. 43 of *Memoirs of the American Philosophical Society* (Philadelphia, 1956), p. 43. The choice of calculus, an obvious case of multiple discovery, makes the comparison less debatable than either universal gravitation or relativity. That choice does serve to remind us of the different levels of creativity in both art and science. As Jacob Bronowski shrewdly remarked, "When the Greek departments produce a Sophocles, or the literature departments produce a Shakespeare, then I shall begin to look in my laboratory for a Newton." (From "The Creative Process" in *Creativity*, ed. by J. D. Roslansky (Amsterdam, 1970), pp. 3–16).

[2] Gunther Stent, "Prematurity and Uniqueness in Scientific Discovery," *Scientific American*, vol. 227 (1972) 84–93.

[3] Norman Campbell, *What is Science* (London, 1921; reprint New York, 1952), p. 102.

[4] Isaac Newton, *Mathematical Principles of Natural Philosophy*, Volume I, Proposition I, Theorem I.

[5] The method is presented in R. P. Feynman, R. B. Leighton, and M. Sands, *The Feynman Lectures on Physics* (Reading, MA, 1963), pp. 9.6–9.9. An analysis is given by Leo Lavatelli, *American Journal of Physics*, vol. 33 (1965) pp. 605–12, and a similar program is explained by Alfred Bork, *FORTRAN for Physics* (Reading, MA, 1967).

[6] The fact that Hilbert submitted the general relativistic field equations for a vacuum for publication a few days before Einstein in no way constitutes an independent discovery nor does it provide evidence that general relativity would have been found in the absence of Einstein. Hilbert presented his solution after Einstein had already given three seminars outlining his theory and *after* Einstein had established the connection between the curvature of space and the presence of matter. See Jagdish Mehra, *Einstein, Hilbert, and the Theory of Gravitation* (Dordrecht and Boston, 1974), especially pp. 82–86.

[7] Transcribed in *The Nature of Scientific Discovery*, edited by Owen Gingerich (Washington D.C., 1975), p. 496.

[8] Reference 3, p. 73.

[9] Reference 3, p. 103.

[10] Since the potential and kinetic energies do not contain time explicitly, this expression is the Hamiltonian function. The method presented here is similar to the formulation by Robert A. Becker, *Introduction to Theoretical Mechanics* (New York, 1954), pp. 230–31.

Epilogue

S ometimes, only half in jest, I explain to my astrophysicist friends that I am the victim of anniversaries. The quadricentennial of Kepler in 1971, the quinquecentennial of Copernicus in 1973—these are the reasons I left the calculation of stellar atmospheres behind and began to devote the majority of my research efforts to the history of astronomy.

As I look over the two dozen papers in this anthology, I realize that the greater part were written for specific occasions. "Johannes Kepler and the New Astronomy" was the 1971 George Darwin Lecture of the Royal Astronomical Society, their most prestigious annual lecture having been devoted to the four-hundredth anniversary of Kepler's birth. " 'Crisis' versus Aesthetic in the Copernican Revolution" was presented at a Copernican symposium that kicked off the Copernican year for the American Association for the Advancement of Science. "Heliocentrism as Model and as Reality" came a few months later when the American Philosophical Society took their turn to honor Copernicus. And "The Astronomy and Cosmology of Copernicus" was the opening address at the International Astronomical Union's Extraordinary General Assembly held in Warsaw in August of 1973.

Others of the papers were commissioned, if not for special occasions, then for specific publications. I count that sixteen of these twenty-five essays were originally presented as lectures or invited as contributions for specific volumes. Most of us historians of science yield to the pressures of the scholarly society in which we are imbedded, which may or may not be a bad thing. Sometimes we are reluctant victims of the calendar, postponing other investigations for the exigencies of the march of time. On the other hand, these festive anniversary occasions may just give us an excuse to polish up research that had intrigued us in any event. Occasionally, as in my case, they help shape a long-term research program.

But several of the papers in the present anthology originated with no external demands to prompt them; they were in turn dedicated to individuals who helped or inspired me as I gradually crossed over from astrophysics to the history of science. "Circumventing Newton" arose out of the Natural Sciences course I have taught for many years, and it seemed appropriate to dedicate it to I. Bernard Cohen, my mentor as a teacher in Harvard's General Education program as well as a distinguished historian of the scientific revolution. "Zoomorphic Astrolabes: Arabic Star Names Enter Europe" was part of a Festschrift for E. S. Kennedy, who was my colleague when I began a teaching career in the 1950s at the American University of Beirut. David King and George Saliba, co-organizers of that Festschrift, are my coauthors on the other paper on astrolabes included in this collection; we researched it together in a seminar (which included Kennedy) that I taught in Beirut in 1971. The paper "The Censorship of Copernicus's *De revolutionibus*," published in the *Annali* of the History of Science Museum in Florence, honors Maria Luisa Bonelli, founder of the journal and ebullient director of the Museum, who was then, alas, in a terminal battle with cancer.

The paper in this collection that elicited the most publicity (but in the computer world, outside history-of-science circles) was "The Computer versus Kepler." The paper showed that an IBM 7094 computer could handle in seconds an iterative computational problem that occupied Kepler for a couple of years, something that proved irresistible for science reporters. It was my first contribution to a History of Science Society meeting, perhaps more fun than serious. I still remember the charming introduction given me there by C. Doris Hellman, whose enthusiasm for Tycho Brahe proved infectious. I also recall that it was Gerald Holton who urged me to publish the essay, certainly an idea I had not originally entertained. If the piece were to carry a dedication, it would be to those two influential colleagues.

The computer versus Kepler paper not only provided a springboard for my entry into the history of science but also taught me an important lesson, probably one that has to be learned several times. It is easy to invent an explanation for how things should have gone in times past, but, in the absence of evidence, the rational reconstruction can be miles off the mark. I imagined that Kepler must have made a large number of computational errors that prevented the convergence of his iterative solution for a circular orbit for Mars. I was curious, however; so, after publishing that conclusion, I asked the Russians for a microfilm of the appropriate volume of Kepler's manuscripts, still preserved in what was then Leningrad and what is once again St. Petersburg.

I repeated my request for a microfilm at intervals of approximately six months for about six years, and then, somewhat to my surprise, the long-awaited reel finally materialized. My reconnaissance of this manuscript legacy showed that the situation was certainly more complex than I had envisioned. There were errors that slowed the convergence and that greatly frustrated Kepler, but they were mistakes in copying the original data rather than in the computations themselves. This fresh insight gave rise to another paper, presented at Kepler's birthplace, Weil der Stadt, at one of the several international meetings held in his anniversary year. In it I showed that, contrary to the received opinion, Kepler's *Astronomia nova* was not a simple chronological record of his researches but a highly reorganized presentation designed to convince his readers that he had tried hard enough on the orbit of Mars. If I were writing the paper today, I would emphasize this point even more: that the *Astronomia nova*, besides being the first account in which we can see a scientist wrestling with contradictory, error-laden data, was also a highly organized rhetorical presentation.

Meanwhile, unpublished researches by D. T. Whiteside, using a copy of my microfilm, have shown that I was, after all, not entirely astray in saying that computational errors prevented the convergence of Kepler's iterations. Whiteside has found that Kepler generally did not carry enough places in the values of trigonometrical functions to guarantee convergence.

I became even more conscious of the minefield of rational reconstructions in another incident, in my ongoing research into the provenances of copies of Copernicus's *De revolutionibus*. One of the most remarkable exemplars carries on its flyleaf a poem in Greek by the Leipzig humanist Joachim Camerarius. Eventually I discovered, in the catalog of the Biblioteca Palatina in Parma, a copy described as carrying such a poem. The book was missing, so naturally I concluded that the copy I had examined must have been stolen from Parma, despite the fact that the book in question contained no physical evidence whatsoever of a library ownership or shelf mark. Because I had examined 240 copies without finding another to match the description, the circumstantial evidence seemed enough to convict. My rationally reconstructed scenario was tossed into a cocked hat when I later saw,

under highly unusual circumstances, the photocopy of another manuscript version of the Greek poem in a clandestine first edition of *De
revolutionibus*. Dare I conclude that this second copy of the Camerarius poem belonged to the Parma library?

Still another deduction that failed was my early attribution to Tycho
Brahe of the remarkable annotations and manuscript pages found in a
copy of Copernicus's *De revolutionibus* in the Vatican Library. I
reached this conclusion because the handwriting in the Vatican volume
matched perfectly with the marginalia in another Copernicus book, in
Prague, that were long attributed to Tycho Brahe. I announced the
discovery in Warsaw at the Congress of the International Astronomical Union, described it in *Scientific American*, and alluded to it in an
article in the *Dictionary of Scientific Biography*. The moment of truth
came five years later when Robert Westman and I finally realized that
neither volume contained Tycho's hand, though we eventually proved
that he had owned both copies. The annotator was Paul Wittich, as is
correctly stated in several articles in this collection written after 1978.

Two of the questions that have greatly exercised historians of science over the past several decade are "Was Ptolemy a fraud?" and
"Why did Copernicus adopt a heliocentric cosmology?" Any answers
must necessarily be rational reconstructions and therefore necessarily
veiled in uncertainty. Did Ptolemy fudge his data or "launder" it?
And, if so, was this unusual in the science of the time? Is it not
probable that a scientist will forge his material into what he under-
stands to be a self-consistent whole, with no intention to deceive? I
strongly believe this to be the case, and we see Copernicus, Kepler, and
Newton doing the same (not to mention others cited in my article
"Ptolemy Revisited"). R. R. Newton did not agree, and he objected to
my initial defense in his response in the *Quarterly Journal of the Royal
Astronomical Society*, vol. 21 (1980), pp. 388–99, which in turn elicited my "Ptolemy Revisited" article.

As for the adoption of the heliocentric cosmology, I have addressed
this tantalizing question in several essays in the middle section of this
anthology. It is perhaps easier to demolish reasons that did *not* moti-
vate Copernicus, though that means challenging a pervasive mythology, such as the existence of a creaky system of epicycles on epicycles.
The secondary literature abounds with statements to the effect that,
by the Renaissance, the Ptolemaic system was ready to collapse of its
own weight. The essay " 'Crisis' versus Aesthetic in the Copernican
Revolution" may be the most important in the collection; it has been
widely admired, but it has not yet stamped out the notion that multiple

epicycles had brought astronomy to a crisis state at the time of Copernicus.

The present selection of papers focuses on comparatively technical aspects of the transformation of astronomy from the ancient geocentric scheme to Kepler's remolding of the Copernican arrangement. In no sense does this group represent an *opera omnia*, even for the topics included here. (And we have chosen to exclude the papers on Galileo, for example.) As mentioned in the Preface, the monograph by me and Robert S. Westman, *The Wittich Connection: Priority and Conflict in Late Sixteenth-Century Cosmology*, complements what is found here. Various articles in this anthology contain analyses of the accuracy of early ephemerides, but an even more extensive investigation is found in "The Accuracy of Historical Ephemerides," by me and Barbara L. Welther, in *Memoirs of the American Philosophical Society* 59S (1983). A much wider historical sweep is found in a second anthology, published by Cambridge University Press under the title *The Great Copernicus Chase*, which contains a number of my more popular articles.

My English colleague Michael Hoskin often uses, as a criterion for judging the importance of a paper, the query "Does it force us to rewrite our lecture notes?" In reflecting on this collection of articles, I have asked myself the same question. The demonstration that Kepler's *Astronomia nova* was not a simple chronological account of his researches might qualify. My attack on the multiple-epicycle mythology might be another. But mostly the themes are more subtle; the papers examine in several ways the idea that Copernicus's intuitive leap to a heliocentric cosmology was a decision of mind's eye, against the apparent evidence of the senses and proposed in the absence of any physical proof. In other words, they furnish a sustained argument against an empiricist view of the astronomical revolution of the sixteenth century. My discovery and identification of Erasmus Reinhold's copy of Copernicus's *De revolutionibus* shows how the Wittenberg professor and leading mathematical astronomer at the mid–sixteenth century considered the new cosmology to be merely a mathematical hypothesis. This evidence became a linchpin in understanding what Robert Westman has called "the Wittenberg interpretation" of the heliocentric doctrine. I elaborate on this theme in the paper "Heliocentrism as Model and as Reality," which won the John F. Lewis Prize of the American Philosophical Society, awarded for "some truth" presented to the Society.

In retrospect, what would I have done differently in preparing this set of papers? Very little, actually—mostly, I might have curtailed a few of my many other activities in order to have completed several papers that would have been appropriate additions to this collection. I have found two copies of the Regiomontanus *Epitome of the Almagest* with manuscript annotations by Erasmus Reinhold, and these will further elucidate his views on astronomical hypotheses. I am convinced that Michael Maestlin had an ingenious (but ultimately unworkable) program for testing the Copernican idea, but my paper on this subject is only partially written. A paper explaining a remarkable refractive sundial bowl by Christopher Schissler is nearly ready for publication, as is an article on a unique "lunarium" by Georg Erlinger. And I would like to return in more detail to one of my earliest historical researches, on the computational framework of Kepler's *Rudolphine Tables* and their aftermath. With luck and persistence, I will be able to finish them before I retire.

Cambridge, Massachusetts
June 1991

Acknowledgments

PTOLEMY, COPERNICUS, AND KEPLER was originally published in *The Great Ideas Today 1983*, edited by Mortimer J. Adler and John Van Doren (Chicago, 1983), pages 137–180.

WAS PTOLEMY A FRAUD? was originally published in *Quarterly Journal of the Royal Astronomical Society*, Vol. 21 (1980), pages 253–266. The article was based on a paper given before the Royal Astronomical Society on January 13, 1978.

PTOLEMY REVISITED was originally published as "Ptolemy Revisited: A Reply to R. R. Newton" in *Quarterly Journal of the Royal Astronomical Society*, Vol. 22 (1981), pages 40–44.

ZOOMORPHIC ASTROLABES: ARABIC STAR NAMES ENTER EUROPE was originally published as "Zoomorphic Astrolabes and the Introduction of Arabic Star Names into Europe" in *From Deferent to Equant: A Volume of Studies in the History of Science in the Ancient and Medieval Near East in Honor of E. S. Kennedy*, edited by David King and George Saliba, *Annals of the New York Academy of Sciences*, Vol. 500 (1987), pages 89–104. The author would like to thank Roderick and Madge Webster, Francis Maddison, Stuart Malin, Carole Stott, Michael Hoskin, and Jon Darius for helpful discussions and for assistance in examining various astrolabes. John North, Gerald Turner, Paul Kunitzsch, Emmanuel Poulle, and Sharon Gibbs also provided stimulating input concerning the problems discussed here. The Trustees of the British Museum, the Cambridge University Library, the

Germanisches Museum in Nuremberg, and the Smithsonian Institution have graciously given permission to reproduce these plates.

THE 'ABD AL-A'IMMA ASTROLABE FORGERIES was originally published in *Journal for the History of Astronomy*, Vol. 3 (1972), pages 188–199.

ALFONSO AS A PATRON OF ASTRONOMY was originally published in Alfonso X of Castile, the Learned King (1221–1284), edited by Francisco Márquez-Villanueva and Carlos Alberto Vega, Harvard Studies in Romance Languages, volume 43 (1987), pages 30–45.

THE 1582 "THEORICA ORBIUM" OF HIERONYMUS VULPARIUS was originally published in *Journal for the History of Astronomy*, Vol. 8 (1977), pages 38–43. I should like to thank Mr. and Mrs. Roderick S. Webster, curators of the Antique Instrument Collection at the Adler Planetarium, for their continuing interest and assistance with this research, and also the many persons who commented on the first draft of this paper, including the Websters, Derek de Solla Price, E. Poulle, Noel Swerdlow, Derek Howse, Alain Brieux, and Silvio Bedini.

THE SEARCH FOR A PLENUM UNIVERSE was originally published as "The Aethereal Sky: Man's Search for a Plenum Universe" in *The Great Ideas Today 1979*, edited by M. J. Adler and J. Van Doren (Chicago, 1979), pages 68–86.

THE ASTRONOMY AND COSMOLOGY OF COPERNICUS was an invited lecture marking the Copernican quinquecentennial at the Extraordinary General Session of the International Astronomical Union and was published in their *Highlights in Astronomy*, edited by G. Contopoulos, Vol. 3 (1974), pages 67–85.

DID COPERNICUS OWE A DEBT TO ARISTARCHUS? was originally published in the *Journal for the History of Astronomy*, Vol. 16 (1985), pages 36–42. The article was based on a paper given at the Aristarchus Symposium held on Samos in 1980.

"CRISIS" VERSUS AESTHETIC IN THE COPERNICAN REVOLUTION was originally published in *Vistas In Astronomy*, edited by Arthur Beer and K. AA. Strand, Vol. 17 (1975), pages 85–95. The volume contains the papers given at a conference held in conjunction with the meeting of the American Association for the Advancement of Science in Washington, D.C. in December, 1972.

EARLY COPERNICAN EPHEMERIDES was originally published by the Polish Academy of Sciences, The Institute for the History of Science, Education, and Technology, in *Science and History: Studies in Honor of Edward Rosen, Studia Copernicana*, Vol. 16 (1978), pages 403–417. This research was partially supported by the Johnson Fund of the American Philosophical Society. The author would like to thank Jerzy Dobrzycki, John North, Peter Huber, and Barbara Welther for providing essential materials.

ERASMUS REINHOLD AND THE DISSEMINATION OF THE COPERNICAN REVOLUTION was originally published by the Polish Academy of Sciences Institute for the History of Science in *Colloquia Copernicana II*: Études Sur l'Audience de la Théorie Héliocentrique, Conférences du Symposium de l'Union Internationale d'Historie et de Philosophie des Sciences, Comite Nicolas Copernic, Toruń 1973, *Studia Copernicana*, Vol. 6, pages 43–62 and 123–125.

DE REVOLUTIONIBUS: AN EXAMPLE OF RENAISSANCE SCIENTIFIC PRINTING was presented as one of four principal addresses at the Inaugural Conference of the Center for Renaissance and Baroque Studies of the University of Maryland in 1982, and was originally published in *Print and Culture in the Renaissance*, edited by Gerald P. Tyson and Sylvia S. Wagonheim, Newark (University of Delaware Press, 1986)and London and Toronto (Associated University Presses), pages 55–73. The research reported in this paper has been generously supported by the Copernicus Society of America.

THE CENSORSHIP OF COPERNICUS'S DE REVOLUTIONIBUS was originally published in *Annali dell'Istituto e Museo di Storia della Scienza di Firenze*, anno 6, fascicolo 2 (1981), pages 44–61. An abridged preliminary version was read before the American Astronomical Society, San Francisco, 15 January 1980.

HELIOCENTRISM AS MODEL AND AS REALITY was originally published in the *Proceedings of the American Philosophical Society*, Vol. 117 (1973), pages 513–522. It was read at the Symposium on Copernicus, 20 April 1973. The author thanks F. L. Whipple, Director of the Smithsonian Astrophysical Observatory, and S. D. Ripley, Secretary of the Smithsonian Institution, for their continued support of this work and especially for the sabbatical year abroad, during which he discovered the Reinhold copy of *De revolutionibus* as well as the Wittenberg astronomy notes. The author also thanks Gonville and Caius College for permission to reproduce several pages of the Wittenberg manuscript.

JOHANNES KEPLER AND THE NEW ASTRONOMY was the George Darwin Lecture delivered on 10 December 1971 to the Royal Astronomical Society. It was later published in the *Quarterly Journal of the Royal Astronomical Society*, Vol. 13 (1972), pages 346–360. The author would like to thank Dr. D. T. Whiteside for numerous provocative discussions about the mathematics of Kepler's warfare on Mars, and to recognize the invaluable assistance given by the Kepler translations of Ann W. Brinkley, made possible by the support of the American Philosophical Society and the Smithsonian Astrophysical Observatory.

KEPLER AS A COPERNICAN was originally published in *Johannes Kepler, Werk und Leistung*, Katalog des Oberösterreiches Landesmuseums, Nr. 74 (Linz, 1971), pages 109–114.

KEPLER'S PLACE IN ASTRONOMY was originally published in *Vistas in Astronomy*, edited by Arthur Beer and Peter Beer, Vol. 18 (1975), pages 261–278. It had been presented to Quadricentennial Symposium at the Franklin Institute in Philadelphia on 27 December 1971, the very anniversary of Kepler's birth.

THE ORIGIN OF KEPLER'S THIRD LAW was originally published in *Vistas in Astronomy*, edited by Arthur Beer and Peter Beer, Vol. 18 (1975), pages 595–601.

THE COMPUTER VERSUS KEPLER was presented to the History of Science Society, in Philadelphia on 29 December 1963. It was later published in *American Scientist*, Vol. 52 (1964), number 2, pages 218–216.

THE COMPUTER VERSUS KEPLER REVISITED was originally published as "Kepler's Treatment of Redundant Observations or, the Computer versus Kepler Revisited" in *Proceedings of the Internationales Kepler-Symposium, Weil der Stadt 1971*, edited by F. Krafft, K. Meyer, and B. Sticker, (Hildesheim, 1973), pages 307–314.

THE MERCURY THEORY FROM ANTIQUITY TO KEPLER was originally published in *Actes du XII Congrés International D'Histoire des Sciences, Paris 1968, Vol. 3A (Science et Philosophie: Antiquité-Moyen Age-Renaissance) (Paris, 1971)*, pages 57–64.

KEPLER, GALILEI, AND THE HARMONY OF THE WORLD was presented at a symposium organized by V. Coelho at the University of Calgary in April, 1989. The symposium papers have been published under Coelho's editorship by Kluwer (Dordrecht, 1992) in a volume entitled *Music and Science in the Age of Galileo,* pages 45–63.

CIRCUMVENTING NEWTON: A STUDY IN SCIENTIFIC CREATIVITY was originally published in *American Journal of Physics*, Vol. 46 (1978), pages 202–206.

Index

'Abd al-A'imma, 102–114
'Abd al-'Ali, 105
'Abd al-'Ali, 111
Aberdeen, 163
Abraham the doctor, 118
accuracy of Kepler's longitudes, 311
accuracy of predictions, 168
Adams, H. M., *Catalogue*, 250, 257, 267
Adler Planetarium, 86, 88, 103–104, 106, 112, 129, 133
Aeneas, 278–279
aesthetic, 200
aethereal spheres, 136, 173
aether, 151, 155
Aetius of Antioch, 188–189
Aga-Oğlu, Mehmet, 114
Aiton, E. J., 347, 405–406
Albatani, 122, 237
Alcabitius, 139
Alfonsine Tables, 31, 115, 122–125, 223; accuracy for Mercury, 381, 383–384; agreement with observations, 168–173, 327; compared to *Prutenic Tables*, 224–226, 231–232; Copernicus' copy, 28, 145, 163, 170–171, 200; ephemerides, 198, 213–214; parameters, 128; Ptolemaic basis, 197; superseded, 172
Alfonso X, 26, 115f, 172, 203
Alhabor (Sirius), 81, 83, 87–88
Alhazen, 140
Allen Memorial Art Museum, Oberlin, 109, 110, 114
Almagest (Ptolemy), 4f, 55f, 276; I,12, 186; IV,19, 185; IX,3, 75; IX,4, 9, 13, 75–76; X,8, 12, 14, 16; X,9, 76; XII, 25; XIII, 25; few observations, 171, 375; "the greatest," 138; 1515 edition, 33, 145, 186, 166, 200; criticisms of, 27; double-entry tables, 324; eclipses, 23; latitudes, 25; lunar theory, 22; Mercury, 24, 379–380; observations in, 171; planetary parameters, 16; scaffolding dismantled, 310; star catalog, 24, 77, 79
Almagestum novum (Riccioli), 148
Altair, 98–99
ambrosia, 4, 55
American Association for the Advancement of Science, 345, 419
American Astronomical Society, 427
Amsterdam, *see De revolutionibus*, 1617
American Philosophical Society, 419, 423, 428; Johnson Fund, 427; Penrose Fund, 347, 376
Ancient Planetary Observations (Newton), 78
angular momentum, 410–412, 415–417
anniversaries, 419
annual equation, 331
anomaly, 13, 15
Antwerp, 256
Apian, Philip, 264
Apianus, Petrus, 60–61, 253, 257, 293
apogee, 13
Apollonius, 7–9, 24–25
archetypes, 49, 182, 349, 353–354, 389, 393
Archimedes, 185
area law. *See* law of areas
Argellata, Petrus de, 163
Argoli, Andrea, 385
Aristarchus, 185–192, 243, 286–287, 300
Aristotelian spheres, 26–27, 173
Aristotle, 137, 186–187, 390, 392

Aristyllus, 186
armillary sphere, 133
Arnaldez, Roger, 156
Arras, 263
astrolabes, 81–114
Astrolabes of the World (Gunther), 84f
astrology, 116
Astronomia nova (Kepler), 41, 309f,
    328–329; title page verso, 265, 300;
    introduction, 33, 272; ch. 16–18, 366,
    368; ch. 24, 338; ch. 26–28, 371; ch. 37,
    317; ch. 39, 351; ch. 40, 43; ch. 52, 334;
    ch. 57, picture, 343; ch. 58, 44; ch. 7, 11,
    39, 334, 339; Book 3, 340; highly
    organized, 421; truly new astronomy, 6,
    306, 344; vicarious orbit, 42, 357–363,
    367–375
Astronomia pars optica (Kepler), 314, 331
Astronomicum Caesareum (Apianus),
    60–61, 257
astrophysicist, Kepler as, 321
Aubrey, John, 266
August, Duke, 266
Aurifaber, Andreas, 167
Austria, censorship, 266
Averroës, 140
axioms of astronomy, 176, 179, 240, 291,
    347
Azarquiel, 118–122

Babylonian astronomy, 7–8, 11; parameters,
    20; system A solar theory, 21, 57; goal
    year texts, 75
Bach, J. S., 414
Bailly, J-S., 116
ballet of planets, 172
Balduinus, Johannes, 292
Bamberg notebook, 262
Barberiniano, 275f
Baronius, Cardinal, 273
Barwiński, E., 204
Basel, 177
Becker, R. A., 418
Bedini, Silvio, 426
Beer, A., 202, 251
Beethoven, 407–408, 413; *Opus 111*,
    413–414
Beham, Hans Sebald, 254
Bellarmine, Robert Cardinal, 273
Bentley, Richard, 151
Benzing, Josef, 266
Berlin Staatsbibliothek, 241, 291
Bernaldo the Arab, 118

Bianchi, Vinzenz, 285
Bibliothèque Arsenal (Paris), 122
Biermann, Ludwig, 192
Billmeier, J. A., 84, 108
Birkenmajer, Alexander, 213, 220, 222,
    232, 235, 241, 301
Birkenmajer, L. A., 184, 204, 213, 235, 237,
    241, 249, 268, 301–302
Birkman, Arnold, 222
Biskup, Marian, 267
al-Bitruji, 140
Blakene astrolabe, 87
boatman, 343
Bodleian Library, *De revolutionibus*, 262
Bohemia, 334
Boll, Franz, 71
Bonelli, Maria Luisa, 134, 420
"book nobody read," 297
Books of Nature and Scripture, 274, 284
Bork, Alfred, 418
Bossong, Georg, 127
Boston Museum of Fine Arts, 107–112
Bott, Gerhard, 101
Brachvogel, Eugen, 192
Brahe. *See* Tycho Brahe
Brieux, Alain, 113, 426
Brinkley, A. W., 347, 387, 428
British Museum, 90, 85f, 301
Britton, John, 56–57
Brocius, Johannes, 214, 299
Brodetsky, S., 280, 285
Bronowski, Jacob, 418
Brożek, Jan, 214, 299
Brück, H. A., 249
Bruckner, Anton, 414
Bruno, Giordano, 266
Buchdahl, G., 355
Budapest, 177
Buffalo Society of Natural Sciences, 113
Buonamici, G., 285
Burmeister, K. H., 267, 301
Butsch, Albert F., 267

Cárdenas, Anthony, 115, 119
Caetani, Bonifacio Cardinal, 275
Calcutta, Indian Museum, 103
Calendarium Romanum magnum
    (Stoeffler), 163
Callippus, 143
Calvisius, Seth, 201
Cambridge, England, 177, 262; University
    Library, 81
Camden, William, 266

Camerarius, J. R., 201
Camerarius, Joachim, 421–422
Campbell, M. S., 248
Campbell, Norman, 408, 414
Canute, King, 284
Carello, J. B., 232, 249
Carter, Elliott, 405
Caspar, Max, 46, 268, 355, 375, 378, 402, 406
Castilian, 116
Catania, 258
Catherine the Great, 321, 365
celestial physics, 321, 328, 333, 344
Censorinus, 191
censorship, 265, 269–285
Charliat, G., Paris, 109–111
Chaucer, Geoffrey, 81–99, 118
Chenakal, V. L., 321, 346, 376
Cherniss, Harold, 192
Chicago, *De revolutionibus*, 282
China, 269, 283
Chrisman, M. U., 250
Christie's, 99
Christina, Queen, 177–178
Cicero, 163, 189
Clagett, Marshall, 191
Clavius, Christopher, 176, 178, 271, 285; as annotator of *De revolutionibus*, 235
Cleanthes, 189
Clessius, J., 250
clockwork, 321, 339, 343
Cohen, I. B., 248, 267, 413, 420
Collegium Maius, 161–163
Collins, Jack, 267
Columbus, 200
Comet of 1577, 147, 181, 306
comet path, 150
commensurability, 31, 34, 204
Commentaria in novas theoricas planetarum (Schreckenfuchs), 130
Commentariolus (Copernicus), 28–29, 33, 145, 163, 168, 172–173, 203
Commentarius in opus Revolutionum Copernici (Reinhold), 239–241
Commentary on the Motion of Mars (Kepler), 299, 375. *See Astronomia nova*
Commentary on the Sphere of Sacrobosco (Clavius), 178, 270
computer, 357–365
conservation laws, 410f
Conversation with the Sidereal Messenger (Kepler), 395
Copenhagen, 177; great fire, 262

Copernican system, 148, 314, 390; accuracy, 63; parameters, 215–216
Copernicus, 125–126, 143–147, 161f, 310; birth, 27; "mind's eye," 32
Copernicus Society of America, 427
Counter Reformation, 394
Côte, Claudius (Lyon), 113
Cracow, 161, 163, 186, 200, 214, 290, 299; Copernicus as undergraduate, 190; longitude, 213
Crawford Collection, 139, 176, 235–237, 291
Cremona, Biblioteca Statale, 270, 278
Cremonensis, 198
Crime of Claudius Ptolemy (Newton), 18, 58
crisis, 193, 200
crystalline spheres, 29, 31; shattered, 38
Curtze, M., 220
Czartoryski, P., 192, 204, 249, 266

Darwin Lecture, 419
Daspa, Guillen Arremon, 117
Dasypodius, C., 242
David, E. S., 113
Davis, C. P., 113
de Berg, Robert, 222
De caelo (Aristotle), 137, 187
De cometis libelli tres (Kepler), 331
De dimensione terrae (Peucer), 293
De expetendis et fugiendis rebus opus (Valla), 189
De humani corporis (Vesalius), 257
De motu cordis (Harvey), 257
De revolutionibus, 34f, 168f, 252–285; annotators, 176; census, 257, 281; copy putatively Tycho's, 179; Galileo's copy, 281–282; importance, 345; inadequacies, 324; initials, 254; Kepler's copy, 261, 325; manuscript, 29, 164–165; number of circles, 197; observational accuracy, 36; price, 261–262, 266; printing, 37; prohibited, 275, 297; permitted in Spain, 283; reception, 37f; size of edition, 166, 256; size of paper, 253; Reinhold's copy, 235–237, 240, 255; type fonts, 254; watermarks, 253
De revolutionibus contents: title page, 4, 34, 167, 221, 265, 290; preface, 264, 323; I,3, 189; I,10, 240, 327, 350; III,17, 241; III,20, 217; IV,3, 35; V,14, 241; pictures, 146, 254–255, 259, 270; epicyclets 175; lunar theory, 35; percent cosmological, 34; precession, 35; 1717 years, 35

De revolutionibus editions: Nuremberg 1543, 218–219, census, 257; Basel 1566, 38, 269, census, 281, in Italy, 262; Amsterdam 1617, 284; Toruń 1873, 215, 218–219; Munich edition, 215, 218–219

De stella nova (Kepler), 331

De triangulis (Regiomontanus), 253, 262

de Vaucouleurs, Gerard, 194

Dee, John, 230, 262

Delambre, J-B. J., 56, 357–359, 364

DeLuccia, M. R., 79

Denmark, 394

Descartes, 136, 149–150, 257, 318, 344

Detroit Institute of Arts, 109, 114

Deutsches Museum (Munich), 134

di oti/to oti, 392

Dialogo della musica antica (Galilei), 396

Dialogue Concerning Two Great World Systems (Galileo), 50, 272, 329

Diels, Hermann, 192

Digges, Thomas, 266

Dineley, M. (London), 109

Dionysian year, 11

Dioptrice (Kepler), 331

Discorsi (Galileo), 329

distance law, 316

divine handiwork ... so vast, 182

Divine Providence, 298, 394

Dobrzycki, Jerzy, 171, 177–178, 192, 249, 285, 427

Dobson, J. F., 280, 285

Doggett, L. E., 79

Dolan, Edmund, 285

Dominicans, 282

donec corrigatur, 274–275, 297

Donner, Georg, 268

Dostoevsky, Fedor, 414

Douai, 263

double-entry table, 12

Drake, Stillman, 272, 285, 405

Dresden, 263

Dreyer, J. L. E., 127, 156, 203, 248–249, 251, 345, 347

Dürer, Albrecht, 254, 322

Duhem, Pierre, 235, 248–250, 285, 301

Dukas, Helen, 73

Duncan, A. M., 192, 404–406

Durret, Noël, 346

Eames, Charles, 165

earth a lazy body unfit for motion, 33, 181, 146

eccentric, 9–11, 16

eccentric anomaly, 14

eclipse, 172

Eclipsium omnium (Leovitius), 231

Eddington, Arthur S., 70, 345

Efemeridi (Moleti), 232

Effemeridi (Carello), 232

Egmont, Count, 266

Egyptian months, 11; year, 12, 76

Egyptians, 403

8' error, 42–43, 312, 316, 319–320, 365, 395. *See also* accuracy

80–34 syndrome, 197

Einstein, Albert, 70–71, 136, 153–155, 200, 328, 407, 413–414

Eisenstein, E. L., 204, 285

Ekphantus, 287

Elementa doctrinae de circulis coelestibus (Peucer), 293

Elements (Euclid), 137, 163

elliptical orbit, 44, 49, 300, 318, 320, 340–344, 349, 370, 410, 412, 416–417

Elskamp, Max, 86, 88, 97

Encyclopaedia Britannica, 125, 152, 154, 196

Engelhart, Valentine, 265

Engelmann, Max, 100, 134

Ephemerides (Magini), 202, 232, 324, 327

Ephemerides (Origanus), 387

Ephemerides (Regiomontanus), 200

Ephemerides anni 1557 (Feild), 230

Ephemerides duorum annorum (Reinhold), 206–207

Ephemerides exactissimae (Argoli), 387

Ephemerides novae (Maestlin), 202, 232, 387

Ephemerides novae (Rheticus), 206–207

Ephemerides novae (Stadius), 202, 231, 297, 387

Ephemerides trium annorum (Feild), 230

Ephemeridum novum (Leovitius), 202, 231

epicycle, 9–11, 15–16; additional, 26

epicycles on epicycles, 125, 172–173, 193–194, 196–198, 422–423

epicyclet, 36, 175, 179, 183–184, 229, 339, 342; Reinhold's use, 241

Epitome astronomiae (Maestlin), 294, 306, 324

Epitome astronomiae Copernicanae (Kepler), 6, 326, 350, 352, 354, 388, 399; Book 5, 49

Epitome doctrinae de primo motu (Strigel), 293

Epitome of Ptolemy's Almagest (Regiomontanus), 47f, 164, 200, 235,

261; with Reinhold's annotations, 250, 424

epoch, 13

equant, 10–11, 15; criticism, 27, 29, 138–140, 143, 175; Copernicus' criticism, 36, 143–144, 199, 228–229, 342; lack in solar orbit, 339; Reinhold's use, 241, 291; relation to law of areas, 174; Kepler's use, 42, 311, 314, 368

equatorial ring, 20, 56

Equatorie of the Planetis (Price), 122

equinox, 20–21, 56–57

Eratosthenes, 19, 186

Erler, Georg, 268

Erlinger, Georg, 424

error patterns, 194–196, 209–212, 233, 332, 385

ether, 154. *See* aether

Ettinghausen, Richard, 98, 101, 113

Euclid, 5, 7, 17, 137, 162, 163, 166

Eudoxus, 8, 137, 143, 149

Euler, Leonard, 56

Evans, James, 79

evection 22, 59, 62

Fabricius, David, 346, 378

fate, 374

Favaro, A., 275

Feild, John, 230

Feisenberger, H. A., 258

Feldkirch, 255

Felli, Marcello, 134

Ferrara, 162

Feynman, R. P., 418

Field, J. V., 405–406

fire of London, 262

Fladt, K., 347

Flamsteed, John, 70

Florence, Biblioteca Nazionale, 281–282

Florence, History of Science Museum, 133

Fogg Art Museum (Harvard), 111

Forbes, Eric, 71

Forstemann, K. Ch., 267

Forster, Johann, 267

FORTRAN, 360–362

Foscarini, P. A., 273–274

Fourier series, 197

France, censorship, 266

Frankfurt, 263

Franklin and Newton (Cohen), 413

Free, John, 220

Freer Gallery of Art, 106, 109–110

Frombork, 28, 252, 288

Fugger, John Jakob, 266

Fusoris, Jean, 97

Galilei, Vincenzo, 396–398

Galileo, 271–274, 298, 329, 414; as annotator of *De revolutionibus*, 235, 266, 281; calls Kepler's lunar theory occult, 149; Copernican, 394–395; *Dialogue*, 50; *Letter to Christina*, 278; trial, 284

Galileo Opere, 275

gallic acid, 282

Garcaeus, J., 250, 294

Gaskell, Philip, 267

Gasser, Achilles Permin, 261

Gdansk, 252

Gebler, Karl von, 285

Gemma Frisius, 231, 297

geo-heliocentric system, 30–31, 38, 45, 147

Geographiae Ptolemaei (Werner), 261

geometry is coeternal..., 401

George II, 266

Gerhard of Cremona, 186

Giambullari, Pietro Francesco, 266

Gibbs, Sharon, 84, 91, 100, 134, 425

Giese, Bishop Tiedemann, 267

Gilbert, William, 316, 320, 344

Giovanni di Paolo, 142

Görlitz, 258

God, 346; as architect, 353

golden chain, 32, 190, 288

Goldstein, B., 71, 156

Gonville and Caius College, 250, 292, 295–296, 324, 428

Gonzaga, Saint Aloysius, 266

Gothus, Laurentius Paulinus, 130–131

Gould, S. J., 69

Grant, Robert, 387

Grasshoff, Gerd, 79

gravitation, universal, 407, 412

Graz, 370, 372, 390, 394

great conjunction, 307–308, 327; of 1563, 232

Greenwich. *See* National Maritime Museum

Gregorian calendar reform, 176

Gregory XIII, Pope, 276

Grendler, Paul F., 268

Grolier, Jean, 258

Gruppenbach, O. and G., 222

Guadalajara Public Library, 284

Gunther, R. T., 84, 97, 103, 105, 110, 126

Gustavus Adolphus, 177

Guttenberg *Bible*, 257

Haebler, Konrad, 237
Hafenrefer, Matthias, 325
Hall, A. R. and M. B., 156
Halley, Edmond, 288
Hamburger, Paul, 394
Handy Tables (Ptolemy), 221
Hanson, N. R., 347
harmonic law, 47, 403, 409, 411, 417
Harmonice mundi (Kepler), 6, 45–47, 326,
    345, 351, 401–403; musical harmonies,
    328–329; V,introduction, 349; V,4, 352
harmony, musical, 388f
Hartmann, Georg, 97
Hartner, Willy, 127, 135, 381
Harvard College Observatory, 365
Harvard Computing Center, 358
Harvard Library fire, 262
Harvard University, 408; Houghton Library,
    260–261
Harvey, William, 257
Heath, Thomas L., 156, 191
Heinrich Petri, Sebastian, 260–261
Heisenberg, Werner, 413
heliostatic system, 333
Hellman, C. Doris, 355, 404, 420
Helmbold, W. C., 192
Henderson, Janice A., 100, 134
Henry II, 266
Hentschel, Klaus, 73
Herakleides, 287
Hermitage, 105, 108
Herodotus, 163
Herrera, Juan de, 266
Herwart von Hohenburg, 343, 372
Hesiod, 163
Hevelius, Johannes, 321
Hicetas, 189
Hilary of Wiślica, 213–214, 220
Hilbert, 418
Hilgers, Joseph, 285
Hipparchus, 7, 9, 11, 24, 56–60, 137–138;
    lunar distance, 185; obliquity, 19; solar
    theory, 20–22; star catalog, 77
Histoire de l'astronomie moderne
    (Delambre), 358
History of Science Society, 420
Hoffman Collection, 103, 198
Holstein, F. W. H., 267
Holton, Gerald, 69, 204, 347, 420
Honeyman, R. B. and M. S., 228, 250
Horblit, Harrison, 289
Horowitz, Michael, 267
Horský, Zdeněk, 135

Hoskin, Michael, 423, 425
Houzeau and Lancaster, *Bibliographie*
    générale, 220, 248–250
Houghton Library, Harvard, 260–261
Howse, Derek, 426
Huber, Peter, 207, 427
Hunter, Joseph, 248
Huygens, Christiaan, 136, 151–152, 257
Hven, 181
hypotheses, 288, 290, 297, 300
Hypotheses astronomicae (Peucer), 293
Hypotyposes astronomicae (Peucer),
    242–244, 294
Hypotyposes orbium coelestium (Reinhold/
    Peucer), 242, 294

IBM, 168
IBM-7094, 358, 361, 369, 420
Ibn al-Haytham, 140, 145
Ibn ash-Shatir, 141, 145–146, 175
Ibn Rushd, 140
Ibn Sam'h, 120
Ibn Sid, Isaac, 117–125
Iliad (Homer), 414
Imsser, Phillip, 201
In Job (Zuñiga), 275
Index librorum prohibitorum, 265
Indian Museum, Calcutta, 103
Inquisition, 265, 269f
Instrumentum primi mobilis (Apianus), 253
Instrumentus sinuum (Apianus), 253
International Astronomical Union, 419
Isfahan school, 105
Iskander, Albert Z., 156
Islamic astronomy, 10
Islamic Astrolabists (Mayer), 103
Italy, censorship, 282–284
iterative methods, 357f, 421

Jacobeius, Stanislaus, 214
Jacobus Cremonensis, 191
Jagiellonian Library (Cracow), 164–165,
    186, 213, 290
Jaipur A astrolabe, 98
Jamal ad-Din, 99
James, M. R., 301
Jarzębowski, L., 204
Jefferson, Thomas, 262
Jeffreys, Harold, 184
Jeronimo de Zurita, 128
Jesuits, 283; libraries, 282
John of Holywood. *See* Sacrobosco

Jones, Christine, 156
Josten, C. H., 100
Jupiter, 196, 227; satellites, 352
just intonation, 398, 400

Kaye, G. R., 101
Keller, John Esten, 126
Kennedy, E. S., 420
Kensington Science Museum, 86, 99, 105
Kepler, Johannes, 39f, 75; 182, 298;
    accuracy, 50; annual calendars, 263; as
    annotator of De revolutionibus, 176, 235;
    becomes Copernican, 306; bet with
    Longomontanus, 361, 394; birth, 39;
    conception, 39, 305; De revolutionibus,
    167, 177, 261, 264–265; grades at
    Tübingen, 306, 325; lunar theory, 48,
    331, 352; on Prutenic Tables, 223, 234;
    optics, 314; prayer, 50; relations with
    Galileo, 149, 271; shape of Mercury's
    orbit, 121; speculations, 47, 349; superior
    to Copernicus in mathematics, 201; to
    Graz, 206; use of epicycle, 317–319
Kepler (Caspar), 46
Kepler's laws, 409–412, 414–417. See also
    law of areas, ellipse, harmonic law
Keynes, Geoffrey, 267
Khalil Muhammad, 102, 105, 108
King, David, 101, 102–114, 156, 420
King Lear (Shakespeare), 408
King list, 13
Klug, Joseph, 253
Königsberg, 206; longitude, 213
Koestler, Arthur, 203, 256, 296
Koyré, A., 184, 357
Kržiž, A., 114
Kugel, Alexis, 89
Kugler, F. X., 71
Kuhn, Thomas, 193, 197–198
Kulikovsky, P. G., 321, 346, 376
Kunitzsch, Paul, 71, 88, 97, 100–101, 126,
    425

labyrinth, 366
Lactantius, 277
Landau Collection, Paris, 108
Langevin, Paul, 413
Lansberg, Philip van, 56, 346
Larmor, Joseph, 152
latitudes, 339, 373
Lavoisier, 414
law of areas, 174, 409, 411

Lawson, Vivienne, 249
Lehman Collection, 142
Leibnitz, G. W., 407
Leighton, R. B., 418
Leipzig, 167, 262, 264, 289, 421
Leningrad, 177, 312–315, 334–337, 365, 370.
    See Hermitage, 105
Leonardo da Vinci, 162
Leoninus, Albertus, 250
Leovitius, Cyprian, 195, 202–203, 231–233
"Letter from Lysis," 187
Letter to the Grand Duchess Christina
    (Galileo), 273, 278
Lettera sopra l'opinione del Copernico
    (Foscarini), 274
Lewis Prize, American Philosophical
    Society, 423
Libro del Saber de Astrologia (Alfonso),
    116, 121
Liddell, Duncan, 272
Liechtenstein, 201
Liège Musee de la Vie Wallonne, 86
Lincei, De revolutionibus, 282
List, Martha, 268, 376
Liverpool, 258
Lodge, Oliver, 154–155
London, great fire, 262
London. See Kensington Science Museum,
    105
longitude, 23
Longomontanus, 41, 75, 339; Kepler's bet
    with, 314, 394; error in Mercury
    prediction, 386
Lorentz, H. A., 73, 153
Lorenz Kirche, 255
Los, J., 204
Louvain, 194, 263
Lubar, S., 89
Lucius, 189
Lufft, Hans, 253
luminiferous aether, 136, 152–153, 155
lunar eclipses, 21
Luther, Martin, 269, 200
Lysis, 187

Maccagni, Carlo, 134
Macomber, Henry P., 267
Maddison, Francis, 89, 100, 110, 113, 425
Madrid, University Library, 121; National
    Library, 122
Maestlin, Michael, as annotator of De
    revolutionibus, 176–177, 235, 262;
    complaint about physical causes, 328;

editor of *Prutenic Tables*, 222;
Ephemerides, 194, 202, 232, 327, 385;
Epitome astronomiae, 294–295, 324;
Kepler's teacher, 39, 271, 298, 306–307,
324–327, 389–390; testing Copernican
hypothesis, 424
Maeyama, Y., 71
Magini, G. A., 202, 232, 250, 377, 387; as
annotator of *De revolutionibus*, 235;
Ephemerides, 194
magnetism, 43–44, 149, 320, 339, 343
Maistrov, L. E., 113
Malin, Stuart, 425
Manitius, K., 71
Mannheimer, E., 109, 114
Manning, Kenneth, 387
Maragha school, 27, 141
Mars, 8f; Copernican positions, 63;
triangulation, 315, 317, 338; errors, 195;
warfare on, 75, 43f, 300, 310–314,
335–340, 357f
Mass in B Minor (Bach), 414
Maxwell, James Clerk, 152–153
Mayer, L. A., 101, 103, 108, 114
Mayer, Tobias, 56
Mediterranean environment, 263, 282
Mehra, Jagdish, 418
Melanchthon, Phillip, 293
Menelaus theorem, 18–19
Mensing Collection, 84, 88, 129
Merchant of Venice (Shakespeare), 388
Mercier, Raymond, 71
Mercury, 379–387; Copernican predictions,
207, 386; errors in longitude, 68;
visibility, 78, 383
Mercury model, 24, 131–132, 144, 340–342;
Copernican, 333, 382f; double perigee,
381; shape of orbit, 121
Merton College astrolabe, 90
Messahalla, 118
Messekunst Archimedis (Kepler), 354
Metonic cycle and year, 22
Metropolitan Museum of Art (New York),
142
Mexico, 284
Meyerhof Collection, 113
Michel, Henri, 100, 110, 114
Michelangelo, 162
Millikan, Robert A., 69
Milton, John, 284
mind boggled, 17
Minkowski, Hermann, 413
Mischler, Chaninal, 285

Moleti, G., 232, 249
monochord, 392
Morhard, Ulrich, 222
Morton, Samuel George, 69
Muhammad Amin, 102–103, 108
Muhammad Baqir, 102
Muhammad ibn Abi Bakr, 98
Muhammad ibn Khidr, 111
Muhammad Tahir, 102, 108
Muhammad Zaman, 110
Mulholland, J. D., 79
Munich, 263
Murdoch Smith, R., 114
Musée Alaoui, 113
Mysterium cosmographicum (Kepler),
40–41, 45–46, 49, 271, 298–299, 307–310,
325–328, 334, 345, 349–351; picture of
polyhedra, 309, 391; wrong, 393
mysticism, 346

Nabonassar, 12
Narratio prima (Rheticus), 166, 176, 235,
252, 288–289; censored, 283
National Maritime Museum, 103–104, 108
National Union Catalog, 257
nature abhors a vacuum, 137
Nelson, Benjamin, 200
Neugebauer, O., 59, 75, 191, 225, 247, 251
Newton, Isaac, 51, 70–71, 136, 151–152,
284, 353, 407–410. *See also Principia*
Newton, R. R., 5, 18, 58–59, 74–79, 422
Nicholas of Cusa, 191
an-Nisaburi, 96, 98
Nixon, Howard, 249, 258
Nobis, Heribert, 191
Nordenmark, N. V. E., 134
North, John, 127, 425, 427
Nova of 1572, 306
Nova of 1604, 331
Novae ephemerides (Origanus), 232
Novae motuum coelestium (Durret), 346
Novae questiones spherae (Peucer), 293
Novara, Domenico Maria, 235
Nuremberg, publication of *De
revolutionibus*, 37, 167, 255, 264–265
Nuremberg Chronicle (Schedel), 142, 161

O'Connell, D. J. K., 178
O'More, Haven, 267
Oberlin College, 109–111
obliquity, 19
On Sizes and Distances (Aristarchus), 188

On the Face in the Orb of the Moon
    (Plutarch), 188
Opera mathematica (Schöner), 130
Opinions of the Philosophers (Aetius),
    188–189
Optica (Witelo), 253
Optics (Newton), 151
orbarium, 133
Oriel College astrolabe, 90
Origanus, David, 194, 232, 332, 385
Osiander, Andreas, 37–38, 167–168, 177,
    179, 182, 235, 255, 264, 270, 273, 289,
    290–291, 297, 300, 323
Otto Heinrich, Elector, 266
oval, 370
ovoid orbit, 44, 318–320, 340, 342
Oxford, 262; History of Science Museum,
    84, 89, 91, 105, 108–112

Padua, 271
Painswick astrolabe, 86, 90
Palermo, Biblioteca Nazionale, 282
Palter, Robert, 128
Pannekoek, A., 71
Paralipomena in Vitellionem (Kepler), 351
parallax, 22
Paris, 342
Parma, Biblioteca Palatina, 421–422
Paul III, Pope, 38, 187, 223, 265, 286
Pauli, Gustav, 267
Peace of 1466, 161
Pedersen, Olaf, 128
Peking National Library, 283
penetration of orbs, 147
Perugia, Museo Archeologico, 134
Petersen, Viggo, 59
Petreius, Johannes, 166, 252–253, 261, 265,
    270, 289
Peucer, Caspar, 156, 242–244, 293
Peurbach, Theoricae novae planetarum, 28,
    129, 131–132, 143, 145, 173, 222, 376;
    Reinhold's commentary on, 243;
    Tractatus super propositiones Ptolemaei,
    253, 261
Pflaum, Jacob, 194
Phaenomenon singulare (Kepler), 331
Philip II, 266
Philolaus, 186, 286–287
Philosophiae naturalis principia mathematica
    (Newton), 151. See Principia
physical causes. See celestial physics
Piero di Puccio, 141
Piñeda, Juan de, 283

pirates, 263
Pisa, 272
planetary harmonies, 399–402
Planetary Hypotheses (Ptolemy), 25–26,
    138, 143
planetary parameters, 16
planetary souls, 49
Plantin-Moretus Press, 256
Plato, 7, 163, 186, 286; numerology, 182;
    uniform circular motion, 174
Platonic Questions (Plutarch), 188
Platonic solids, 334, 345. See polyhedra
Plutarch, 188, 287
Poincaré, Henri, 413
polygon, 389
polyhedra, 40, 298, 307, 309, 334, 339, 345,
    349, 390–392, 393, 395, 399, 403
polyphony, 46, 396, 402
Pommersfelden, 214
Pontus de Tyard, 266
Portugal, non-censorship, 266, 281–283
positions, accuracy of predictions, 169
Posterior Analytics (Aristotle), 392
Poulle, Emmanuel, 127, 203, 381, 425–426
Praetorius, Johannes, 265
Prague, 259, 370, 372, 394; De
    revolutionibus facsimile, 178–179;
    Kepler's arrival, 342; National Technical
    Museum, 132
prayer, 329
precession, 13, 22, 26, 58
Price, Derek de Solla, 84, 100, 113, 122,
    134, 426
Principia (Newton), 51, 151, 256–257, 288,
    331, 345, 353, 407–408
Principia philosophiae (Descartes), 150–151
Procter, Evelyn S., 127
Progymnasmata (Tycho), 43, 276
prosthaphaereses, 15, 215, 225, 245
Prutenic Tables (Reinhold), 171–172, 202,
    205–206, 208–218, 221–237, 240, 253,
    290–292, 294, 296, 332, 384–385; errors,
    327; Maestlin's 1571 edition, 324, 327;
    name, 177; sample calculation, 224–226,
    245–247; with Reinhold inscription, 249
Psalms 19, 272
Pseudo-Plutarch = Aetius, 188, 198
Ptolemaic system, 123, 138–139, 181, 314,
    333; accuracy, 66f; as monster, 34, 199;
    Clavius' opinion, 271; inferior to
    Copernican, 298, 390; lunar theory, 59,
    62; Mercury errors, 386; Mercury theory,
    379–382; number of circles, 197

Ptolemy, Claudius, accepted by Peucer, 243; *Atlas*, 163; equant considered cheating, 175; fraud, 5, 16f, 55f, 422; superior to Copernicus in mathematics, 201. *See Almagest, Planetary Hypotheses*
Pythagoras, 174, 186–187, 407

qibla, 103
Questiones novae in Theoricas novas planetarum (Urstitius), 130–131
quintessence, 136

Rabbi Saq, 117. *See* Ibn Sid
Ramus, Petrus, 300, 346
Rankghe, Laurentius, 292
Raphael, 162
Ratdolt, Erhard, 163, 201
Ravetz, Jerome, 176, 184, 204, 235
Rawlins, Dennis, 77
refraction, 351
Regiomontanus, Archimedes manuscript, 191; *De triangulis*, 253, 262, 293; *Ephemerides*, 194, 200; *Epitome of Ptolemy's Almagest*, 164, 235, 261, 424; publisher of Peurbach, 131–132; *Tractatus contra Cremonensis*, 198
Reinhold, Erasmus, 63, 221–247, 323; annotations in Regiomontanus' *Epitome*, 424; commentary on Peurbach, 341; *De revolutionibus*, 176–177, 179, 235f, 255, 257, 263, 265, 270–271, 347; Dean and Rector at Wittenberg, 234, 287; depth of plague, 287; *Ephemerides* 202, 205–208; handwriting samples, 238–239; premature death, 244; *Prutenic Tables*, 171, 209–218, 253, 290–292, 294, 296, 332
relativity, general, 154
relativity, special, 136
Remus Quietanus, Johannes, 353–354
retrograde motion, 7–10, 29, 138, 182
Rhaw, Georg, 253
Rheticus, 32, 145, 164, 166, 287–290; as annotator of *De revolutionibus*, 176, 235; *Ephemerides*, 202, 205–208, 213, 218–219; publication of *De revolutionibus*, 37, 187, 252–255, 264
Riccioli, J. B., 56, 148, 386
Richeliene Tables (Durret), 346
Rico y Sinobas, Manuel, 115–116, 118–122
Ripley, S. D., 427
Roberts, Julian, 268
Rodin, Auguste, 414

Roguski, Sylvestrus, 214
Rome, 178–179; *De revolutionibus*, 282
Rose, Valentin, 241
Rosen, Edward, 128, 156, 184, 189, 192, 202, 205, 248–249, 277, 280, 285, 301, 302, 405
Rosenthal-Schneider, Ilse, 73
Rostock, 272
Royal Astronomical Society, 419
Royal Observatory Edinburgh, 139, 176, 235, 255, 257, 291
Royal Society, 51
Rudolf II, 41, 45, 310, 395
Rudolphine Tables, 321, 328–329, 369, 386, 396, 424; accuracy, 50; preface, 327
Rufus, Carl, 345
Russian Academy of Science, 365
Ruttelius, A, 202
Rybka, E., 302

Sacrobosco, 28, 162, 291–292
St. Louis, City Art Museum, 103, 108
St. Petersburg Kepler Archive, 321, 396, 421
Saliba, George, 91, 101, 102–114, 156, 420
Sambucus, Johannes, 266
Sand-reckoner (Archimedes), 185
Sands, M., 418
Sarton, George, 126
Saturn, 196, 224–226
Schaffhausen, 177
Schedel, Hartmann, 161
Scheibel, Johann Ephraim, 250
Schickhard, William, 386
Schissler, Christopher, 424
Schöner, Johann, 130, 176, 220, 294, 296, 383–384; as annotator of *De revolutionibus*, 235
Schönborn, Bartholomew, 292–293
Schönbornsche Bibliothek, 214
Schofield, Christine, 248
Schradi, Johann, 267
Schreckenfuchs, E. O., 130
Schwarzschild radius, 412
Scorriggio, Lazzaro, 275
Scultetus, Bartholomew, 268
Sédillot, L. A., 113
Seidelmann, P. K., 79
sexagesimal notation, viii, 225–226, 245–247
Shakespeare, vii, 388, 408, 418; first folios, 257
Shaykh Muhammad 'Iraqi, 111
Sherburne, Edward, 201

Shevchenko, M., 79
Shipman, Joseph C., 267
ash-Shirazi, Qutb ad-Din, 141
sidereal period, 32
sidereal year, 20, 22, 24, 76
Sidereus nuncius (Galileo), 395
Sigismund II Augustus, 266
Simocatta, Theophylactus, 162
Simplicius, 137
size of moon, 23; of universe, 48
Skeat, Walter W., 83, 88, 97
Skopova, Otilic, 135
sleep walking, 319
Sleepwalkers (Koestler), 203, 256, 296
Sloane astrolabe, 86, 88, 97
Small, R., 360, 363
Smithsonian, 91, 103, 106, 108, 112, 291, 428
Smithsonian astrolabe, 81–82
solstice, 19, 57
Somnium (Kepler), 331
Sophocles, 418
Soriano Viguera, José, 128
Sotheby's, 89, 114, 258
Spain, non-censorship, 266, 281–283
Sphaera (Sacrobosco), 28, 142, 292
spheres, crystal, 27. *See* aether
Stadius, Johannes, 194–196, 202, 222–223, 231, 233, 297, 384–385, 387; as annotator of *De revolutionibus*, 235
Stahlman, William, 360
Steinschneider, Moritz, 116
Stockholm, *Commentariolus*, 163; *De revolutionibus*, 282
Stoeffler, Johannes, 163, 194–195, 197–198, 202, 204, 213–214
Stott, Carole, 425
Straker, Stephen, 322
Strasbourg, 263, 294
Strigel, Victorin, 293
Strubius, Caspar, 222
Structure of Scientific Revolutions (Kuhn), 193
Sturge, Mrs. George, 285
as-Sufi, 116
sun, errors, 233; sun on royal throne, 181
sunspot, 331
suspiciendo despicio, 4
Swerdlow, Noel, 79, 156, 191–192, 267, 426
symmetriam, 204, 299
Symphony No. 5 (Beethoven), 407, 414
Syntaxis (Ptolemy), 5, 58, 138
systema diapason duplex, 389, 398

Table Talk (Luther), 269
Tabulae Bergensis (Stadius), 202, 222
Tabulae Rudolphinae (Kepler). *See Rudolphine Tables*
Tadhkira (at-Tusi), 140
Taisnier, Johannes, 316
Taliafero, R. C., 71
Teheran, 112
telescope, 331, 353
Tertullian, 273
tetrachord, 397–398
Teutonic Knights, 161
textbooks, cheap printed, 292
Theodoricus Winshemius, Sebastian, 292
Theodosius, 249
theorica orbium, 129–135
Theorica planetarum, 142–143, 198
Theoricae novae planetarum (Peurbach), 28, 129, 132, 143, 145, 173, 222, 376, picture, 341
Thirty Years' War, 163, 262, 395
Thorn, 250
Thorndike, Lynn, 202
Thoth 1, 12–13
Timon (Shakespeare), 408
Tokyo, *De revolutionibus*, 282
Toledan Tables, 115, 125
Toledo, longitude, 213
Toledo Museum of Art (Ohio), 111
Tomba, Tullio, 84, 91
Toomer, Gerard, 12
Toruń, 179
Tractatus brevis (Garcaeus), 294
Tractatus super propositiones Ptolemaei (Peurbach), 253, 261
Tracy, Melvin, 285
transit of Mercury, 331, 342, 385–386
treadmill, 45, 396
Treatise on the Astrolabe (Chaucer), 81–83
trepidation, 26–27, 35
Trinity College, Cambridge, 48, 257, 326; *De revolutionibus*, 282
tropical year, 20, 22, 24, 76
Tübingen, 45, 253, 262, 264, 271, 298, 324–325, 332, 389
Tuckerman, Bryant, 168, 171, 202, 207, 220, 233, 332, 381
Turnbull, H. W., 156
Turner, Gerald, 425
at-Tusi, Nasir ad-Din, 140–141
Tycho Brahe 172; attitude toward Copernicus, 32, 38, 146–147; complaint re: tables, 327; copy of *Prutenic Tables*,

249; crystal spheres, 30, 149; diligent observer, 394–395; discovers variation, 60, lunar theory, 331; geo-heliocentric system (see also geo-heliocentric), 31; heirs, 45; invites Kepler, 41–43, 310, 334, 342; Kepler's poem to, 312; re: Leovitius, 231; motto, 4; observations, 276, 340, 403, of Mars, 336, 344, 358, of Mercury, 385, of conjunction, 196; papers in Vienna, 249; putative annotator of *De revolutionibus*, 176–181, 259; reviews accuracy of predictions, 202; solar positions, 332; use of ephemerides, 194; visits Wittenberg, 244

Tychonic system, 148, 179, 181, 190, 241
Typen von Sternzeichnissen (Kunitzsch), 97

Uppsala University Library, 163, 170, 289, 301
Uraniborg, 41, 181
al-'Urdi, 146
Urstitius, Christian, 130–131

Valgriso, Vincenzo, 232
Valla, Giorgio, 185, 189, 235
variation, 60, 62, 210
Varmia, 161
Vatican, Library, 177–178, 180; Museum, 134
Vega, 98–99, 109
Venus, 66; perigee, 63–64
Venus de Milo, 414
Vesalius, Andreas, 257
via buccosa, 320
vicarious hypothesis, 310–313, 316, 367f
Victoria and Albert Museum, 103, 108, 258–259
Vienna Nationalbibliothek, 163, 378
Vieta, François, 366
Virgil, 278–279
Virginia, University library fire, 262
Vistula, 383
Voet, Leon, 267
Volpaia, Girolamo Della, 129–135
vortex theory, 149–150
Vossius, Isaac, 201
"votes and ballots," 44
Vulparius, Hieronymus, 129–135

Wacker von Wackenfels, 405

Wagner, Richard, 414
Walderman, William, 202, 248
Walker, D. P., 404
Wall, B. W., 192
Wallis, C. G., 330
Walther, Bernard, 191, 211–212, 220, 383–384
Ward, F. A. B., 100
warfare on Mars, 306, 310, 342, 349
Watzenrode, Lucas, 161–162
Webster, R. and M., 113, 425–426
Wegener, Alfred, 126
Weibel, Adele Coulin, 114
Weil der Stadt, 389, 421
Weil, Ernst, 258, 267
Welack, Matthew, 222
Welther, Barbara, 79, 168, 195–196, 202, 233, 332, 387, 427
Werner, Johannes, 235, 261
West Point Military Academy Library, 223
Westman, R. S., 156, 404, 423
Whipple Museum, 88
Whipple, F. L., 427
Whiteside, D. T., 421, 428
Wilson, Curtis, 375
witchcraft, 396
Witelo, 253
Wittenberg, 166–167, 179, 206, 242, 245, 257, 265, 270–272, 287; astronomy notes, 250, 291, 324
Wittenberg interpretation, 423
Wittich, Paul, viii, 184, 259, 423
Wlodarczyk, Jaroslaw, 79
Wolfenbüttel, 266
woodblocks, 253
World of Learning, 258
Württemberg, 396
Wursteisen, Chris, 272

Yale University, 290; Medical Historical Library, 111
Yehuda ben Moses Cohen, 116–117, 123, 125

Zarlino, Giuseppe, 266, 397–398
al-Zarqali, 118, 120
Zeller, F. and C., 220, 300
Zinner, Ernst, 135, 203, 213, 235, 248–249, 268, 301
zodiac, 393
Zuñiga, Diego de, 274–275